Practical Robotics

Canadian Cataloging-in-Publication Data

Davies, Bill, 1932-
Practical Robotics : Principles and Applications
Includes bibliographical references and index
ISBN 0-9681830-0-X
1. Robotics. I. Title

TJ211.D384 1997   629.8'92   C97-930284-6

Published by WERD Technology Inc.
Unit 35B Suite 155
10520 Yonge Street
Richmond Hill, Ontario L4C 3C7

Text reproduction from camera-ready WordPerfect electronic files. Line drawings are by the author using Corel Draw 4, except Figure 5-9 kindly provided by Measurements group, Raleigh NC.

Printed in Canada by University of Toronto Press

PRACTICAL ROBOTICS

PRINCIPLES AND APPLICATIONS

Bill Davies

Published by WERD Technology Inc.
Unit 35B Suite 155
10520 Yonge Street
Richmond Hill, Ontario L4C 3C7

Contents

Acknowledgements

Students also teach instructors by asking basic but often very profound questions. I have learned a great deal by trying to find answers to such queries.

Jake Unger, a friend and colleague for thirty years kindly read Chapters 1 through 4 despite his busy schedule. Thankfully my understanding wife and family have never complained about my unusual working hours, because a pleasant home environment is essential for tinkering and writing.

Warnings and Disclaimer

Making mechanical items requires exceptional care to avoid injury. Working with power line potentials and high voltages is hazardous, and can produce lethal shocks. Death can also occur in any situation where more than 6 milliamps is conducted through body tissues.

This book is intended merely as a guide and information source for readers. While efforts have been made to provide accurate information, errors may be present in this text and this must be taken into consideration when constructing apparatus described herein.

Both author and publisher disclaim any responsibility for accidents or damages resulting from use of this book's contents.

If you do not wish to accept conditions stated above, this book may be returned to the publisher for a full refund.

Getting started

Building Sophisticated Robotic Projects

This book is a response to requests from second year students enrolled in the Engineering Design course at the University of Toronto. They have found information on robotics applications is widely dispersed, often fragmented and seldom practical in nature. Their observations are not unique since even experienced personnel sometimes have difficulty finding such information.

Addressed primarily to students at the high school through third year university levels, this text will be a useful companion for the hobbyist, robotics enthusiast, and those who love to experiment. Mathematics has been used only where essential and a knowledge of Ohm's law and the Basic programming language are the only prerequisite for using this text.

During ten years of teaching design courses I have been amazed at the capabilities of young people, and know from experience that given the tools and a little guidance their achievements are truly remarkable. At the beginning of each term I used several three-hour lectures to explain fundamentals of electronics and robotics to a typical class of 100 students - a select cadre of the best engineering prospects. I showed them videos of projects their compatriots completed in previous years. These screenings were accompanied by grunts of amazement and incredulity that such wonderful projects could be completed in a period of 10 weeks, with no prior training in electronics or computer hardware experience. Each year I told them several of the current year's projects would surpass the quality and sophistication level of their predecessors. They always fulfilled this prediction.

A magic elixir?

What is the secret formula to their success? Intelligence, dedication, perseverance, synergistic group activities, high tech support equipment, adequate guidance, moral support, expensive tools and hardware? Well, some portion of each of these ingredients is necessary, except the expensive items.

Students work in groups of two to four, choose their own project, make all hardware elements they require, and also shop and pay with their own money for components they need. Selecting appropriate devices, shopping wisely and using such purchases with prudence, is an invaluable learning experience for budding engineers. Working as part of a team prepares them for real world engineering.

Students' interests are eclectic, their projects spanning the gamut from humanitarian: *(Head Controlled Mouse, Electromyographically Controlled Wheelchair)*, musical: *(Trumpet and Articulated Piano Players)*, sports: *(NHL Air Hockey, Ball Catcher)*, marine: *(Robot Submarine, Squid Tracker)*, pastimes: *(Computer Controlled Mechanical Rubik's Cube Solver, Chess Player)*. These are not singular successes, each year 90% of all projects met their objectives and over half were chosen for special recommendation. Many projects were at the fourth year bachelor's thesis level, complemented by comprehensive and erudite final reports, portraying understanding and insights well beyond their authors' years.

Why is this so? I believe young people are charged with innovation, intelligence, a will to succeed and that wonderful youthful innocence of not knowing that something is 'impossible'. Such latent potential is just itching to be released. Yet perhaps the most important element is that when pursuing their own ideas a project becomes a labor of love that is 'theirs'.

This may give the impression that the instructor is an inconsequential part of the control loop. Not quite, since students need a source of experience and inspiration when things are not going the way they should. They need to know which multiplexer to use, "How can I bend acrylic sheet?", "Is there a simple inexpensive electronic level sensor?", "Can I tell which way this motor is turning from my encoder disk output?", "How do I switch 40 amps at 12 volts?", "Will I be able to control this 60 psi air flow remotely using an IR link?", "Can this laser beam be occluded in 60 microseconds?". Answers to such questions are currently an instructor's domain, but comprehensive texts should make things easier for students.

Hopefully this book will be of interest to others too. Teachers and students will benefit from techniques and protocols suggested here. Hobbyists, experimenters and those who love to tinker will find something of consequence. Because unique solutions are described that have never appeared before in print, there may be something of interest for many. I have spent endless hours solving problems for students, but the real fun for me has always been just fiddling with hardware.

Because the robotics field spans so many disciplines, introducing neophytes to every robotic technique required during a single course is not easy. In this book a phenomenological interpretation of the function and utility of devices is stressed, and practical applications are given in many cases. Playing with the systems described, promotes a depth of understanding and familiarity that can be obtained in no other manner.

"How much does it cost to build a robot?"

An important parameter for an individual experimenter is cost. Budget limitations place restrictions on size and complexity when building a project, and demand circumspection in spending habits. The question of "how much will it cost?" has been addressed, and prices (in 1997 US$) are given for many items, also sources where these may be purchased.

Although time changes the absolute cost of everything, estimates of up-to-date values may be obtained by inter-comparing current prices of different items. Such relative scaling works well over many decades, because individual prices escalate in lockstep, until items are obsolete.

Information sources for beginners

My own experiences have been culled from many sources, and some of these are listed in Appendix B. It is important that both teachers and students have a knowledge of 'what's out there' and where this information can be obtained. It should not be necessary for a novice to complete a literature search before beginning a project.

Classics such as Don Lancaster's Cookbooks and his columns in popular magazines, are milestones in their fields and important for both teachers and students alike. Forrest Mims' notes are readable and satisfyingly practical, making even a tyro immediately comfortable. Unfortunately only university libraries now have large selections of technical books, but these establishments are in many cases open to everyone. It is a sad commentary on current societal interests that 621.x, 629.x, 681.x book sections in most public libraries are minuscule fractions of those devoted to fiction, cooking or diet!

SI versus American units

Readers will find a mix of units is used in this book, unfortunately this is a necessary practicality. In North America most items are still sold by the foot and pound. Being familiar with this system is necessary for our future engineers, because structures and devices based on this measurement system will be around for decades. Most materials and instrumentation sold by surplus houses have their specifications based on US dimensions, and many offshore suppliers of new equipment still tailor their products to satisfy the US market. Conversion to SI is straightforward, and equivalents are given where confusion may arise.

Electronic Equipment for the Roboticist

Robotics tool kit, electronic test equipment, classroom workstations

Sophisticated robotic projects can be built with only a modest investment in tools and equipment. Total parts' costs for most course projects are about $300, or $100 per student for a three-person group. This is only possible by purchasing items such as motors, solenoids, transducers, power supplies, dc-dc converters etc. at surplus prices. For example, a 555-timer chip may be sold for 35¢ by a surplus supply house, but will cost three times as much in a blister pack from a general electronics store in a shopping mall. New solenoid prices start at around $20 from most dealers, but used prices for the same item can be as low as $3. It is easy to spend several hundred dollars on a flashy set of tools, but the kit recommended below is adequate for building even complicated electronic systems.

A classroom workshop should be equipped with a drill press, bench, vise and basic hand tools for student use. Sheet metal work is much easier if a shear/bender/roller is available, and a 30" machine with these features retails for about $600 (Figure 8-2).

Students seldom need encouragement to work at home on an interesting project of their own choosing, but will find it difficult if they have no suitable tools. Essential articles listed in Table 1-1 include the most important tools for building and testing electronic circuits, or making small mechanical items. One kit per group is usually adequate, and this can be lent against a refundable deposit to ensure all elements are returned in good working order. Hobbyists will find the listed items are necessary too, if they wish to do some serious robotics work. Prices (in US dollars), and sources for kit items have been quoted for comparison purposes, or where difficulties may be encountered in finding parts locally.

Table 1-1. Robotics tool kit for hobbyists and students

Qty	Item	cost	model # and *supplier*
1	Power supply[1]	$20	+5V@18A, -5V@0.3A, +12V@3.8A, -12V@0.3A
1	DMM (digital multimeter)	$20	#990087 *Kelvin*
1	Side cutters	$3	ST-1 *C&S Sales*
1	Pliers	$3	ST-2 *C&S Sales*
1	Wire strippers	$3	ST-3 *C&S Sales*
1	Screwdriver set (6pc)	$2	TL-8 *C&S Sales*
1	Soldering tool kit[2]	$10	ST-12 *C&S Sales*
1	Soldering iron stand/sponge	$5	SH-1 *C&S Sales*
1	Iron controller	$5	dimmer switch (Figure 7-8)
	Copper wire[3]		24 AWG solid core wire for breadboards
	Solderless breadboard	$6	(6" x 2") #42-103 *Hosfelt*
	Resistor selection[4]	$5	approx values 1,10,100,1k,5k,10k,100k,1M(Ω) 5% ¼W
	Capacitor selection[5]	$5	0.1µF ceramic, 1µF, 10µF, 100µF electrolytic (25V)
10	Alligator test leads	$2	#52-100 *Hosfelt*
2	Miniclip leads	$1	#80-211 (black) #80-212 (red) *Hosfelt*
2	Banana plugs	$1	#2349B 2349R *Hosfelt*
1	Hand power drill & bits	$30	sale price from various sources (AC line cord model)
1	Hacksaw and blades	$5	
	Total cost	$126	

[1] Power supplies from used PCs are available from various sources, (Figures 1-1, 1-7, 4-1)
[2] Kit includes soldering iron, solder, desoldering pump and solder wick
[3] Solid core telephone wire is ideal for solderless breadboards
[4,5] Resistor kits are ~$45, and capacitor kits start at ~$70 (see Chapter 3).

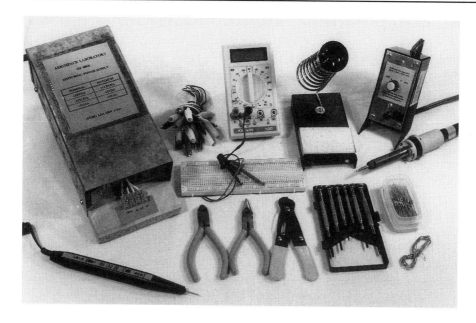

Figure 1-1. Many items listed in Table 1-1 are illustrated in this photograph. A DMM is shown with miniclip leads described in the text. An open frame switching supply has been mounted on a wooden base with a terminal block for attaching power leads. Galvanized 20 gage/0.0313" sheet steel is used for the power supply cover that is spot welded and provided with ventilation holes. Power indicator, line cord and a switch have been added. A soldering iron controller is essential for good soldering, and is shown at upper right. This figure shows a logic probe, but hand drill and hacksaw are omitted.

Power supplies and electronic test equipment are required for any serious robotic work, but does not have to be expensive. As shown below, an investment of about $1000 will cover everything needed for a good start. With an oscilloscope ($325), function generator ($225) and the robotics tool kit ($126) a hobbyist will be well equipped, and still have several hundred dollars left to buy electronic parts, materials and a few extra hand tools.

Oscilloscopes :   S-1325 analog oscilloscope 25MHz $325
Hitachi V-212 analog oscilloscope 20MHz $449
DS-303 storage oscilloscope 20MHz 10M samples/sec $1095
(All prices are *C&S Sales*)

An S-1325 25 MHz analog oscilloscope at $325 is adequate for home or classroom use, and is part of a classroom workstation, shown in Figure 1-8.

A storage scope is only occasionally required so one instrument is sufficient for even a large class. All oscilloscopes listed above are sold with two probes.

Function generators: GF-8026 $225 *C&S Sales*, #9600 $29.95 *C&S Sales*

A GF-8026 function generator in Figure 1-8 produces sine, square, triangle, pulse and ramp waveforms from 0.02Hz to 2MHz. Outputs are variable to 10V peak to peak into 50Ω, with DC offset control. An integral digital frequency meter can also be used to monitor external sources up to 10MHz.

A #9600 generator blox illustrated in Figure 1-2, is an inexpensive and compact alternative for a GF-8026 function generator. This unit provides variable amplitude sine, square and triangle waveforms (1Hz to 1MHz) which can be am or fm modulated. Although ±5V is indicated on the specification sheet provided with a #9600, this unit can be safely operated from a ±12V switching supply.

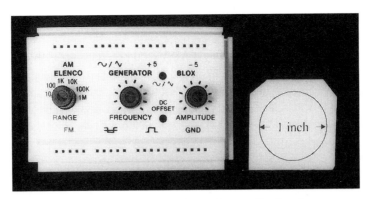

Figure 1-2.   A #9600 generator blox is an inexpensive and versatile function generator that can be purchased assembled as shown, or in kit form for $28.95. It may also be built on a solderless breadboard from components purchased separately. This generator produces sine, square, and triangle waveforms from 0.01Hz to 1MHz and can be operated from a ±12V switching power supply.

Digital multimeter and miniclip probes

Test probes supplied with most new DMMs are inconvenient for circuit testing, and troubleshooting is more efficient using miniclip test leads. When buying a DMM make sure it is equipped with banana input sockets, so that miniclip probes can be connected to the meter. Miniclip probes can be assembled from 2" miniclips and banana plugs listed in Table 1-1. Each probe consists of a 12" length of flexible 22 gage stranded wire (eg. Belden 8525), with a miniclip at one end and a banana plug at the other. Leads should be soldered to both miniclips and banana plugs, for reliability and long service life.

Digital multimeters are widely available, many with a dazzling array of features. A 3½ digit model shown in Figure 1-3 and listed in Table 1-1, meets most requirements. If extra funds are available a DMM with capacitor, transistor and frequency measuring capabilities is very handy. Transistor checkers on most DMMs are useful but can usually only test regular transistors, not Darlingtons.

Figure 1-3. Miniclips shown in this figure are more convenient for troubleshooting. If a short length of wire is attached to the V/Ω clip, it can be inserted quickly into any breadboard location. The common clip is usually hooked onto a short loop of wire inserted into the ground/common rail.

Digital voltmeter, inductance, capacitance and transistor checker: LCM-1850 $75 *C&S Sales*

This 10-function meter is more expensive than a DMM recommended in Table 1-1, but it can measure inductance to 20H, capacitance to 20μF plus the normal ranges of current, voltage and resistance. Transistor and diode checking features are also built into this versatile low-priced instrument.

Inductance can always be determined by measuring the resonant frequency of an LC circuit, formed by a known capacitor and the unknown inductor. Therefore if a hobbyist buys an oscilloscope, purchase of an inductance meter can be postponed until funds are available.

Logic probe: LP-550 $14.95 *C&S Sales*

Status checks on static or slowly varying signals, or monitoring CMOS/TTL logic levels is possible using only a DMM. High speed pulses cannot be detected in the same way. They require either an oscilloscope or a logic probe such as the LP-550.

An audio output provided with an LP-550 shown in Figure 1-4 is particularly useful, eliminating the need to look at the probe's visual indicator when troubleshooting. A high tone indicates a logic 'high' (usually +5V) and a low tone represents a 'low', or power ground potential.

Logic probes are usually connected to the power lines of the circuit being tested. If the probe is powered from a separate supply, the probe's ground clip must also be connected to circuit ground or common.

Logic pulser: LP-600 $21.95 *C&S Sales*

Digital fault finding can be tackled in a systematic manner using a logic pulser shown in Figure 1-4. Pulses are injected at appropriate circuit points and their effects evaluated with a logic probe, oscilloscope or logic analyser. Removing chips from a circuit when using the LP-600 is not necessary, so the pulser is particularly useful when checking components soldered to a printed circuit board. This technique works well for a singular fault, but is time consuming if a problem is due to failure of several chips.

IC test clips: TC-16 $6.50, TC-40 $16.95 *C&S Sales*

IC test clips in Figure 1-4 are useful for monitoring ICs on a PC board but are not essential for hobbyists, or most project work. Test clips make it easier to interrogate IC pins or inject tracer signals.

Figure 1-4. Most hobby projects do not need IC test clips or a logic pulser for troubleshooting. IC clips are handy for checking chips mounted on a PCB (printed circuit board). Some IC clips (top right), are equipped with LEDs that show each pin's status. This type of clip is powered from the chip being tested. Logic probe LP-550 is shown at top left, and a Radio Shack logic pulser is at lower right in this figure.

IC extractor tools, pin straighteners

Extractor tools in Figure 1-5, are seldom necessary for a hobbyist. A slim screwdriver blade inserted carefully under ends of a chip is a good tool for removing ICs from sockets or solderless breadboards. When removing large chips, such as a 40 pin DIP (**D**ual **I**n line **P**ackage), pry successively from both ends of the chip.

DIP IC pin straighteners are convenient for quickly reforming buckled pins, but a pair of flat faced pliers can be just as effective although a bit slower. Pins on a brand new DIP should initially be bent against a flat surface so they are perpendicular to the chip's body, this makes them easier to insert into a socket or breadboard.

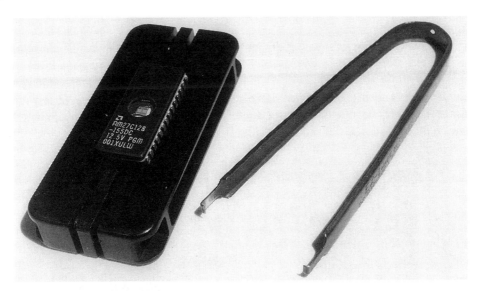

Figure 1-5.   An inexpensive steel spring chip extractor removes all ICs from sockets or breadboards (including 40 pin chips). A slim blade screwdriver is just as effective, but more awkward to use if chips are very close together. A pin straightener shown at left handles both 0.25" and 0.5" width ICs.

Mini vise with vacuum base

Small parts can be soldered more conveniently if they are securely held in a mini vise, as in Figure 1-6. Vise shown is *Radio Shack* 64-2094, $4.59. Modelling clay, Plasticine, or Panduit's duct seal are all useful for positioning and temporarily supporting awkwardly shaped objects, for gluing or soldering.

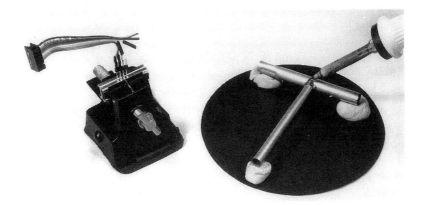

Figure 1-6.   Soldering wires to a connector is much easier if it is held in a small vise, as shown in this photograph. Don't forget to slip heat shrink tubing over a wire *before* soldering wires to the connector. Precise positioning of parts for soldering is simpler if modelling clay is used for support, as shown in this figure.

Power supplies

Switching power supplies, regulated DC linear supplies

A power supply from an old PC is excellent for project work. Such supplies are well packaged and equipped with a power switch, fan and overload protection. Even a small computer power supply is adequate for most project requirements. Prices vary, and sometimes buying an old computer with a working power supply is cheaper than purchasing a power supply separately. Switching supplies are extremely rugged and will tolerate a surprising amount of abuse. They seldom fail in computers, and even with rough treatment meted out in a project course, power supply failures are rare. Information on selecting an appropriate load resistor for a switching supply is given in chapter 4.

An open frame switching power supply is the most economical purchase in large quantities, and this type of supply is issued to groups at the University of Toronto (Figure 1-1). A very good open frame switching supply can be purchased for about $20.

Exposed high voltages are present on open frame supplies and they must be encased in a suitable enclosure for safety reasons. Open frame supplies also require installation of a line cord, power switch, load resistor, power distribution block, and a power indicator, (usually an LED on the +5V power rail).

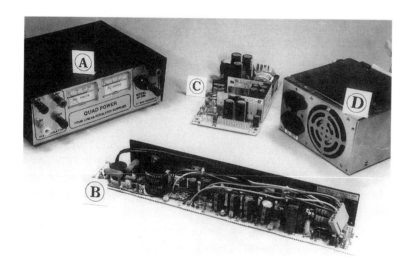

Figure 1-7. A linear power supply will always be larger, heavier and less efficient than a well designed switching power supply having the same power capabilities. [A] XP-580 linear supply. [B] and [C] ±12 V +5V open frame switching supplies. [D] This 140W switching supply from an old PC is equipped with a line switch and dc output power leads. It has the following ratings: +5V/18A, -5V/0.3A, +12V/3.8A, -12V/0.3A

Linear dc supplies fill an important niche, since they are the only convenient method for obtaining continuously variable dc voltages. A high current variable dc supply is invaluable for testing high power motors and electromagnets before running such devices from batteries. A fully adjustable 0 to 15V dc supply sold by *Hosfelt* (#40-166 $200), is rated at 25A with a surge capability of 30A. It is imperative that a reversed biased power diode is placed across a power supply's output terminals before testing any large inductive device, otherwise the supply may be damaged.

Noise on output lines of a switching power supply can be troublesome, for some special projects. Switching noise is most noticeable when pushing the limit on low level signal detection, and can be significantly reduced by using a linear power supply. An XP-580 linear supply sold by *C & S Sales* $70, provides a variable output from 2 to 20V @ 2.5A and three fixed outputs, +5V @3A, -5V @ 0.5A, 12V @ 1A. This well regulated linear dc power supply with short circuit protection and panel meters, is shown in Figure 1-7. The XP-580 is suitable for low noise work and other general robotic applications.

A classroom electronic workstation

A classroom electronic workstation shown in Figure 1-8, consists of a ± 12V, ± 5V dc switching power supply, S-1325 oscilloscope, and GF-8026 function generator. Students also use their robotics kits during scheduled class times. Nine workstations are shared by fifty students during a normal three hour session, this has proven satisfactory because not all groups require workstation facilities at the same time. Too many workstations can also be a disadvantage since this decreases space available for other activities.

All test equipment listed in this chapter has been used by second and third year students at the University of Toronto for project design courses. Items were selected for their cost/performance benefits and continuing availability. No failures have occurred since they were purchased several years ago.

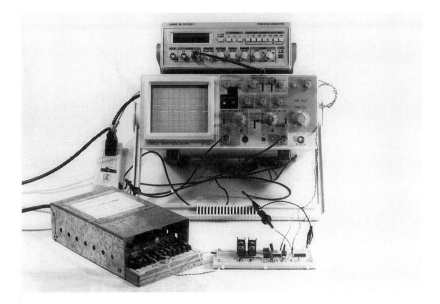

Figure 1-8. A classroom electronic workstation is shown in this figure. An S-1325 oscilloscope shown has two switchable probes, (x1 and x10) sensitivity. One ±5V ±12V switching power supply is provided, and students also use their robotics kits at workstations. A twin lead cable with #20 gage solid wire ends, is used with the GF-8026 function generator. This cable makes circuit testing easier, because solid wire ends can be inserted directly into breadboard sockets. A multiple outlet ac power bar is also provided at each station.

Buying and using an oscilloscope

An oscilloscope is one of the most useful diagnostic tools, and probably the most expensive instrument a hobbyist will buy, so it is wise to give this purchase some thought. If possible, try to test a scope to see if it suits your requirements, ask friends or co-workers for their advice and experiences too.

Digital Storage Oscilloscopes (DSO)

Menu driven DSOs have a learning curve that becomes steeper with scope sophistication. If you enjoy pressing buttons and have a photographic memory, you will probably like them. Signal brightness does not change with sweep speed so single event high speed sweeps are easy to see on a DSO, unlike single-shot traces with an analog scope.

If a DSO's sampling rate is low compared with the event time, it may be difficult to tell what the 'true' signal is from a sample of disparate dots scattered across the screen. Linear or sine wave interpolation (join the dots processing), produces a nice looking output but this may bear little resemblance to the true signal. Nowhere is this more confusing than on inexpensive low sample rate DSOs, with poor internal diagnostic capabilities.

Aliasing and signal sampling rate

Figure 1-9 shows how a 'true signal' can be distorted when sampled by a DSO at discrete times, indicated by squares on the diagram. A dot-connect routine draws a smooth curve through the squares and displays the dotted line or 'aliased signal' on the oscilloscope screen.

Unfortunately, such displays are so satisfyingly precise it is tempting to accept the information as factual, although it is grossly inaccurate. Most expensive DSOs have adequate smart diagnostic routines to avoid this problem, but cheaper scopes are often confused.

To avoid aliasing, a signal of frequency f must always be sampled at a frequency greater than $2f$ (this is known at the Nyquist criterion). Very high sampling rates are required for monitoring high frequency signals accurately, and this capability usually increases an oscilloscope's price.

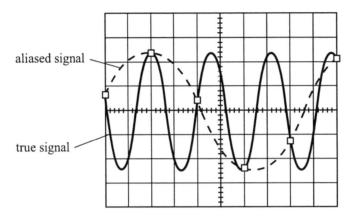

Figure 1-9. A 'true' signal (solid line) is interpreted as the dotted line signal if it is sampled at a repetitive rate indicated by the squares. To avoid this aliasing problem, any signal of frequency f must be sampled at a rate $2f$ or faster.

Real time signal monitoring

Real-time displays of jitter or hash on a cheap DSO look like snow storms, and even expensive instruments cannot compete with an analog scope for jitter measurement. For raw speed an analog scope is also tops. Special analog scopes with tiny [6] screens capture one-shot real time traces up to 1GHz, and analog sampling scopes can do the same for repetitive signals on a full size screen.

Although digital storage scopes can reduce random noise by time averaging, this is not always an advantage for a novice. Clean signal traces look great, but when jitter and noise are eliminated critical information may be lost.

[6] Deflecting an electron beam at very high frequencies requires powerful magnetic or electric fields. Deflection rates can be faster with a smaller raster because the electrons travel shorter distances.

Some key benefits of digital storage scopes

For accuracy in frequency, period and voltage measurements, a DSO is the best choice. Tricky triggering problems on one-shot signals of unknown amplitude are also easier with a DSO. On many models 'pre-trigger' trace information can also be displayed. This allows recovery of vital information, ordinarily lost forever when an analog scope sweep is accidentally triggered by noise

Some DSOs provide disk storage for recording trace data, and a printer can be used for hard copies. Contrast these conveniences with the hassles of photographing an analog oscilloscope screen. On more expensive DSOs, Fast Fourier Transforms (FFTs) turn a scope into a sequential frame spectrum analyser. RS232 and GPIB are often available as standard features. Replacement cathode ray tubes for a DSO are inexpensive compared with variable persistence phosphor screens on some older analog scopes.

Feature-wise DSOs can only get better, and each year they capture more of the scope market, although they may never displace analog oscilloscopes entirely.

Analog oscilloscopes

Analog scopes are reliable, easy to use, and usually better buys at prices below about $5000. Although typical 8 bit DSOs and analog scopes have the same vertical sensitivity (1 or 2mV/div.), an analog scope often gives better real time performance at these lower levels.

No aliasing, intuitive controls and lower prices for an analog scope, are definite advantages for a hobbyist, who rarely needs measurement accuracy better that 2%. Replacing an inexpensive analog scope also involves far less pain if it is accidentally 'blown up'.

Using oscilloscopes

All diagnostic tools should be easy to use, and oscilloscopes are really very good. Apart from misreading sweep speed or voltage sensitivity settings, there are only a few areas where one can go astray.

Frequency response

Measuring very high frequencies with a low bandwidth scope is not possible. This is why a 200MHz instrument costs more than a 20MHz scope. If a 100MHz signal is monitored with a 20MHz scope, all frequencies above 20MHz will be attenuated, introducing amplitude errors in the displayed signal. For example, if signal attenuation is -3dB at 25MHz a real signal will appear with only 70% of its true amplitude, and this error increases with frequency.

Input signal coupling (ac and dc)

AC coupling is useful when small ac fluctuations sitting on top of a much larger dc component, must be examined in detail. For example, it is difficult to make an accurate analysis of the 200mV sine wave modulating a 1 volt dc signal in (a) of Figure 1-10, because increased signal amplification will send the trace off screen.

However, this dc component can be removed by ac coupling, allowing the signal to be amplified and displayed as shown in Figure 1-10 (b).

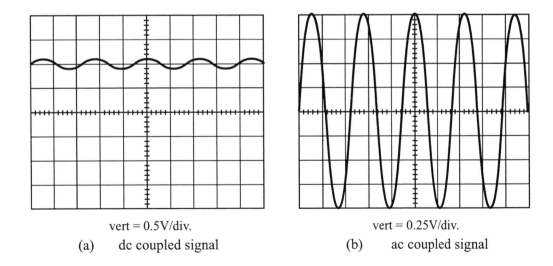

vert = 0.5V/div.	vert = 0.25V/div.
(a) dc coupled signal	(b) ac coupled signal

Figure 1-10.   Small ac components of a large dc signal in (a) are easier to amplify for closer examination if they are ac coupled on an oscilloscope as shown in (b).

Oscilloscope probe sensitivity

Many oscilloscope probes have switchable sensitivity settings and the following information may be helpful for first time users.

[a] A x1 (times one, or 1:1) probe position, means an oscilloscope will display the unattenuated input signal amplitude as indicated by the oscilloscope's vertical sensitivity setting (volts/div). Input impedance is ~ 10MΩ shunted by 20pF, when using a probe on its x1 switch position.

[b] Using a x10 or 10:1 setting attenuates an input signal by a factor of ten. Consequently, a 5V signal will be displayed as 0.5V. Input impedance is reduced to 1MΩ when a probe is used at its x10 position.

[c] Probes with no switchable sensitivity option are by default x10 probes. Therefore signals will appear at only one tenth their true value on an oscilloscope screen, when a x10 probe is used.

Probe compensation

When measuring signals with an oscilloscope, maximum frequency response is obtained using a x10 probe setting. If a x10 probe is used, it must be trimmed by the user to give optimum performance, and most scopes have a front panel test point for this purpose. A 1kHz square wave is normally provided at these test points, and probes are adjusted to give a trace similar to (a) in Figure 1-11.

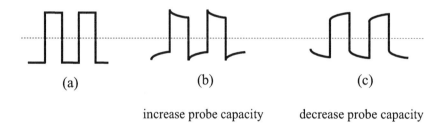

(a)   (b)   (c)

increase probe capacity   decrease probe capacity

Figure 1-11.   A test waveform from an oscilloscope's test point shown in (a) is usually a 1kHz square wave. Optimum high frequency response is achieved when a probe's x10 setting is used. Adjusting a probe's compensation trimming capacitor (by turning a small screw head on the probe's body), results in waveforms (b) and (c). Note that compensation adjustment is ineffective for a probe on its x1 setting.

Curing some common signal distortions

Sometimes during normal circuit testing, a square wave signal may be appear distorted as in Figure 1-11b,c. If the scope probe has already been properly adjusted then distortion may be corrected as follows.

[1] A circuit signal similar to (b) in Figure 1-11, may suggest a larger external coupling capacitor is required. If no external capacitor is being used, then the scope must be dc coupled to reduce signal distortion.

[2] Integrated waveforms as in (c), are often produced when a circuit signal is monitored through a large external resistor. A poor contact or high impedance to the probe produces similar results.

Measuring signals without a probe

Ordinary wire can be used for monitoring large amplitude low frequency signals on an oscilloscope, but more care is necessary when measuring fast rise waveforms and low level signals. Twisted pairs of wires are better than loose wires or parallel flex cable, but coaxial cable is even better. More information on this topic is provided in Chapter 7, and typical signal distortions are shown in Figure 7-6.

Signal reflections in an unterminated cable occur because electrons travel at about 0.7c (70% of the speed of light), producing pretty but undesirable 'ringing' on fast rise/fall waveforms. A 10-foot length of coax cable can produce a delayed echo easily seen on a fast oscilloscope.

Signal retardation in a long cable has its uses too. Pulse delay lines are sometimes made by sending signals along a cable of known length, producing a predictable transit time.

Cable terminations

Terminating a cable by its characteristic impedance allows current to travel down the line, then dump all its energy into a resistor, eliminating reflections. Series or parallel load resistors are used but the latter is more common and easier to install, because a resistor can be soldered to a BNC 'Tee' connector at the scope's input. Some oscilloscopes have a 50Ω resistor that can be switched on demand. However, not all coax cables have 50Ω impedances, so this is not a 'universal' termination. A good 50Ω coax cable for hobby use is RG174U. This cable is only 0.1" diameter and very flexible. It will not pull protoboards off the bench, unlike many heavier and stiffer cables.

Curing oscillations caused by signal sensing cables

Because most coaxial cables have a distributed capacity of ~30pf /foot, a coax line can easily resonate with circuit components being monitored. Two common solutions to this problem are described below.

[1] Attach a 100Ω or 1kΩ resistor to the cable's center core conductor (the end remote from the scope), then use this resistor's tip as a probe.

[2] Duplicate the features of a regular scope probe by using a 9MΩ series resistor shunted by a 20pF trimmer capacitor. Put these components in a shielded box at the end of the cable furthest from the scope, then adjust the capacitor for optimum waveshape with a 1kHz square wave.

Oscilloscope and probe suppliers

C & S Sales carry Hitachi digital storage oscilloscopes from $1895 to $5995 and Hitachi analog scopes from $425 to $1069.

A two-channel 25MHz analog scope is sold by *C & S Sales* for $325 and an analog storage scope for $1095 (these two items were discussed earlier in this chapter).

60 MHz probes with x1 and x10 sensitivity are available from *Hosfelt* PP-80 $15.95, and a 100MHz PP-150, x10 probe is $29.95.

Electronics Refresher and Quiz

Electronics and mechanics play important roles in robotics. Although one can get by with just a little knowledge in these fields, the operating principles are not difficult to learn. Rough estimates of circuit conditions are generally adequate to see what makes things tick, and most mechanical problems require only high school physics.

It is not necessary to understand the detailed operation of circuits you wish to use in robotics projects, since any circuit in this book may be used simply as a black box and incorporated into your designs. However, learning by using equipment, building and testing circuits, then applying your efforts to real life problems can be very satisfying. As you gain confidence and understanding, you will be able to modify circuits to suit your specific requirements.

This book assumes a basic knowledge of electronics, but if you are apprehensive or have no previous electronics experience, the following user friendly books are recommended.

Getting Started in Electronics Forrest Mims III *Radio Shack* 276-5003 $4.99

Engineer's notebooks: (*Timer ICs, Op amps, Optoelectronics, Basic semiconductor circuits, Digital logic circuits, Communications projects, Formulas tables and basic circuits, Schematic symbols*). Forrest Mims III *Radio Shack* 266-5010, 5011, 5012, 5013, 5014, 5015, 5016, 5017 $1.99 each

A first year university physics text is an excellent source of information on electricity and magnetism. These texts explain capacitance, inductance, reactance, resonance, mechanics, linear and rotary motions, and other robotics related topics.

You can test your current knowledge on the following problems covering some typical difficulties experienced by students starting in electronics. Some questions look disarmingly simple, but contain subtleties that may provoke "oh yeah I knew that!", after reading the answer.

Answers are given after the last question

Question 1

(a) What are the approximate currents i_1 and i_2 in *Figure 2-1* ?
(b) What is the total current drawn from the battery?
(c) Estimate the power rating of each resistor.
(d) Guess the size of each resistor, (inches or mm).

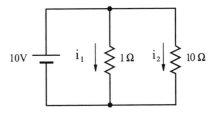

Figure 2-1. (*Question 1*)

Question 2

(a) What voltage will a 3½ digit DVM show when connected across AB of *Figure 2-2* ?
(b) Give an estimate for i_1
(c) Estimate i_2

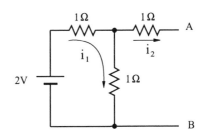

Figure 2-2. (*Question 2*)

Question 3

An inexpensive 3½ digit DVM is used to measure a divider voltage in *Figure 2-3*, what is V_o?

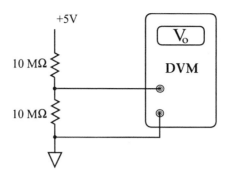

Figure 2-3. *(Question 3)*

Question 4

The tabled instruments are used to measure V_o in *Question 3*. Can you fill in the last column of *Table 2-1?*

Table 2-1. *(Question 4)*

test instrument	input impedance	Vo (measured)
expensive DVM	100,000MΩ	?
inexpensive DVM	10MΩ	?
oscilloscope	1MΩ 30pF	?
analog meter*	20kΩ/volt	?

* input impedance = 100kΩ if a full scale reading of 5V is assumed.

Question 5

How would you measure the input impedance of a DVM?

Question 6

A DVM can sometimes influence the operation of a circuit it is monitoring. Do you know how?

Question 7

If wipers in (a) and (b) of *Figure 2-4* are positioned as shown and then connected to ground, what are the consequences? Both potentiometers are rated at 0.5W and have linear resistive elements.

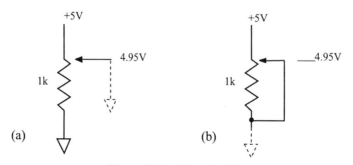

Figure 2-4. *(Question 7)*

Question 8

Potentiometers are frequently used with extra resistors or a capacitor, (a) (b) (c) of *Figure 2-5.*
A potentiometer is still the only variable element in each example, so are there any advantages in adding these additional components?

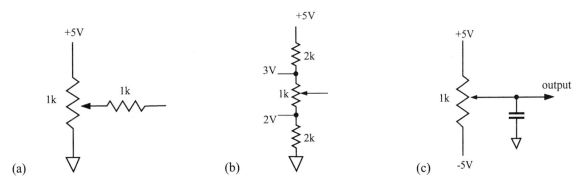

(a)   (b)   (c)

Figure 2-5. *(Question 8)*

Question 9

Not all resistive devices used in robotics are shown in *Figure 2-6.* Can you name some others?

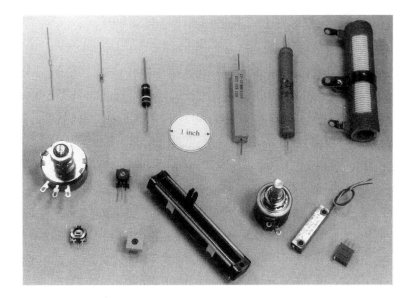

Figure 2-6. *(Question 9)* Resistors in top row from left: ⅛W ¼W 2W 10W 20W, the last item is a 50W rheostat. Bottom row (potentiometers): Potentiometers come in many shapes and sizes. A slide resistor (bottom center) and a multiturn rotary potentiometer to its right, are often used for sensing and controlling mechanical robot movements

Question 10

Why is it extremely difficult to electrocute yourself by grabbing hold of the terminals on a 12V auto battery with both hands?

Question 11

True or false? 115V ac is used in North America because 230V ac is much more dangerous.

Question 12

True or false? Getting 230V from 115V household electrical outlets in the US is not possible?

Question 13

True or false? One of the safest places to be during a lightning storm is inside an automobile. If true, why?

Question 14

Is there any connection between *Question 13* and shielding circuits from noise in robotic electronics?

Question 15

Size isn't everything! Which item in *Figure 2-7* has the largest capacitance?

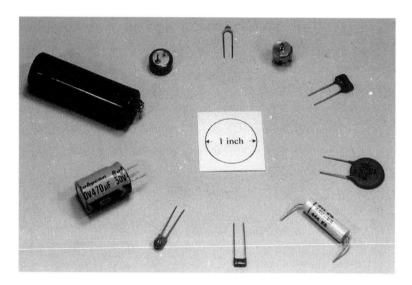

Figure 2-7. *(Question 15)* Clockwise from top: 10pF 100V ceramic plate, 3 to 80pF mica trimmer, 43pF 300V silver mica (rf), 0.01μF 1kV ceramic, 0.063μF 63V polystyrene, 0.1μF 100V mylar, 0.1μF 16V tantalum, 470μF 50V electrolytic, 0.5μF 400V high voltage, 100,000μF 5V supercap.

Question 16

Sketch the approximate output waveforms V_o for both circuits in *Figure 2-8.*
Do you know what these circuits are called?
Can you name some uses for these circuits.

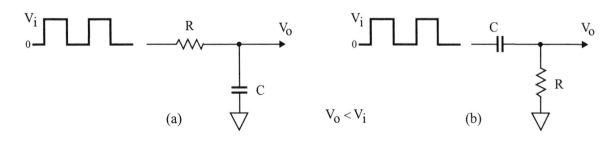

Figure 2-8. *(Question 16)*

Question 17

Your friend gives you a black box, too small to hold a battery and with only three wires. A noisy signal is magically smoothed as shown in *Figure 2-9*. What's inside the box?

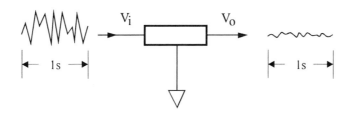

Figure 2-9. *(Question 17)*

Question 18

(a) A bubble detector design *Figure 2-10 (a)*, is proposed for measuring the presence of bubbles in a water flow. It is supposed to work because air bubbles reduce capacitance between two separated copper plates that cup the glass tube.
Would you invest in this concept?

(b) A friend says he will build a similar device with copper pipe outside the glass tube, *Figure 2-10 (b)*
Is this a better idea?

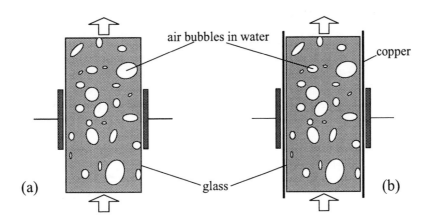

Figure 2-10. *(Question 18)*

Question 19

A charged capacitor looks harmless but can pack a potent wallop. If a 5lb sledgehammer falls from knee height onto your toe, the energy transferred to your foot is about 10J (joule). The same energy, applied to the heart by touching high voltage terminals of an equivalently charged capacitor, does far more damage.

(a) How many microfarads are required to store 10J at 1kV?
(b) Guess the size of a 100μF 450V capacitor, (inches or mm).
(c) Your friend designs a robotic beer can lifter to prevent arm strain when the World Series is on TV. He will use energy stored in a 100μF capacitor charged to 100V to lift a full 12oz can.

Is this a viable plan?

$$E = mgh \quad (m = 0.35kg, \quad g = 9.8 \text{ ms}^{-2}, \quad 1 \text{ Joule} = 1 \text{ Newton meter})$$

Question 20

Two problems illustrated in *Figures 2-11, 2-12, 2-13* are at least 40 years old but still raise eyebrows.

A 100µF capacitor in *Figure 2-11* is charged to 10V
then connected to a similar uncharged capacitor.
Initial energy stored in the first capacitor is given in *Eqn. 2.1*.

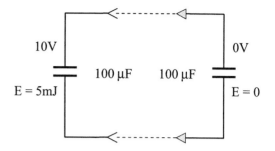

$$E = \frac{1}{2}CV^2 = 0.5 \; x \; 100 \; x 10^{-6} \; x \; 10^2 = 0.005 J \quad (2.1)$$

Figure 2-11. *(Question 20)*

Capacitors add in parallel and charge is also
conserved, so we now have a 200µF capacitor
charged to 5V as shown in *Figure 2-12*.

Energy stored in the 200µF capacitor is shown in *Eqn. 2.2*.

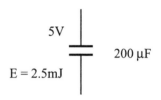

$$E = \frac{1}{2}CV^2 = 0.5 \; x \; 200 \; x 10^{-6} \; x \; 5^2 = 0.0025 J \quad (2.2)$$

Figure 2-12. *(Question 20)*

2.5mJ of energy has disappeared, where has it gone? This is not a mathematical trick. Get out your DVM, a battery, a couple of 100µF electrolytics and convince yourself!

A mechanical version of the same effect using potential energy stored in water tanks, and with similar resultant energy loss is illustrated in *Figure 2-13*.

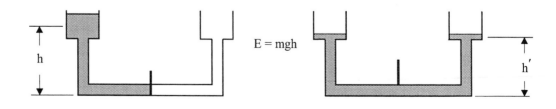

Figure 2-13. *(Question 20)*

Question 21

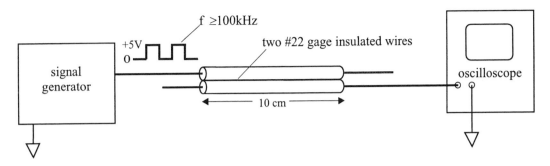

Figure 2-14. *(Question 21)*

Two 4 inch lengths of #22 gage insulated wire are run close together as shown in *Figure 2-14.* The conductors *DO NOT* make direct electrical contact with each other.

(a) If a 5V square wave ($\geq$100kHz) is injected into one wire what would you expect to see on the scope?
(b) Will the scope signal change if the input frequency is reduced to 1kHz?
(c) Can you explain what is happening? (Assume the scope has 1MΩ and 20pF input).
(d) Has this any practical application?

Question 22

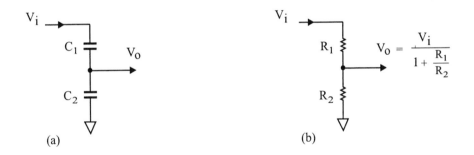

Figure 2-15. *(Question 22)*

(a) Does (a) in *Figure 2-15* depict the ac equivalent of a dc resistive divider?
(b) A resistive divider formed by R_1 and R_2 in (b) of *Figure 2-15* also gives proportional outputs for ac or dc input signals, so why use capacitors?

Question 23

(a) Which types of capacitors are polarized?
(b) What happens if a polarized capacitor is connected to the wrong polarities?

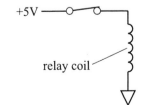

Question 24

Would you use a computer (with some auxiliary circuitry)
to open the switch in *Figure 2-16* ?

Figure 2-16. *(Question 24)*

Question 25

A large turns ratio transformer is depicted in *Figure 2-17*

(a) What is V_{AB} with the switch in a closed position?
(b) What is V_{AB} immediately after the switch is opened?
(c) Is the capacitor necessary?
(d) Where is this type of circuit extensively used?

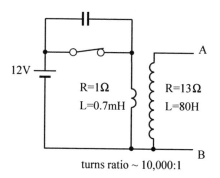

Figure 2-17. *(Question 25)*

Question 26

(a) If the 1kΩ resistor is replaced by 100kΩ what is the approximate voltage drop across each diode of *Figure 2-18?*

(b) If the 1kΩ resistor is replaced by 10Ω what is the approximate voltage drop across each diode of *Figure 2-18?*

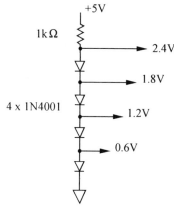

Figure 2-18. *(Question 26)*

Question 27

10kHz signals are applied to the circuits in *Figure 2-19* and outputs monitored with a 10MΩ oscilloscope probe. If 1N4148 signal diodes are used, can you sketch the output waveforms V_o and show any change in these levels due to diode voltage drops?

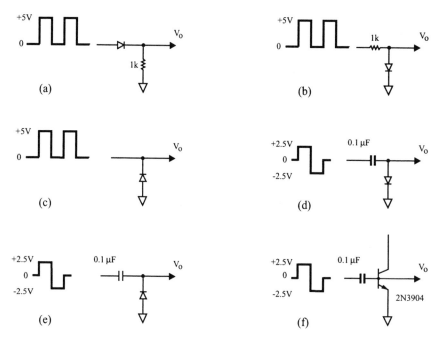

Figure 2-19. *(Question 27)*

Question 28

When you plug your bedside lamp into an extension cord borrowed from a friend, you notice the lamp is quite dim. He made the cord himself and likes playing tricks. From a cursory examination, end connectors and cord appear normal. Neither cord nor connectors get hot.
What has he done?

Question 29

(a) Is the lamp ON or OFF in *Figure 2-20*?
(b) What happens if the transistor's base is connected directly to +5V without going through the 270Ω?

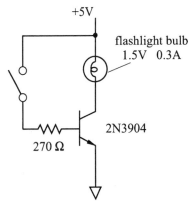

Figure 2-20. *(Question 29)*

Question 30

Is the lamp ON or OFF in *Figure 2-21?*

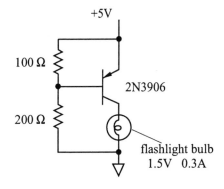

Figure 2-21. *(Question 30)*

Question 31

V_o rises in circuit (a) of *Figure 2-22* when the photodiode is illuminated, but falls in (b). Why is this so?

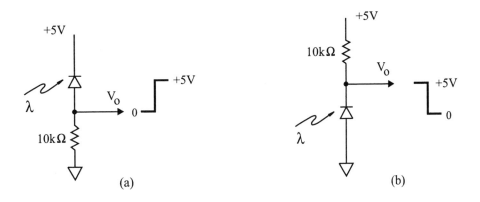

Figure 2-22. *(Question 31)*

Question 32

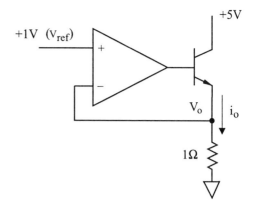

(a) What is the magnitude of i_o in *Figure 2-23?*
(b) If V_{ref} is increased to +2V what is i_o ?

Figure 2-23. *(Question 32)*

Question 33

Inductors pass dc but resist sudden current changes. Capacitors pass ac but block dc. Both circuits shown in *Figure 2-24* act as frequency filters. Can you tell which type they are?

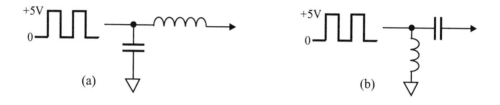

Figure 2-24. *(Question 33)*

Question 34

What are the differences between a regular transformer, an autotransformer or Variac® and a dimmer switch?

Question 35

How would you make a soft iron electromagnet with two south poles as indicated in *Figure 2-25?*

Figure 2-25. (*Question 35)*

Question 36

(a) Can you sketch the magnetic field lines between
north and south poles in (a) of *Figure 2-26?*

(b) Can you sketch the magnetic field lines between
poles and inside the soft iron slug in (b) of *Figure 2-26?*

(c) Is a magnetic compass reading reliable when taken
inside a building constructed with steel girders?

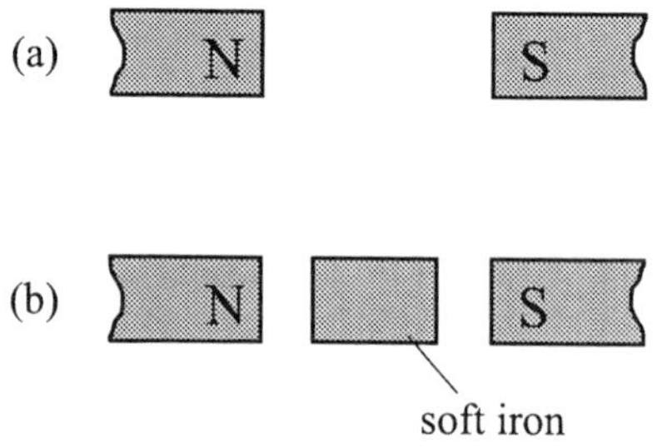

soft iron

Figure 2-26. *(Question 36)*

Question 37

Current flowing through the coil of *Figure 2-27* creates a magnetic field.

(a) Energy is stored by the electric field in a capacitor. Does an inductor also store energy in its magnetic field?

(b) What happens to the magnetic field when the switch is opened?

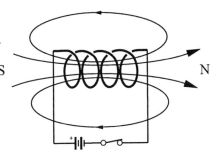

Figure 2-27. *(Question 37)*

Question 38

For safety reasons every roboticist should know outlet details and wire colors for a 115V ac receptacle, *Figure 2-28.*

(a) Can you state the function of each pin socket?
(b) What are the wire colors for each pin?

Figure 2-28. *(Question 38)*

Question 39

Does the oscilloscope trace in Figure 2-29 depict the true signal when the switch is closed?

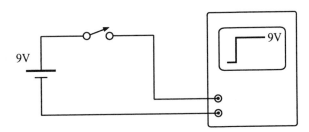

Figure 2-29. *(Question 39)*

Question 40

(a) It is 68°F today and the forecast calls for a temperature twice as hot tomorrow, what will tomorrow's temperature be? (68°F=20°C).

(b) Does this question have any connection to practical electronics?

Answer 1

(a) $i_1 = 10$ amps, $i_2 = 1$ amp
(b) A guess of 10amps is close enough.
     True current is 11amps (usually written as 11A).
(c) $1\Omega \equiv 100W$ and $10\Omega \equiv 10W$, from *Eqn. 2.3.*
(d) 100W (6"x 1"dia),   10W (2" x 0.25" dia).

$$(2.3) \qquad Power \;=\; P = \frac{V^2}{R}$$

See *Figure 2-6* in *Question 9* for examples of 10W and 50W resistors.

Answer 2

(a) $V_{AB} = 1$volt,   see (c) below for the reason
(b) $i_1 \approx 1A$
(c) $i_2 \approx 0$ (An inexpensive 3½ digit DVM has an input impedance of 10MΩ so i_2 is less than 1µA)

Answer 3

A DVM in *Figure 2-30* will not show the anticipated divider
voltage of 2.5 volts. It will instead display **1.667V**

The divider is 'loaded' when a meter is attached to
the circuit because a small portion of the divider current
can now flow through the 10MΩ meter resistance.
A smaller divider current consequently produces a
smaller voltage for measurement by the meter.

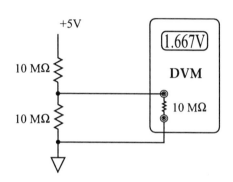

Figure 2-30 *(Answer 3)*

Answer 4

Table 2-2. *(Answer 4)*

test instrument	input impedance	Vo (measured)
expensive DVM	100,000MΩ	2.500V
inexpensive DVM	10MΩ	1.667V
oscilloscope	1MΩ 30pF	0.417V
analog meter	20kΩ/volt	0.049V*

* Analog meter needle barely moves for this 1/100 of full scale deflection

An analog meter requires an appreciable current to deflect the needle and a 10MΩ divider only provides sufficient
current for 0.01 full scale deflection. Low resistance dividers are called 'stiff' because their midpoint voltage is minimally
affected by loading.

Answer 5

(i)   Adjust R_L in *Figure 2-31* until DVM[†] indicates half
the no-load battery voltage (ie.V/2). Battery must be
disconnected from other circuit elements to measure
this no-load voltage.

(ii) Measure R_L, this is equal to the DVM's input impedance.
† Method also works for an oscilloscope, VOM etc.

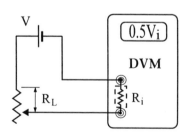

Figure 2-31. *(Answer 5)*

Answer 6

DVM's have internal switching circuitry that can cause op amps and other devices to break into oscillation. If you are sure everything in your own circuit is ok and yet still have a weird DVM reading, try an oscilloscope or a VOM instead and see if the problem disappears. Analog to digital converters used in DMM's employ fast pulses for timing and these are easily coupled to other devices, (see *Question 21*).

Answer 7

Grounding wipers in both (a) and (b) of *Figure 2-4* will burn out top portions of the resistive elements. Both potentiometers have a 0.5W rating and this specification applies to dissipation over the whole resistive element. Remaining intact portions of each potentiometer can still be used as variable resistors.

An output voltage of 4.95V means that only 10Ω of resistive element isolates the wiper from +5V. If the wiper is taken to ground, 2.5 watts will be dissipated in that 10Ω segment, as shown in *Eqn. 2.4*. This error is common for beginners and the term 'toast' means more than bread in a classroom setting

Measuring a potentiometer over its entire range with a DVM is safe since the test current will not exceed 100µA.

$$(2.4) \quad Power \ = \ \frac{V^2}{R} \ = \ \frac{5^2}{10} \ = \ 2.5W$$

Answer 8

(a) and (b)   Excessive power dissipation described in *Question 7* can be avoided by adding 'current limiting resistors'. These extra resistors restrict wiper current to a safe level. For safety, make limiting resistors the same value or larger than the potentiometer resistance.

(b) Fixed 2kΩ resistors provide protection for the potentiometer and in this example also give finer wiper control from 2V to 3V.

(c) A 0.1µF capacitor reduces wiper noise when monitoring robotic motion. It is good practice to use a capacitor in this way whenever voltage jitter can cause problems. Use a scope to pick an optimum capacitor value for difficult conditions

Answer 9

Non-linear potentiometers are available, these include log, sine, cosine and special tapers. Log potentiometers are widely used for audio applications, because human hearing (like many other human responses), is approximately logarithmic. Linear potentiometers are most commonly used for robotic sensing and circuit design.
    Continuous-turn potentiometers are sometimes handy. They have a dead band of 10 to 20 degrees on each revolution to provide clearance for contact wiper width. Rotary rheostats can be used for monitoring longitudinal and rotary motion as discussed in Chapter 5.

Answer 10

Safe let-go currents are 9mA for males and 6mA for females. Currents exceeding these levels can promote fibrillation, respiratory inhibition or death. Body resistance through major blood vessels (eg hand to hand) is about 200Ω, because blood behaves electrically like salty water. Hand to hand resistance depends on gripping force and whether hands are moist. Results of tests conducted using two 6" lengths of clean copper pipe connected to an ohmmeter are given in *Table 2-3*. Maximum gripping force was used in each case.

Table 2-3. Hand to hand resistance on ½" copper pipe   *(Answer 10)*

test conditions	resistance (hand to hand)
dry hands	≈10kΩ
wet hands	≈3.5kΩ
wet (soapy)	≈2.5kΩ
wet (detergent)	≈1.5kΩ
wet (salty) *	≈1kΩ

* 1tbsp/1 liter water

Ohm's law shows that worst case conditions will only drive 12mA through the body even using large grip areas. We may conclude that grabbing skimpy 12 volt battery posts is safe.

Answer 11

Because electrocution is a current driven effect 230V is just twice as bad as 115V. Many countries use 230V for household outlets since it is more efficient for power distribution and only marginally more hazardous.

Answer 12

Most North American residential electrical service is based on two phase 230V. If you measure the potential between various sockets (hot to hot) you will find some combinations that have 230V between them.

   For safety, always keep plug polarization the same when summing power outputs from more than one outlet, or you may be in for a big surprise.

Answer 13

True, unless you own a Corvette or other auto with an electrically insulating body. Tires do not keep one safe - it is the metal shell. Electrical current always takes the easiest path, it travels through the surrounding metal and then jumps to ground with a big flash, leaving occupants unscathed.

Answer 14

In *Question 13* the metal car body acts as a Faraday cage. In fact every metal box is a Faraday cage and even metal screen or mesh is employed at higher frequencies, (check the perforated screen on your microwave oven door). You may find a Faraday cage helpful if you have a noisy circuit and suspect radiative pick-up.

The quickest way to see if your circuit will benefit from a Faraday cage?

[i]   Slip your circuit into a non conducting plastic bag.
[ii]   Cover the bag with aluminum foil.
[iii]   Clip a grounded lead to the aluminum foil.
[iv]   Cross your fingers

Answer 15

Supercap at 11 o'clock in *Figure 2-7* is the winner by a comfortable margin as shown by information in the photo's caption.

Answer 16

(a) V_o rises slowly in *Figure 2-32* because R limits current supplied to the capacitor ,and V_o falls when V_i changes direction. This circuit conducts dc (through the resistor) but shunts ac to ground (through the capacitor).

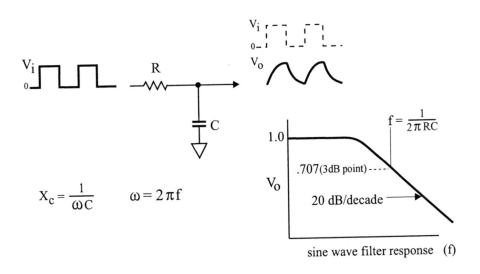

Figure 2-32. passive RC integrator (low pass filter) *(Answer 16)*

(b) V_o in *Figure 2-33* is also governed by changes in the direction of V_i as described in *Answer (a)* above.

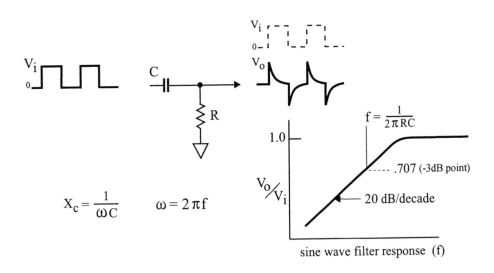

Figure 2-33. passive RC differentiator (high pass filter) *(Answer 16)*

Both circuits are useful as frequency filters and used for power-up resets, soft turn-ons and slow turn-offs. Differentiated outputs are great for pulse production, one-shots etc. In fact RC circuits have dozens of applications, but are often surrounded by so many other components we take their presence for granted.

Answer 17

The box contains an integrator or low pass filter shown in *Figure 2-34*.
If you have a noisy signal try an RC filter - it takes only a few minutes to see if it improves a signal.

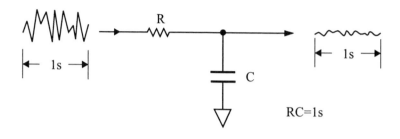

Figure 2-34. *(Answer 17)*

Answer 18

(a) The technique has been employed and works well because capacity between plates is proportional to dielectric volume. Water has one of the highest dielectric constants of any material. At room temperature κ_{water} = 80 and κ_{air} = 1.0 so detecting bubbles is easy. In practice, the bubbly water capacitor forms part of a tank circuit whose resonant frequency scales with the percentage of bubbles.

(b) Your friend didn't read *Question 13*. A conducting pipe acts as a Faraday cage precluding any electronic sensing of its interior.

Answer 19

(a)

$$(2.5) \quad C = \frac{2\,E}{V^2} = \frac{2\,x\,10}{10^6} = 20\,\mu F$$

(b) A 100μF 450volt capacitor is about 15cc, or 1" x 1" diameter.

(c) Calculations in *Eqns 2-6 and 2-7* show the can will only rise 6" so your friend should have very short arms or be prone!

Initial conditions: m = 0.35kg   g = 9.8 ms^{-2} 100 μF charged to 100V

$$(2.6) \quad E = \frac{1}{2}\,C\,V^2 = 0.5\,x\,100\,x\,10^{-6}\,x\,10^4 = 0.5\,J$$

$$(2.7) \quad E = 0.5\,J = mgh \therefore h = \frac{0.5}{0.35\,x\,9.81} = 0.146\,m = 5.7\,^{\prime\prime}$$

Using a capacitor has some practical limitations in this proposed application:

(1) Capacitors deliver an exponentially decaying current into a load.
(2) No energy transfer is 100% efficient, he may be lucky to realize 20% with this system.
(3) The source impedance for a 100μF capacitor may be too high to drive a motor efficiently.
(4) Motive drivers (DC motors etc.), prefer a steady voltage for optimum performance

If he is intent on lifting that bottle, a battery or power supply are better choices than a capacitor.

Answer 20

All capacitors have internal inductance and resistance, and when two capacitors are joined the combination of capacitance and inductance forms a resonant tank circuit. A tank circuit is the basic element in almost every radio transmitter, so initial stored energy is converted to electromagnetic radiation. Some power is also resistively dissipated.

In the second example it is impossible to transfer water from one tank to the other without turbulence or oscillations, even though these may be on a micro scale level. Frictional losses due to viscous forces from fluid/fluid and fluid/vessel interactions, account for energy loss.

Answer 21

(a) A good square wave with amplitude $\sim \pm 75$ mV
(b) Yes. It will now be a differentiated waveform with peaks about ± 150 mV
(c) The two wires form a 4pF capacitor that allows ac signals to pass from wire to wire.
(d) The effect has practical use, because small capacitors are easily made by twisting two insulated wires together. Some experimenters use these 'gimmick' capacitors as trimmers, and their capacity is quickly changed by snipping them to a suitable length.

Be alert for stray pickup whenever high frequency signals are around. In this example fast rising and falling edges of the square wave contain frequency components in the MHz region, even if the pulse repetition rate is low.

A good square wave may have rise and fall times <10ns corresponding to frequency components of 100MHz. We can calculate the capacitive reactance X_C (the impedance or 'resistance') between wires, *Eqn. 2.8*.

$$(2.8) \quad X_C = \frac{1}{\omega C} = \frac{1}{2\pi f C} = \frac{1}{2\pi \times 10^8 \times 4 \times 10^{-12}} \approx 400\,\Omega$$

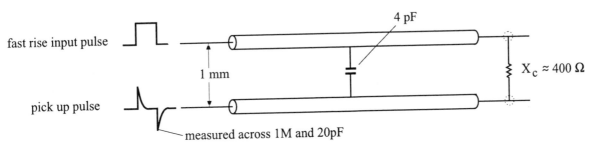

Figure 2-35. *(Answer 21)*

Answer 22

(a) Yes. *Figure 2-36* shows how this division takes place.

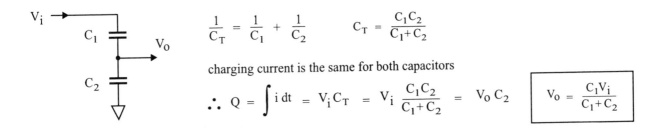

$$\frac{1}{C_T} = \frac{1}{C_1} + \frac{1}{C_2} \qquad C_T = \frac{C_1 C_2}{C_1 + C_2}$$

charging current is the same for both capacitors

$$\therefore\ Q = \int i\,dt = V_i C_T = V_i \frac{C_1 C_2}{C_1 + C_2} = V_0 C_2 \qquad \boxed{V_0 = \frac{C_1 V_i}{C_1 + C_2}}$$

voltage divider for high voltage AC measurements

(b) Capacitive dividers are commonly used for high voltage ac measurements. Because of their low inductance, some high voltage high frequency capacitors present a smaller load to the circuit being measured, than their resistive counterparts.

Answer 23

(a) Electrolytics, tantalum, supercaps and many oil filled high voltage capacitors are polarized.

(b) An incorrectly installed polarized capacitor can burst with explosive force. Even a small back biased electrolytic running at 12V startles a room full of students into temporary silence when it explodes. Imagine the consequences if you are examining your circuit closely when this happens.

Answer 24

Only if you want to buy a new computer! During turn off, the collapsing magnetic field from a 5V relay coil may generate a voltage spike of several hundred volts. This voltage is lethal to most solid state components. Power supplies can also be damaged by merely switching off power from inductive devices.

All switched inductive loads must have a reversed diode clamp or decoupling capacitors. An appropriate voltage clamp for a small relay is shown in *Figure 2-37*. Decoupling capacitors are commonly employed for shunting voltage spikes on brush motors.

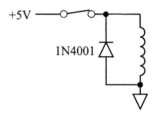

Figure 2-37. *(Answer 24)*

Answer 25

(a) $V_{AB} = 0$ when switch is closed
(b) $V_{AB} \approx 100\text{kV}$ momentarily (under open circuit conditions).
(c) No, but it reduces switch contact arcing by soaking up back EMF energy transferred to the primary coil. A capacitor also shortens break time giving a higher rate of current change.

Fast current decay and large inductance L produce higher secondary voltages as indicated by *Eqn. 2.9.*

$$(2.9) \quad \boxed{V_{AB} = L \frac{di}{dt}}$$

(d) The transformer shown in Figure 2-17 is an automobile ignition coil used on most gasoline powered engines.

Answer 26

Diode drop increases with current through the diode. Measured voltage drops across one 1N4001 for the circuit shown in the question are listed in *Table 2-4.*

Table 2-4. *(Answer 26)*

1MΩ	100kΩ	1kΩ	100Ω	10Ω
0.36V	0.44V	0.63V	0.72V	0.8V

Answer 27

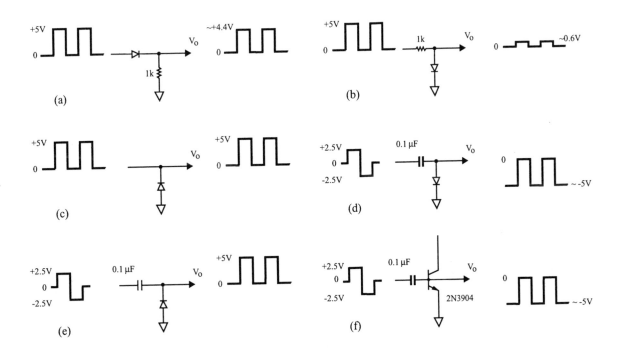

Figure 2-38. *(Answer 27)*

Small output offsets are present in many circuits shown in *Figure 2-38*. These offsets depend on output impedance of the signal source and circuit loading. Although (d) and (e) are widely used in high frequency and logic circuits as dc restorers, they have limited use in robotics. Your projects will usually require exact dc restoration, and a summing amplifier is a better choice to correct offsets and biases.

In (f) the base emitter junction acts as a diode (which it is), and the circuit is electronically equivalent to (e).

Answer 28

He has inserted a 1N4001 diode in one line and concealed it in the plug. A simple modification as shown in *Figure 2-39*. Your lamp only gets half its normal power.

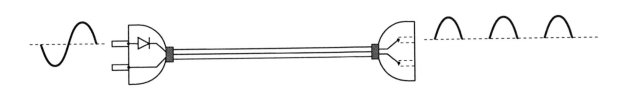

Figure 2-39. *(Answer 28)*

Answer 29

(a) The lamp is OFF. Close the switch and lamp operates at 0.3A.
   A positive base voltage must be applied to turn on an NPN transistor.

(b) Lamp lights briefly then burns out.

A 270Ω resistor shown in the circuit limits base drive current. With limited base drive, the transistor is only partially on and lamp current is restricted to 300mA. Lamp current will rise to 0.6A if full base drive is applied by connecting it directly to +5V.

Cold lamp resistance is 0.66Ω and increases to 3.3Ω when lamp operates at 0.3A. This positive temperature coefficient is a characteristic of all incandescent tungsten filament lamps. It is also the reason that old household lamps near the end of their life sometimes fail with a bright flash when turned on.

Answer 30

Lamp is ON because a PNP transistor turns ON when its base is more than ≈ 0.6V below the emitter voltage.

Answer 31

(a) In low light conditions photodiode (PD) resistance might be 1MΩ, as shown in *Figure 2-40 (a)* and V_0 will be close to ground potential. V_0 is calculated in *Eqn. 2.10* using values from *Figure 2-40 (a)*.

$$(2.10) \quad \boxed{V_O = IR = 5\mu A \times 10k\Omega = 0.05\,V}$$

As the light level increases, photons cause a release of electrons in the PD junction. These electrons increase conduction, reducing the PD's resistance and V_0 rises towards +5V as shown in *Figure 2-40 (b)*. Similar reasoning applies to *Question 31 (b)*.

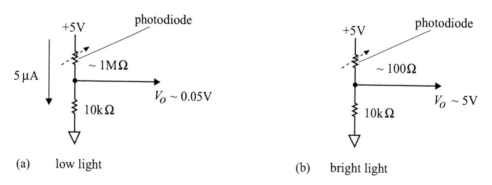

Figure 2-40. *(Answer 31)*

Answer 32

(a) Feedback keeps both + and - op amp inputs at the same level (1volt), *Eqn. 2.11*.

$$(2.11) \quad \boxed{V_0 = 1\,volt \qquad therefore \qquad i_0 = 1\,amp}$$

(b) This circuit is a constant current source and i_0 scales directly with V_{ref} as shown in *Eqn. 2.12*.

$$(2.12) \quad \boxed{V_0 = V_{ref} = 2\,volts \qquad therefore \qquad i_0 = 2\,amps}$$

Answer 33

(a) High frequencies are shorted to ground by the capacitor in *Figure 2-41*. Low frequencies are passed by the inductor. This is a low pass filter or integrator.

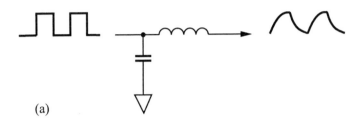

(a)

Figure 2-41. *(Answer 33)*

(b) Low frequencies are shorted to ground by the inductor and high frequencies passed by the capacitor in *Figure 2-42*. This is a high pass filter or differentiator.

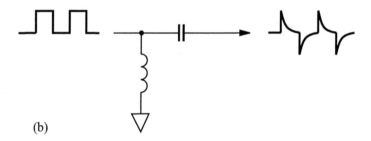

(b)

Figure 2-42 *(Answer 33)*

Answer 34

Secondary windings on a regular transformer are electrically isolated from the primary, but this is not the case for autotransformers, Variacs or solid state dimmer switches.

If a Variac® is supplied with a three-prong line cord do not use a cheater plug to operate the transformer from a two-pin socket. This can be extremely hazardous, especially if the Variac is used with apparatus that is improperly grounded.

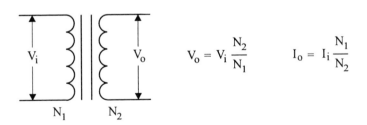

$$V_o = V_i \frac{N_2}{N_1} \qquad I_o = I_i \frac{N_1}{N_2}$$

Figure 2-43. *(Answer 34)*

A regular transformer in *Figure 2-43*, can be operated with primary and secondary functions interchanged. Always check that both primary and secondary windings have adequate voltage and current capacities. Special precautions are necessary when using an autotransformer shown in *Figure 2-44*.

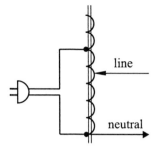

Figure 2-44. variable autotransformer or Variac® *(Answer 34)*

Warning:

Output leads from a variable autotransformer shown in *Figure 2-44*, and Variacs or dimmer switches are *NOT* isolated from the ac power line. An isolation transformer *must* be used in series with such devices if electrical isolation is required. Unshielded leads from any of these devices are personnel hazards.

Solid state dimmer switches allow only part of an ac waveform to go to the load. This creates electronic switching noise that is difficult to eliminate and readily coupled to sensitive circuitry. If such noise is a problem in an application, use a variable transformer instead of a solid state dimmer switch.

Answer 35

Several poles can be produced on a soft iron core using winding protocols illustrated in *Figure 2-45*.

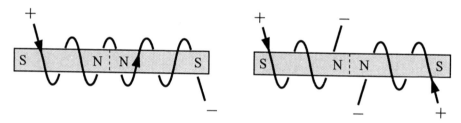

wrap right hand fingers around the iron core following the arrow,
thumb now points to a north pole

Figure 2-45. *(Answer 35)*

Magnetic levitation and maglev train projects require an understanding of magnets and electromagnetism. Coil windings on electromagnets should be placed at pole tips for maximum magnetic pole field strength.

Answer 36

(a)

(b)

soft iron

Figure 2-46. *(Answer 36)*

(c) Magnetic field lines follow the path of least resistance and swerve to enter high permeability materials, as shown by their deviation into soft iron *Figure 2-46 (b)*. Consequently steel girders are easy rides for magnetic fields and distort the earth's magnetic field. Compass readings taken inside a steel structured building should always be viewed with suspicion because of this latter effect. The earth's magnetic field is quite weak, ~ 0.6G (gauss) or 6 x 10^{-5} T (tesla), so residual magnetic fields in steel assemblies are also troublesome for a magnetic compass.

Answer 37

(2.13) $E = \frac{1}{2} L i^2$

(a) Energy is stored in any current carrying coil, *Eqn. 2.13*

(b) The collapsing magnetic field generates a voltage across the coil. This voltage is proportional to coil inductance and current decay rate as indicated by *Eqn. 2.14.*

(2.14) $V = L \dfrac{di}{dt}$

This principle is used to make a Tesla coil or a stun gun producing 100,000V sparks. Install diode clamps on every switched inductor and relay in your circuits, or you may blow up an expensive piece of electronic equipment with a high voltage inductive kick.

Answer 38

Answers to (a) and (b) of *Question 38* are provided in *Figure 2-47*.

Always break the *hot line* going to ac powered equipment when using a SPST power line switch. A switch in the neutral line will also turn off your equipment **but the apparatus will remain hot**.

Neutral is at ground potential but **must never be used** as a signal or power ground. Equipment grounded from the neutral line can be lethal.

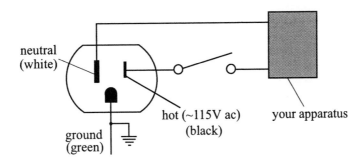

Figure 2-47.   *(Answer 38)*

Answer 39

No.   Because of switch bounce present in all mechanical switches the trace will be more like Figure 2-48.

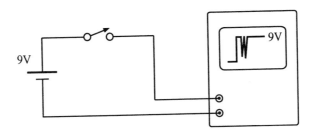

Figure 2-48.   *(Question 39)*

Answer 40

(a) 313.15°C or 595.67°F !!! This is hot enough to melt all soft solders since high temperature solder melts at 300°C/572°F and low temperature solder at 179°C/354°F. Hard or silver solder melts at brazing temperatures but even some of these will melt at 596°C.

The original temperature (20°C) must be converted to degrees Kelvin. ie: $20°C = 293.15K$ because $(T_K = T_C + 273.15)$. You must multiply this value by two then subtract 273.15 to get tomorrow's temperature in Celsius.

(b) Temperature related electronic parameters (and most physical and chemical phenomena), are based on the Kelvin scale. The thermocouple or Seebeck effect, Johnson noise, semiconductor junction effects, thermal radiation, Big Bang cosmic radiation, and a host of other phenomena are based on the absolute temperature scale or 'degrees Kelvin'.

Electronic Components for Hobbyists and the Classroom

Component standardization is important in a classroom since it reduces total part's inventory. Limiting the range of components also permits students to obtain parts from an instructor, or trade with their colleagues in emergencies. Parts recommended in this book have been selected for their performance/cost ratio and availability. Most items are stocked by major electronic stores and can be obtained at low cost by mail from surplus outlets.

Even sophisticated projects can be constructed with items listed in this chapter, and experimentation is always easier if these components are on hand. Hobbyists can limit their purchases to components for specific circuits, or build a wider stock for future convenience.

Some useful books and data manuals are tabled in this chapter and these will be of help to beginners and advanced experimenters. Additional reference material is listed in Appendix B.

Resistors and capacitors

Resistors, potentiometers and rheostats

For most circuits in this book ¼W 5% resistors are adequate, and a kit similar to an RC100 $45 sold by *C&S Sales*, is suitable for classroom use. This multi drawer cabinet kit with 20 resistors of 60 different values ranging from 1Ω to 1MΩ fulfils most needs, but resistors smaller than 1Ω or larger than 1MΩ are sometimes required. Independent hobbyists can conserve funds by initially buying only resistors listed in Table 1-1 Small 5% resistors (¼ and ½W) can be purchased in quantities of 500 for less than a cent each, single item purchases usually triple that price.

Metal film 1% precision resistors have superior temperature characteristics and must be employed where specified to prevent unacceptable signal drifting. Many are manufactured in different values than composition resistors, and all employ an extra body band for resistance denomination.

Power resistors are also needed occasionally, and a selection of five and ten watt resistors in values of 0.1Ω, 1Ω, 5Ω, 10Ω, 100Ω, 1kΩ, will cover most project needs. Situations may arise where a particular power resistor is not on hand. In such cases an equivalent resistor may be assembled, using series or parallel combinations from stocked resistors.

Small ½W multi-turn trimpots (~$1 each) are adequate for most circuits in this book. Buy the style with a top adjuster screw shown in Figure 3-1, these are easier to use on solderless breadboards, and are available from 10Ω to 5MΩ in single-turn to twenty-five turn styles. The most commonly used trimpots are 100Ω, 1k, 10k, 100k, 1MΩ, and these are adequate for a starter kit.

Try to remember that a potentiometer can easily be destroyed by applying just a small voltage across part of its track. For example, if a 1kΩ ½W potentiometer is set to a value of 1Ω, then this section is connected across a 5 volt power supply, a current of 5 amps will be drawn instantaneously. That 1Ω section is required to dissipate 25W and will be vaporized! Questions 7 and 8 in Chapter 2 give more details and some advice on avoiding this very common mistake.

"Shall I use a potentiometer, rheostat or a trimpot?"

For projects, the main difference between these elements is size, power rating and resolution. In most circuits you will find it best to use trimpots, they take up less space on a breadboard and are easier to mount and adjust. Multi-turn trimpots provide finer control and are useful for setting offset trims and reference voltages. Trying to set a voltage precisely with a single turn pot can be frustrating at times.

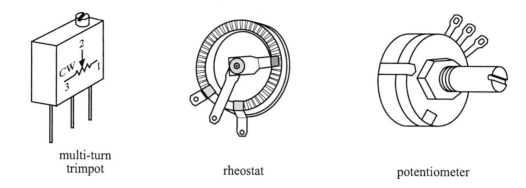

multi-turn
trimpot   rheostat   potentiometer

Figure 3-1. Variable resistors come in a variety of sizes but multi-turn trimpots (10 or 20 turns), are items used most frequently in robotics circuits. Rheostats are very useful for high power applications, but standard potentiometers are used less frequently in robotic circuits because of their larger size.

Rheostats or high power potentiometers are usually required only for test purposes. They are especially useful for checking the output impedance of a battery or power supply, simulating a dynamic load, determining total power drain in a circuit, or measuring high currents. Rotary rheostat values of 1Ω @ 20A, 25Ω @ 5A and 1k @ 1A are those most commonly used for projects, and adequate for an initial purchase because these items are expensive. Figures 2-6 and 3-2 represent just a small fraction of power resistors, potentiometers and trimpots available for roboticists. Many more are listed in electronic catalogs.

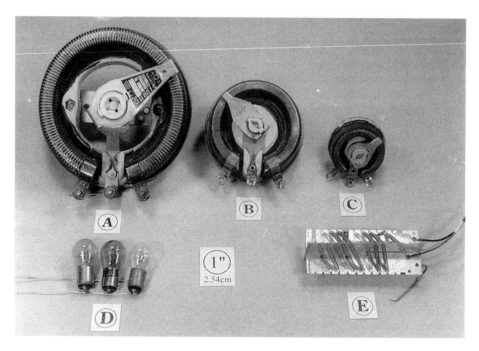

Figure 3-2. Rheostats and power resistors are invaluable for robotics work, and a cache of these parts is always handy. [A] Rheostat 25Ω 3.5amp. [B] Rheostat 4.5kΩ 0.25 amp. [C] Rheostat 6Ω 2.9amp [D] Automobile bulbs make good power resistors. Type 1157 shown here has dual filaments with 0.5Ω and 2Ω cold resistances. Leads can be soldered to these lamps as shown. [E] Hair dryer elements use several gages of heating wire suitable for home made power resistors. Heater wire can be crimped inside small diameter brass hobby tubing, and leads may then be soldered to the tube. More resistive elements are presented in Figure 2-6.

"But I don't have the recommended resistor/capacitor/transistor for the circuit I'm building!"

A difference of 25% in a component's value will not affect the performance of most circuits in this book.

For example, if a 4.7k resistor is recommended but is not available, try 3.9k, 4.3k, 5.6k or 6.8k, any of these will probably be satisfactory. A similar philosophy can be adopted for selecting alternative capacitors and even transistors, but recommended power ratings for components should always be followed to avoid overheating.

Because about 8% of males (<1% females) are color blind, using the wrong resistor in a circuit is not unusual. The problem is exacerbated because some manufacturers use colors that are easily confused, so it is worthwhile checking with a DMM if color coding is not clear. Resistor and capacitor codes are easy to learn, and with a little practice you will be able to pick up a resistor, confidently knowing its value at a glance.

Capacitors

No capacitor kit on the market is targeted toward robot builders' needs. Consequently, commercially available kits are a poor investment because most items will not be used. It is more cost effective and practical to assemble your own kit starting with parts listed below, adding extra items to suit your personal needs.

Ceramic disk	**10pF, 100pF, 0.001µF**	**(Qty 5 of each)**
Ceramic	**0.1µF**	**(Qty 20)**
Electrolytic	**1µF, 10µF, 100µF (50V with radial leads)**	**(Qty 5 of each)**

The most commonly used capacitor for projects is a 0.1µF ceramic. They are used to decouple power supply lines in many circuits, and placed across dc motor leads to limit brush noise. Radial lead capacitors are the best choice for experimenters, and take up less space on solderless breadboards because they can be installed vertically. Most small ceramic capacitors are made with radial leads.

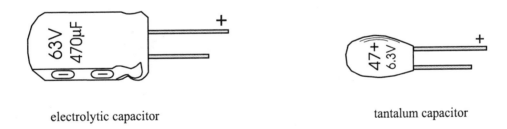

electrolytic capacitor   tantalum capacitor

Figure 3-3. Radial lead capacitors are easiest to use on solderless breadboards and take up less space. Always observe the correct polarity when applying power to a capacitor, or it may explode.

"What caused that explosion?"

Probably an incorrectly installed polarized capacitor! Yes, some capacitors are polarized and must be installed correctly, otherwise they may explode as discussed in Question 23 Chapter 2.

Polarity indicators usually specify the negative pin on electrolytics, and positive pin for tantalum capacitors - these markings are important. On many two-lead semiconductor devices and small capacitors, the positive lead is made slightly longer than the negative lead, as indicated in Figure 3-3.

Always connect the positive lead to the most positive potential in a circuit where the capacitor is installed. For example, if a capacitor is connected between -12V and ground (or common), connect the capacitor's positive lead to ground.

Relays

Even a small relay coil powered from a 5V battery can produce a 100V pulse when the power supply is disconnected, and this principle is used to manufacture 500,000 volt stun guns that can knock down a 250lb human. Questions 24 and 25 in Chapter 2 elaborate on this problem, and stress how cautious we must be when attaching inductive elements to solid state circuitry.

To prevent high voltage spikes from destroying valuable equipment, all relay coils and solenoids must be clamped with a diode as shown in Figure 3-4.

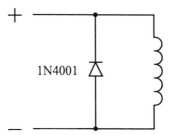

Figure 3-4. Every relay coil or solenoid must have a diode clamp to prevent high voltage inductive kicks from destroying valuable electronic equipment.

Mechanical relays are not widely used in amateur robotics but there are times when solid state substitutes are inadequate, because no solid state switch has an 'ON' resistance as low as a set of relay contacts. A relay is also sometimes the most economical choice for brute force current handling, while small DIP and reed relays fill other special needs.

Low power SPST, SPDT and DPDT solid state relays can be fabricated from 4066 CMOS switches and 4049 inverters. Relays made from 4066 chips have a 'contact' resistance of about 100Ω and are dynamic devices, so test one in your application before building a bank of these devices.

Figure 3-5. Some relays such as [A] look complicated, but all are constructed according to simple principles described in the text. [A] 4PDT 5amp contacts, 12V 1kΩ coil. [B] SPDT 40amp contacts, 12V 90Ω coil. This brute will handle most high current robotics switching needs, (*Hosfelt* 45-287 $3.50). [C] SPDT 2amp contacts, 12V 2.5kΩ coil [D] DPDT 2amp contacts, 5V 200Ω coil. A handy DIP style that plugs directly into a protoboard. [E] This reed relay also fits into protoboard sockets. SPST with one amp contacts, 12V 1kΩ coil. *Radio Shack* 275 233 $2.49. SPDT reed relays are also manufactured.

Using surplus or unclassified relays

If you have a relay without pin specifications, it is not difficult to find out how it works using a power supply and an ohmmeter, because most relays fall into one of the categories in Figure 3-6. Don't be confused if your relay has more pins than are shown in this figure, even the 4PDT relay in Figure 3-5 is merely four sets of DPDT contacts and a single coil.

Pin allocations and functions for an unclassified relay can be determined as follows:

[1]   Find the coil pins first - these can be identified by their resistance, typically 100Ω to 10k, so set your meter on the 20k range. Reverse the meter leads once you have found the coil pins to check if an internal suppression diode is installed across the coil. Protective snubber diodes should be installed *at the relay,* as shown in Figure 3-4. Do not use long leads to place diodes at a more convenient location.

[2]   Each set of NC contacts will have approximately zero resistance - locate all NC sets.

[3]   Monitor any set of NC contacts while connecting a 5V power supply across the coil (make sure the power supply polarities are connected as in Figure 3-4 to avoid blowing an internal or external diode). If the relay activates there will be an audible click, and all NC resistances will go to >>20MΩ. Simultaneously all NO contacts will switch to zero resistance.

[4]   If no relay activation is observed in [3] then increase the coil voltage to 12V and try [3] again.

Although many surplus 24V relays are available, these are of limited use in mobile projects. They can be used for bench work, if a suitable power supply or ac adaptor is available. Operating voltages are usually marked on 115V ac relays and they can be activated with an opto-coupled driver.

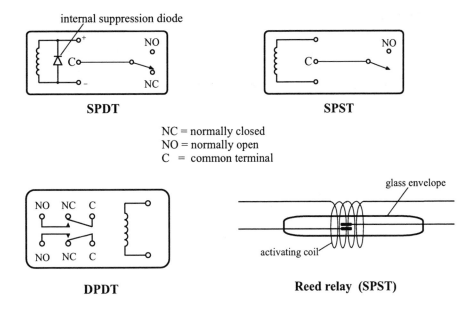

Figure 3-6. Most relays have internal features based on those presented in this diagram. It is straightforward to determine how all common relays operate, by following instructions given in the text in conjunction with this figure.

Computer control of relays and low power devices

Activating devices from low power signals with a computer is a common robotic requirement and some of these needs are satisfied by the circuit in Figure 3-7, (where R represents a load). Very large relay coils, and medium sized dc motors or lamps, can be controlled from logic signals using this circuit. If a 2N3906 is used instead of the TIP115, we have a smaller 200 mA driver suitable for small relays, motors etc.

       The key advantage to this type of driver is its ability to operate devices running at higher voltages than a normal logic signal level. For example if V^+ is 24 volts it can still be switched by a 5V logic signal, this is not possible using a single transistor driver shown in Questions 29 and 30 of Chapter 2. Drivers are important elements for robotic control and are discussed several times in this book.

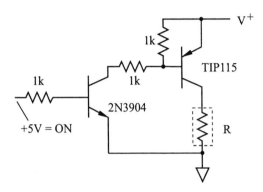

Figure 3-7. This 3 amp logic controlled switch can be operated directly from computer commands. Use heat sinks if transistors feel hot. A smaller current device capable of switching about 200mA can be constructed by replacing the TIP115 with a 2N3906.

Diodes, transistors and power MOSFETs

Checking diodes with a DMM

Diodes can be quickly tested using a DMM's 'diode check' mode, but more information is obtained if diodes are tested as shown in Figure 3-8, and this method can also be used to test Zener diodes. Additional information on diodes and their characteristics is given in Question 26 and 27 of Chapter 2.

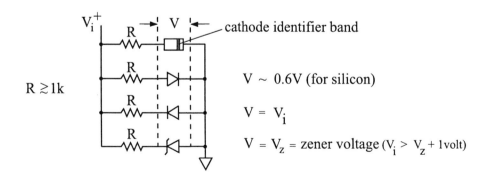

Figure 3-8. Some meters have a special feature for checking diodes, but any DMM can be used to test diodes by measuring their forward voltage drop as shown in this figure.

Diodes and bridge rectifier modules

1N4001 rectifier diodes and 1N4148 or 1N914 signal diodes are used most frequently for projects, and hobbyists can start their collection with the following items.

1N4001	**(Qty 10)**
1N4148 or 1N914	**(Qty 5)**

Table 3-1. Low power diode rectifiers

Part #	Peak reverse voltage	Current rating	Description	Source	Price
1N4148	100V	10mA	fast switching/signal diode DO-35 glass case	*Hosfelt*	5¢
1N914	100V	10mA	fast switching/signal diode DO-35 glass case	*Hosfelt*	5¢
1N4001	50V	1A	general purpose rectifier epoxy DO-41 case	*Hosfelt*	4¢

Three most commonly used silicon diode rectifiers for robotic projects are listed here. 1N4001 and either 1N4148 or 1N914 components should be kept on hand. Both 1N4148 and 1N914 have similar characteristics and can be used interchangeably.

Although full-wave power rectification is always possible with discrete diodes, bridge modules have many advantages. They are more compact and neater than four diodes, and can be replaced quickly with less chance of making a wiring error. High power bridge modules such as a KBPC1004 from Table 3-2 (and shown in Figure 3-9), are designed for easy mounting to a suitable heat sink.

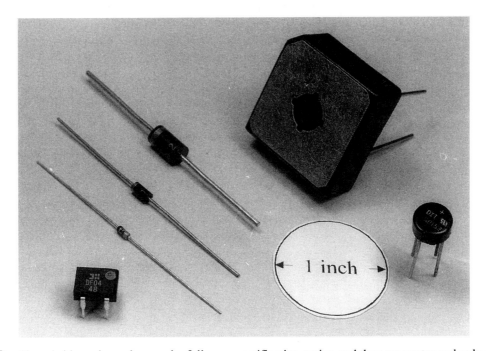

Figure 3-9. Three bridges shown here make full wave rectification easier, and they are neater and only marginally more expensive than using individual diodes. Clockwise from bottom left: DF04 1 amp 400V ac bridge. 1N4148 signal diode. 1N4001 1 amp 50V power diode. 1N5401 3 amp 100V power diode. KBPC1004 10 amp 400V bridge. RB152 1.5 amp 100V bridge.

45

Table 3-2. Bridge rectifying modules

Part #	Peak reverse voltage	Current rating	Description	Source	Price
DF10	1000V	1A	DB-1 case 4 pin DIP, 50amp surge	*Hosfelt*	50¢
KBPC1004	400V	10A	1.125"x1.125"x.5", 200amp surge	*Hosfelt*	$2.45

Specifications on two full wave bridge rectifiers illustrated in Figure 3-9 are listed in this table.

Bipolar transistors and power MOSFETs

General purpose bipolar transistors

Peripheral devices such as motors, lamps etc. must sometimes be controlled from a computer, but the latter has insufficient current drive to do this directly. A computer's signal current must first be amplified, and transistors are frequently used for this purpose. Circuits providing a computer or logic level interface will be given in a future book. This later text also deals with drivers and controllers for DC brush and stepper motors, showing how these may be controlled from a computer.

Transistors are often employed as power switches, such as the TIP115 in Figure 3-7. When used in this mode a transistor should be driven so it is either fully ON or fully OFF, because more heat is generated in a transistor when operated in its linear region, (ie. partially ON or OFF). You can be sure a transistor is operating in its nonlinear or saturated mode, by applying the appropriate base bias current. This is done by operating a transistor switch at a gain of 10, or Darlington switch at a gain of 100.

Solderless breadboards are ok for TO-220 transistors, but TO-3 transistors may draw too much current for contacts on these boards, which are rated at 4amps max. Large transistors can be wired and supported from soldered connections, using # 16 AWG bus wire as shown in Figure 7-16 - not very elegant but practical and simple. More information on solderless breadboards is given in text following Figure 7-15.

Table 3-3. Commonly used transistors for robotic applications

item	type	power	f_TMHz	V_{CEO}	I_C max	substitute	h_{FE}	case	price	source
2N3904	NPN	350mW	300	40V	200mA	2N2222	30/100	TO-92	6¢	*Hosfelt*
2N3906	PNP	350mW	250	40V	200mA	2N2907	30/100	TO-92	6¢	*Hosfelt*
TIP29	NPN	30W	3	40V	1A	TIP29A/B/C	15/75	TO220	45¢	*Hosfelt*
TIP30	PNP	30W	3	40V	1A	TIP30A/B/C	15/75	TO220	39¢	*Hosfelt*
2N3055	NPN	115W	2.5	60V	15A		20/70	TO-3	59¢	*Hosfelt*
MJ2955	PNP	115W	2.5	60V	15A		20/70	TO-3	$2.30	*Hosfelt*

A few basic transistors listed here will cover most project needs. Items in this table can handle high frequency and medium to high power requirements.

Power Darlington transistors

Darlingtons are made from two separate transistors and have about twice the voltage drop of a single transistor. Consequently, motors will have their top speed reduced a little when operated from Darlington drivers. Less base current is required to turn on a Darlington, so when replacing a regular transistor with a Darlington increase the original base drive resistor by a factor of 10 to compensate for higher gain.

Table 3-4. Darlington power transistors

item	type	power	f_TMHz	V_{CEO}	I_C max	substitute	h_{FE}	case	price	source
TIP110	NPN	50W	25	60V	2A	TIP111/112	500	TO-220	45¢	*Hosfelt*
TIP115	PNP	50W	25	60V	2A	TIP116/117	500	TO-220	59¢	*Hosfelt*

As shown in this table, power Darlington transistors have higher gains than regular transistors. They also drop more voltage, but usually this cannot be avoided if only small drive currents are available. Assign a gain of 100 to a Darlington when it is used as a simple ON/OFF switch.

Testing transistors

Bipolar junction transistors in TO-92 and TO-220 case styles can be tested on a DMM's transistor test socket. Larger transistors may be checked this way too if their leads are extended with bus wire. DMM transistor test sockets usually do not work with Darlingtons, but the following procedures will check all transistors and are handy for those hobbyists using a utility DMM.

Equivalent diode junction schematics for NPN and PNP transistors are shown in Figure 3-10 and, (for testing purposes only), these diagrams are valid for all types of transistors - signal, power and Darlington devices.

A memory aid: For *NPN* transistors the arrow is *N*ot *P*ointing i*N*

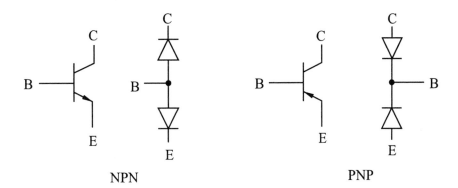

NPN   PNP

Figure 3-10. Regular transistors behave as simple diode junctions when using test procedures described in the text.

Procedures for testing an NPN transistor using a DMM

[1]   Set meter to the diode check position
[2]   Connect V/Ω lead to base B (Figure 3-11)
[3]   Connect COM lead to collector C
[4]   Meter will now show junction's forward voltage drop ($V_f \sim$ 0.5V to $\sim$ 0.75V)
[5]   Short B to emitter E
[6]   Meter reading should drop slightly ($\sim$ 0.004V to $\sim$ 0.1V)
[7]   Connect COM to E, (leave V/Ω attached to B)
[8]   Meter shows V_f as in [4]
[9]   Short B to C
[10]   Meter reading should decrease as in [6]

Darlington transistors show slightly larger drops in [6] and [10]
PNP transistors are tested using the same procedures but with COM and V/Ω leads interchanged.

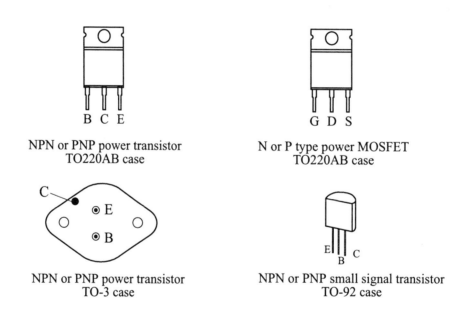

NPN or PNP power transistor
TO220AB case

N or P type power MOSFET
TO220AB case

NPN or PNP power transistor
TO-3 case

NPN or PNP small signal transistor
TO-92 case

Figure 3-11. Lead designations and case styles for transistors most commonly used in projects are shown here. Pinouts for TO-92 style shown here, only apply to transistors with their pins in a straight line.

Power MOSFETs

Power MOSFETs simplify high current switching, they are fast (nanosecond operation), and can be used for both analog and digital applications. However, for a roboticist their real forté is their extremely low 'ON' resistance $R_{DS(on)}$, and ability to control high currents with only a small steering current. A small 'ON' resistance ensures low power dissipation, and minimal heating in the MOSFET itself.

A high current switch that can be turned 'ON' from a low current logic signal, is depicted in Fig 3-12. This circuit can operate continuously at 20 amps, using an IRFZ44 attached to a TO-3 heat sink from Figure 3-13. Natural convection is adequate for this arrangement, but power handling can be increased substantially if the sink is fan cooled.

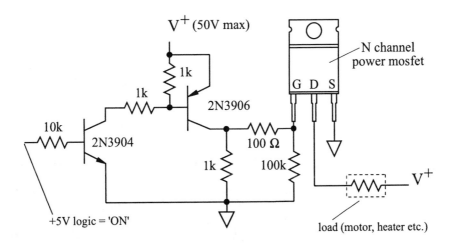

IRFZ44 'on' resistance ~ 0.028Ω i_{max} = 35A ($2.75)

Figure 3-12. Using this circuit an IRFZ44 power MOSFET can switch 35 amps with a low current logic signal from a computer. A TO-3 heat sink shown in Figure 3-13 is adequate at 25 amps, but a larger heat sink or fan cooling is necessary for continuous operation at 35 amps.

Table 3-5. Power MOSFETs

item	type	$V_{BR (DSS)}$	$R_{DS (on)}$	power	$I_{D(cont)}$	case	price	supplier
IRFZ44	Nchannel	60V	0.028Ω	150W	50A	TO-220	$2.75	Hosfelt
IRFZ42	Nchannel	50V	0.028Ω	125W	50A	TO-220	$2.15	Hosfelt
IRFZ530	Nchannel	100V	0.18Ω	75W	14A	TO-220	$1.65	Hosfelt

N-channel power MOSFET prices are very reasonable for the performance they provide. Unfortunately P-channel versions still cost too much for most hobby applications. Items in this table will handle most project requirements

Heat sinks

The necessity for a heat sink can be determined by checking a transistor's operating temperature *very quickly* with moistened finger tips - if too hot to hold for more than a few seconds then attach a sink. A very hot transistor will iron out fingerprints in less than a second, so take care.

Always leave adequate clearance between adjacent heat sinks, because transistors may blow if two sinks are shorted together.

Some commercial heat sinks are shown in Figure 3-13 but any good thermal conductor can be used if necessary. A coat of flat black paint raises the cooling efficiency of a smooth metal surface, because it increases both radiative surface area (it isn't as smooth as metal), and emissivity (its ability to radiate heat).

Figure 3-13. Heat sinks for transistors listed in this chapter are available commercially but any good thermal conductor can be used. *Bottom row, from left:* 2N3904 TO-92 case. IRFZ44 TO-220 case. 2N2222 TO-18 case. 2N3055 TO-3 case. *Upper row:* Heat sink size scales with power handling capability. A small 'clip on' (far left) is appropriate for many jobs and a TO-3 sink (far right) will handle about 5 watts, more if fan cooled.

Integrated circuits

Most ICs used in projects are *DIP*s (*D*ual *I*n line *P*ackages) from *TTL* (*T*ransistor *T*ransistor *L*ogic) or *CMOS* (*C*omplementary *M*etal *O*xide *S*urface) families. Integrated circuits or chips come in various sizes but all use a pin pattern shown in Figure 3-14. Location of pin #1 is important and usually denoted by a dot, depression, or by its proximity to a semi circular cut out.

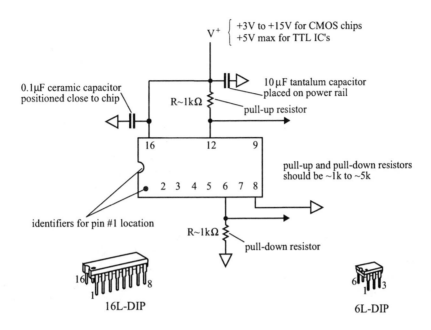

Figure 3-14. DIP (Dual In Line Package) conventions and some standard techniques for noise suppression and logic level controls are shown in this figure. A pull up/down resistor's value is not critical but must be 'stiff' enough to prevent the logic level from being influenced by other circuitry. If voltage drop across a resistor does not change appreciably when loaded, it is said to be 'stiff'.

Chip tips

Manufacturers usually omit pull-up and pull down requirements from their device descriptions, but these resistors are essential for proper chip operation. If a chip does not function as it should, try adding pull up or pull down resistors on those pins being interrogated as shown in Figure 3-14, (though not necessarily on pins 6 and 12).

If chip behavior is erratic, it may be due to pick up, and the following procedures are sometimes helpful and a good general policy.

[a] Decouple power rails with 10μF tantalum capacitors to ground as in Figure 3-14, keep these leads short and take them directly to ground/common.

[b] Place an additional 0.1μF ceramic capacitor in parallel with the 10μF tantalum capacitor of [a]. Always use the shortest possible path to ground for decoupling capacitors because their effectiveness is diminished by added inductance from long leads.

[c] Connect all unused pins either to ground or the positive power rail. Because many operating quirks can be traced to 'floating' pins, this is a very important protocol.

Table 3-6 lists integrated circuits commonly used in projects and stock items for classrooms, they are inexpensive especially if purchased by mail. The utility for many of these items will become apparent as you read this text and see how components are used. Hobbyists can conserve funds by selecting specific items for their immediate needs, adding extra components as required.

Table 3-6. Commonly used IC's for robotic projects

CMOS			TTL		
4001	Quad NOR, 2 input	22¢	74LS00	Quad NAND, 2 input	29¢
4011	Quad NAND, 2 input	22¢	74LS02	Quad NOR, 2 input	29¢
4013	Dual D flip flop	35¢	74LS04	Hex inverting buffer	29¢
4027	Dual JK flip flop	39¢	7407	Hex noninverting buffer	$1.04*
4043	Quad NOR RS flip flop	39¢	74LS08	Quad AND, 2 input	35¢
4044	Quad NAND RS flip flop	59¢	74LS14	Hex Schmitt, inverting	39¢
4046	Phase locked loop	59¢	74LS73	Dual JK flip flop	35¢
4049	Hex inverting buffer	25¢	74LS74	Dual D flip flop	35¢
4050	Hex noninverting buffer	25¢	74LS86	Quad excl OR	35¢
4051	8 Ch. mux/demux	55¢	74LS138	3 to 8 line decoder	39¢
4066	Quad switch	32¢	74LS148	3 to 8 line encoder	79¢
4067	16 Ch. Mux/demux	$1.35	74150	16 Ch. Mux	$1.59
4070	Quad excl OR	29¢	74LS151	8 Ch. Mux	39¢
4071	Quad OR, 2 input	26¢	74LS154	16 Ch. demux	$1.29
4081	Quad AND, 2 input	25¢	74LS240	Octal inverting buffer	49¢
4538	Dual analog timer	69¢	74LS244	Octal noninverting buffer	59¢
4553	3 digit BCD counter	$2.92	74LS245	Octal bus transceiver	59¢
4584	Hex Schmitt trigger	55¢	74LS373	Octal D type latch	65¢

(Prices from *Hosfelt*, except *=*Newark*)

This table contains components most often required for robotic projects. They should be kept on hand if possible, because they are inexpensive and used in so many applications.

Because of its immunity to noise and low power dissipation CMOS should always be your first choice, followed by LS (Low power Schottky) TTL, then regular TTL. Low power Schottky can be recognized by its 'LS' label, for example 74LS373 is a lower power equivalent of a 74373 TTL latch.

"What's the difference between CMOS and TTL?"

Regular CMOS requires less power, has greater noise immunity, is more tolerant of power supply fluctuations and is a bit slower than TTL. However hundreds of functions are available only in TTL format so one cannot be too picky. TTL can always be interfaced to CMOS and vice versa, as shown in Figure 3-15. High speed CMOS is seldom used for student robotic projects.

Useful operating parameters and some differences between CMOS and TTL, are outlined in Table 3-7.

Table 3-7.   TTL and CMOS operating parameters

	TTL	**CMOS**
'HIGH' trigger level	2.4 to 5V	> 0.8 of power supply
'LOW' trigger level	0 to 0.8V	< 0.2 of power supply
Power supply	5V ± 0.5V	3 to 15V
Chip power	High	Low
Frequency response	High	Medium
Noise immunity	Fair	Excellent

CMOS and TTL components have some important operating differences as outlined in this table, but they can be interfaced to work with each other as shown in Figure 3-15.

Power consumption is an important consideration in battery powered designs. It is surprising how quickly power escalates when using many TTL chips. Using CMOS or low power Schottky TTL where possible, will significantly reduce power requirements and keep equipment cooler, because most power dissipation reappears as heat.

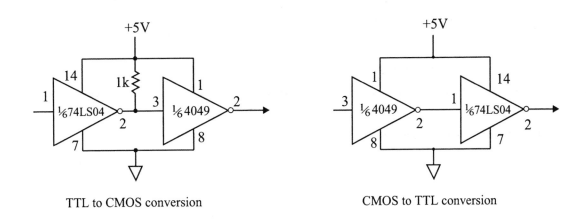

TTL to CMOS conversion   CMOS to TTL conversion

Figure 3-15.   TTL can be interfaced to work with CMOS and vice versa as shown here.

"Where can I get more information on this integrated circuit?"

Unfortunately no single source contains comprehensive data for even the few devices cited in this book. Photocopies of specifications are usually obtainable from any large electronic component store, such as Future Electronics/Active Electronic Components. These companies are mostly very cooperative, especially if one offers to pay for photocopying or buy parts under consideration

Useful electronic books

Books in Table 3-8 are among the best on the market. They are practical texts covering important areas in electronics and provide pin allocations and other parameters for many devices. Hobbyists can preview these books in a large bookstore such as *Barnes and Noble*. Major electronic component stores such as *Active Components* stock useful electronic texts, but technical books are no longer found in shopping mall bookstores.

Table 3-8. Selected electronics books for hobbyists and classrooms

title	author	publisher	ISBN #	price
CMOS Cookbook	Don Lancaster	H.W.Sams	0-672-22459-3	$25
TTL Cookbook	Don Lancaster	H.W.Sams	0-672-21035-5	$25
Active-Filter Cookbook	Don Lancaster	H.W.Sams	0-672-21168-8	$25
OP-AMP Circuits and Principles	Howard M.Berlin	H.W.Sams	0-672-22767-3	$25
The Art of Electronics	Horowitz and Hill	Cambridge Univ. Press	0-521-37095-7	$80

These useful electronics texts are among the best from a robotics viewpoint. They are well written user friendly practical books, which should be on every serious experimenter's and classroom bookshelf.

Detailed specifications and parameter data on many linear and digital devices can be found in manufacturers' data books listed in Table 3-9. When using a component for the first time, you will find application notes and circuits accompanying most device descriptions, particularly handy.

Manufacturers' data books are necessary for classrooms but are an expensive investment for hobbyists, because full device specifications are seldom necessary to make a circuit work.

Table 3-9. Electronic manufacturers databooks

title	manufacturer	book #	supplier	price
Linear and Interface Integrated Circuits	Motorola	DL128	Newark	$12.46
Linear Databook	NSC*		Electrosonic	$15.50
Schottky TTL Databook	Motorola	DL121	Newark	$12.46
Logic Databook	NSC*	Vol II	Electrosonic	$12.50
CMOS Logic Databook	Motorola	DL131	Newark	$5.32
Discrete Semiconductor Products Databook	NSC*		Electrosonic	$12.50
Optoelectronics Device Data	Motorola	DL118	Electrosonic	$4
Data Book (covers many optoelectronic items)	OPTEK		Newark	$15.68

* National Semiconductor

Manufacturers' data books not only give component specifications. They frequently have useful application notes and practical circuits showing how their products can be used to best advantage.

Semiconductor replacement guides

Three excellent volumes containing a wealth of useful information, and pin specifications for thousands of ICs and discrete semiconductors are listed below. Transistors, diodes, bridges, transient suppressors, opto-electronics, linear and digital, microprocessors, memories, hardware, TTL and CMOS are cross referenced and often accompanied by sufficient information to use a device. Asian and European equivalent replacement numbers are given for thousands of components, making these books information goldmines.

ECG	Semiconductor Master Replacement Guide	#84F596	$3.81	*Newark*
NTE*	Semiconductors	NTE-CRM7	$4.95	*Hosfelt*
Sams	Semiconductor Cross Reference Book	#61000	$24.95	*Hosfelt*

* available on diskettes

Storing integrated circuit chips

Many ICs can be safely stored in trays made from *non-conductive* 1" rigid Styrofoam™ foamboard. Despite very low humidity high static winter conditions in Toronto, hundreds of chips have been successfully stocked and dispensed for many years without mishap, using techniques outlined in Figure 3-16. Common sense handling procedures are always necessary when working with static sensitive devices, either in the classroom or elsewhere. These very important precautions are reviewed in the Figure's caption.

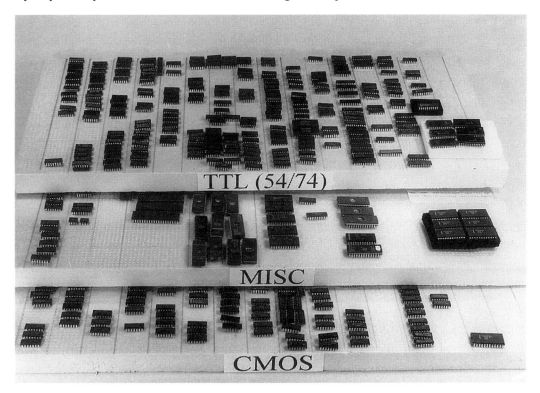

Figure 3-16. Styrofoam™ board is very useful for storing large numbers of chips if simple common sense anti-static handling procedures are followed. Before picking up a semiconductor or working on any electronic equipment, first 'GROUND' your body by touching a metal surface at electrical outlet ground potential. Always place an IC on an insulating surface for another person to pick up. Passing from hand to hand may destroy a chip if handlers are at different electric potentials. Anti-static work stations are unnecessary for the hobbyist or classroom, since static sensitive components can be handled safely if these simple precautions are followed.

Chip storage trays can be kept in an open bookcase with 2" tray separators attached to the bookcase sides. This arrangement allows trays to be slipped in and out, permitting quick access to all components.

Operational amplifiers, buffers and comparators

Op-amps, comparators and buffers are some of the most versatile and useful electronic components in robotics circuitry. For a beginner, op-amp parameters such as current and voltage offsets, biases, gain bandwidth, CMRR and slew rate can be confusing, but it is possible to build working circuits without delving into these areas. If you wish to design your own circuits, it is only necessary to understand a few basic principles to use an op-amp in an intelligent and practical manner. Excellent texts are available which deal with this topic thoroughly, and some are listed in Appendix B.

General purpose operational amplifiers

An LM358 is satisfactory for most op-amp project needs, it is inexpensive (39¢), and pin compatible with several other devices as indicated in Figure 3-17.

When higher speed and greater input impedance is required an LF353, TL072 or TL082 can be substituted for an LM358. These FET amps have an input impedance ~$10^{12}\Omega$, are fast (slew rate 13V/μs), and deliver an output current of 20mA. All four amplifiers have the same pin functions and can be used interchangeably for many circuits in this book. Specifications for these op amps are listed in Table 3-10. Single supply operational amplifiers and chopper stabilized op-amps for monitoring very small signals, such as those produced by thermocouples are available.

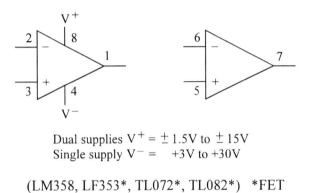

Dual supplies $V^+ = \pm 1.5V$ to $\pm 15V$
Single supply $V^- = \quad +3V$ to $+30V$

(LM358, LF353*, TL072*, TL082*)   *FET

Figure 3-17.   These dual operational amplifiers are 8 pin DIPs with two amplifiers in each package. All four amplifiers quoted are inexpensive, pin compatible, and satisfy most amateur robotic needs.

Table 3-10.   Two versatile operational amplifiers

Part#	Pkg.	Mfr. *	Power supply max (V)	Input offset max (mV)	Input offset max (nA)	Input bias max (nA)	Slew rate (μs)	Description	Source	Price
LM358N	8pin DIP	Mot	±16 or +32	7	50	250	0.6	Low power, dual op amp bipolar or single power supply	*Hosfelt*	39¢
TL082CN	8pin DIP	NSC	±18	15	0.2	0.4	13	Low noise, dual FET op amp high speed, high input impedance	*Hosfelt*	64¢

* Manufacturer: Mot = Motorola, NSC = National Semiconductor.
TL072 and LF353 operational amplifiers have similar performance characteristics to a TL082. The LM358 is an excellent, low priced general purpose amplifier, fulfilling most robotic circuit needs. All these devices are widely available from good electronic sources.

Buffer or voltage follower

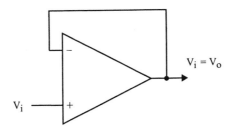

Figure 3-18. A buffer or voltage follower has a very high input impedance and can be used where circuit loading is a problem, (Question 4, Chapter 2). Buffers are also used as isolation elements, reducing interaction between circuit components.

"All my circuits work perfectly except when I connect them to the computer"

This problem is invariably solved by placing 4050 buffers in every signal line attached to the computer.

"Why does this work?"

[1] A buffer increases line drive current to a level where it dominates spurious pick up noise.

[2] Computer interface circuitry (usually a multiplexer, A/D, D/A converter etc.), is forced to definite state levels by the buffer, and cannot 'float' between HI and LO conditions. Gate 'floating' produces equivocal instructions to and from the computer or peripheral device, causing erratic behavior.

Buffers have other uses too. They are used extensively in projects for logic level conversion, and as drivers for LEDs or even small motors.

Hex CMOS buffers 4049 and 4050 shown in Figures 3-19 and 3-20 are among the most commonly used ICs for projects. Both devices handle larger currents in their sink than source mode (a characteristic common to many other buffers and drivers). This factor of four in current drive capability can be very useful at times.

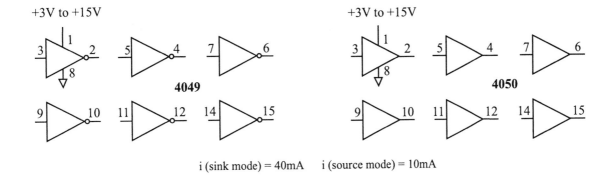

Figure 3-19. Hex inverting CMOS buffer   **Figure 3-20.** Hex non-inverting CMOS buffer

Voltage comparators

A comparator can be fabricated from an operational amplifier running at a high gain, or open loop as indicated in Figure 3-21. However, using a device designed specifically for comparing voltages generally works best. Fortunately several excellent inexpensive components are available for hobbyists, and two recommended comparators are listed in table 3-11.

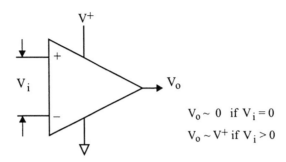

Figure 3-21. A voltage comparator can be built from an operational amplifier but an LM339 has four comparators in a 14-pin package and is usually a better option.

An LM339 is a good choice for robotic projects, it is inexpensive and has four separate comparators on each chip. Pull-up resistors (4.7k in Figure 3-22), must be installed because outputs on both LM339 and LM311 are open collectors of internal transistors - these chips *will not work* without pull-up resistors.

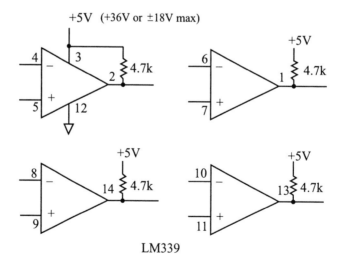

Figure 3-22. Four high speed comparators are contained in this 14-pin quad package. The LM339 is an excellent device and suitable for many robotic tasks. Pull-up resistors shown are essential for proper operation.

Noise immunity is improved on a comparator if a small capacitor is connected between signal input and ground. By smoothing out unimportant voltage fluctuations a capacitor helps to reduce spurious comparator switching. A $0.1 \mu F$ ceramic capacitor is satisfactory for this purpose at frequencies below 1MHz.

Table 3-11. Comparators for projects

Part #	Pkg.	Mfr*	Power supply max (V)	Input offset max (mV)	Input offset max (nA)	Input bias max (µA)	Time resp. (µs)	Description	Source	Price
LM339N	14pin DIP	NSC	± 18 or +36	5	50	0.25	1.3	Low power quad comparator bipolar or single power supply	Hosfelt	39¢
LM311N	8pin DIP	Mot	± 15	7.5	50	0.1	0.2	High speed single comparator bipolar or single power supply	Hosfelt	49¢

* Manufacturer: Mot = Motorola, NSC = National Semiconductor
Both LM311 and LM339 are great chips but usually an LM339 is a better buy. An LM339 has four comparators on one chip, and occupies less breadboard space if more than one comparator is required.

Additional useful Integrated circuits for robotic applications

Active filter: MF10ACN $4 *Newark*

The MF10 is a dual universal switched capacitor filter, with all the capabilities of a regular active filter up to 30kHz. It can be configured as a bandpass, low pass, high pass, or notch filter, all with variable Q. Not only is an MF10 a very versatile filter, its frequency characteristics may be computer controlled, a useful feature in audio frequency spectrum scanning projects.

Each MF10 chip contains two filters that can be used independently, or connected in series to provide enhanced filter performance. All classical filters such as Butterworth, Bessel, Cauer or Chebyshev can be derived with an MF10, and details on using this device are given in a future book.

Timers, multivibrators, waveform generators, BCD counter

Multivibrators can be used to generate single pulses or continuous rectangular waveforms. Such signals are used for event timing or as clock signals for various functions. Practical circuits describing uses for 555, 4538 or XR2206 components have been reserved for another book in this series.

Timer : NE555 35¢ *Hosfelt*

The 555 timer has been one of the most popular chips for many years and is still used in many applications. A dual CMOS version of this versatile IC, sold as TLC556 is also widely used by experimenters.

Non symmetrical waveforms from a 555 are occasionally a problem but its high current drive of 200mA, in both source and sink modes is a valuable asset for operating relays or running small motors directly. It will operate to about 100kHz and many ingenious applications have been devised for this device. An exact 50% duty cycle waveform can be derived by using a frequency divider after the 555 output. A basic circuit for operating a 555 as a square wave generator is shown in Figure 3-23.

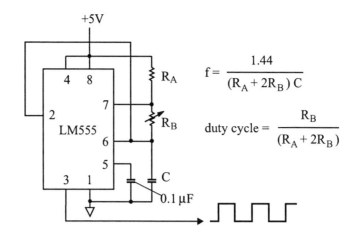

Figure 3-23. A 555 timer does not produce a true square wave output, but it can source or sink 200mA and this is a valuable feature. It can be used to drive small motors, or operate even large relays and other low power devices.

Dual analog timer: CD4538 69¢ *Hosfelt*

A 4538 can produce accurate pulses for precision timing and is a better choice than the 555 for such tasks. This 16-pin CMOS IC is a dual retriggerable and resettable monostable multivibrator that can generate predictable and accurate timing signals from external resistors and capacitors. Basic functions performed by this versatile and useful chip are illustrated in Figure 3-24. Be sure to include pull-up and pull-down resistors shown in this diagram.

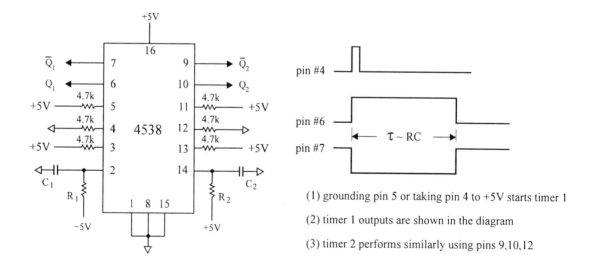

(1) grounding pin 5 or taking pin 4 to +5V starts timer 1

(2) timer 1 outputs are shown in the diagram

(3) timer 2 performs similarly using pins 9,10,12

Figure 3-24. A 4538 dual analog timer can be used to provide timing signals from 10μs to 10 seconds, and is superior to a 555 for event timing operations. With two timers in a 16-pin package a 4538's performance is hard to beat.

Function generator: XR2206 $4 *Active components*

Excellent stability triangle, ramp, square waves and low distortion sine waves from 0.01Hz to about 2MHz can be derived from an XR2206. Its output may be amplitude and frequency modulated from an external voltage source, and a dc output offset is easily added using a summing amplifier. This chip is more versatile than an ICL8038 and produces better sine waves too.

BCD counter: CD4553 $2.92 *Hosfelt*

Precision event timing with 1µs resolution is possible using a CD4553 three-digit BCD counter. Circuit details showing how a 4553 can be used with a computer to record accurate time intervals are provided in a future book.

Variable voltage regulators: LM317 65¢, LM337 $1.25 *Hosfelt*
Fixed voltage regulators: 7805 (+5V), 7905 (-5V), 7812 (+12V), 7912 (-12V) 59¢ each *Hosfelt*
Voltage references: LM317 or LM337 are recommended for use as voltage references

Voltage regulators are used anywhere a stable potential is required, and are available in fixed and variable versions. Small linear dc power supplies with current limiting and good regulation can be constructed from an ac adapter followed by a fixed regulator. More information on these components and their uses is available in Chapter 4.

Precise and stable voltage references from ±1.2V to ±37V can be derived from LM317 or LM337 variable regulators as shown in Figure 3-25. Because these devices provide large currents they need no buffering when used as a voltage reference, and accuracies of about 0.5% are attainable if the desired voltage is set with a 3½ digit DVM.

Although dedicated voltage references are sold, hobbyists will find it better to use an LM317 or LM337 in most cases. These chips are inexpensive, widely available, and serve as multipurpose components in a parts kit.

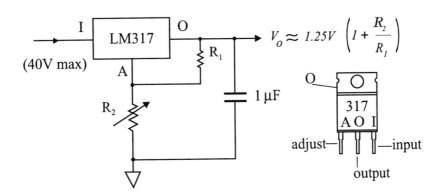

$$V_o \approx 1.25V \left(1 + \frac{R_2}{R_1}\right)$$

Figure 3-25. LM317 and LM337 variable voltage regulators also make good voltage references. They can be set to give a stable voltage from ±1.2V to ±37V and will tolerate considerable loading without their outputs sagging.

Table 3-12. Miscellaneous solid state integrated circuits

DAC08 or MC1408	D/A converter	$1.74	*Hosfelt*
ADC0808N	A/D converter	$5.99	*Newark*
P8255	Programmable Peripheral Interface	$5.99	*Active Components*
MC145026P	Remote control encoder	$2.42	*Newark*
MC145027P	Remote control decoder	$3.42	*Newark*
LM567	Tone decoder PLL	59¢	*Hosfelt*
LM335Z	Temperature sensor	$1.99	*Hosfelt*
LM331	Precision frequency to voltage converter	$6.00	*Electrosonic*
LM2917N-8	Frequency to voltage converter	99¢	*Hosfelt*
IR3C02A	Laser diode driver	$6.00	*Sharp*
2N6342	Triac	$2.24	*Newark*
MOC3011	Triac driver (opto-coupled)	$1.25	*Hosfelt*

Opto-electronics

TLN110 (#25-147)	IR diode, 30mW/Sr, T1¾,	20¢	*Hosfelt*
LD271	IR diode, 15mW/Sr, T1¾	60¢	*Active Components*
(#25-261)	Super bright red LED 47mcd, T1	24¢	*Hosfelt*
SFH205	Photodiode with IR filter 350ns	$1.25	*Active Components*
OP505A	Phototransistor light blue tint filter	$1.00	*Active Components*
MRD370	Photodarlington	$2.47	*Newark*
(276-116)	LDR (Cadmium sulphide photocell)	$3.00	*RadioShack*
(276-137)	IR detector module by Sharp	$3.59	*RadioShack*
OPB804	Slotted optical switch	$2.50	*Active Components*
OPB706A	Reflective assembly	$3.00	*Active Components*
4N26	Opto-isolator 1500V	49¢	*Hosfelt*

BCC52 components

80C52-BASIC	Microprocessor (40 pin or PLCC)	$19	*Micromint*
2764	EPROM	$3.45	*Hosfelt*
6264	SRAM	$5.99	*Active Components*
8255	Programmable Peripheral Interface	$5.99	*Active Components*
MC1488	RS232 driver	39¢	*Hosfelt*
MC1489	RS232 receiver	45¢	*Hosfelt*
74LS14		39¢	*Hosfelt*
74LS373		65¢	*Hosfelt*
74LS08		35¢	*Hosfelt*
7407		$1.75	*Active Components*
74LS138		39¢	*Hosfelt*
74LS245		59¢	*Hosfelt*

Analog to digital converters, peripheral interface adaptors, remote control chips and other useful devices for robotics are listed and priced with suppliers. Additional information and circuits using these items are given in a future book.

Power Supplies, Batteries, Energy and Power

Power supplies

Every roboticist needs a bench power supply that can provide voltages and currents adequate for circuits, motors, and ancillary robotic equipment. Suitable supplies are available at such reasonable prices from surplus outlets that it is a waste of effort, and almost always more expensive to build one. A switching power supply from an old computer is recommended, and Chapter 1 gives sources and relative prices for such supplies.

For those enthusiasts who like to build everything or just learn more about the subject, a handy primer is sold by Radio Shack:

Building power supplies   276-5025 $5.99   *Radio Shack*

Warning - Unlike batteries and ac adapters, power supplies cannot be connected in series or parallel to provide higher voltage or current performance, without some risk. Connecting supplies in this manner can sometimes be a good way to blow up a supply, because of duelling between output regulators.

Switching power supplies for hobbyists

A switching supply can power op amps, motors, optical devices, relays, microphones etc. In only a few special cases is anything more sophisticated required. Switching supplies are available on the surplus market with fixed dc outputs of +5V, -5V, +12V, -12V and +24V. Variable voltages can be derived from such supplies using LM317 or LM337 regulators, described later in this chapter.

Switching supplies convert more than 60% of ac wall plug power into useful dc power, and they are dependable rugged and reliable. High frequency switching noise (about 50mV p/p), is usually present on all dc outputs and some RF radiation occurs. However, these cause no problems except for the most demanding projects.

Many circuits designed to operate at ±15V run satisfactorily on ±12V, and increasingly ±12V and ±5V are becoming de facto voltage standards for much robotic instrumentation. Circuits designed on the bench for ±12V can also be conveniently operated from 12V SLA (Sealed Lead Acid) batteries, or nickel cadmium packs, simplifying the transition to autonomously powered robotic control.

Selecting load resistors for switching power supplies

Most switching power supplies will not function properly unless an appropriate load resistor is first installed. Under some operating conditions this load is supplied by microprocessors, disk drives or other circuitry permanently connected to the supply. However, this is not so when a PC or other switching supply is used for hobby or robotics projects.

Erratic behavior and very high output ripple will be present on most switching power supplies unless a proper load resistor is installed. Using the following procedures you can determine whether your supply already has a load resistor, or the appropriate resistor to use if one is not present.

[1]   Set up your switching power supply, and test equipment as shown in Figure 4-1.

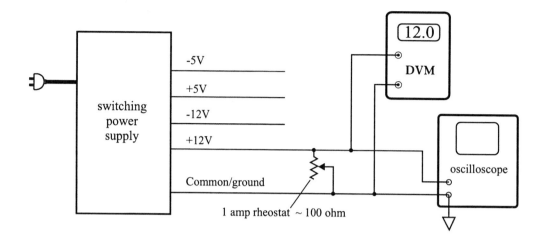

Figure 4-1. Switching supplies need a fixed load on one of the high current output power lines, (usually +12V or +5V). An appropriate resistor value can be determined using this circuit and instructions given in the text.

[2]   Set the rheostat to 100Ω, turn on supply, then monitor the +12V line for output voltage with a DVM, and output ripple with an ac coupled oscilloscope on 1V/div. Power lead designations for PC's or clone power supplies are given in Table 4-1

Table 4-1. IBM and PC clone power lead assignations

color	voltage	current*
red	+5V	18A
black	common/ground	
yellow	+12V	3.8A
white	-5V	0.3A
blue	-12V	0.3A
orange	power good	

* for supply D shown in Figure 1-7

IBM and PC clone power supplies use output power line coding given in this table. Supplies often have many sets of leads of the same color but only one set is required for hobby use.

[3]   If ripple is more than ~200mV p/p at a frequency below 5kHz, a load resistor must be installed.

[4]   Reduce the rheostat's resistance until ripple decreases, ripple frequency will increase at the same time. Measure the rheostat's resistance when ripple starts to drop rapidly. Measure the rheostat's resistance, then install a power resistor with the same value as a permanent load resistor.

Test results on power supply D of Figure 1-7 are listed in Table 4-2.

Table 4-2.   12 volt output ripple vs load for a PC switching power supply.

no load	ripple (p/p)	25 Ω load	ripple (p/p)
11.1V	1V @ 30 Hz	11.7V	40mV @ 100 kHz

These test results using the circuit of Figure 4-1 show +12V output voltage ripple decreases by a factor of 25 when a PC clone switching power supply is loaded with an appropriate resistor, (25Ω for this supply).

A 25Ω load dissipates 5.5W, so a 25Ω 10W power resistor installed on the +12V line is satisfactory. However, further tests detailed in Table 4-3, show it is more efficient to load the +5V line with a 30Ω 1W resistor. This reduces load resistor power loss to 0.9W.

Table 4-3.   5 volt output ripple vs load for a PC switching power supply.

no load	ripple (p/p)	30 Ω load	ripple (p/p)
5.31V	0.4V @ 30 Hz	5.14V	20mV @ 100 kHz

For the PC clone power supply tested +5V line loading is more efficient. However, this is not so for all supplies and it pays to test both +12V and +5V outputs, to see which line is best for loading.

Warning: Never touch components on an open frame switching power supply, even if it has been unplugged. A long time after power has been removed, these boards can deliver a kick you will remember for the rest of the day.

Characteristics of two switching power supplies used for projects are listed in Table 4-4.

Table 4-4.   Characteristics of switching supplies for student use

U of T student supply	U of T workstation supply
+12V 2A	+12V 0.8A
-12V 0.5A	-12V 0.8A
5V 4A	+5V 4A
	-5V 0.5A

Characteristics of two types of power supplies issued to students at the University of Toronto are listed in this table. Surplus switching power supplies with ±12V and ±5V outputs are not always available. However, -5V can be derived when required by using a 7905 negative voltage regulator powered from the -12V supply line.

Finding mating connectors for power supplies and other surplus equipment

Don't waste time or money buying mating connectors for surplus supplies or other surplus equipment. These can be difficult to find, and are sometimes expensive. Cut off connectors and trim unused leads. New connectors or header strips may be attached to power supply leads. However, a terminal block in Figure 7-14 is more versatile for general use.

Linear power supplies

Linear supplies are less efficient, have lower noise, provide variable voltage dc outputs and are sometimes less sophisticated than their switching cousins. A linear supply with similar features to switching supplies listed in Table 4-4 is usually more expensive, even at surplus prices.

An Elenco linear supply with dc outputs at +5V 3A, -5V 0.5A, +12V 1A, and a variable output from +2V to +20V at 2.5A is sold by *Hosfelt* #40-100 for $65. This is a good general purpose supply with short circuit protection, power jacks and voltage/current meters, ('A' in Figure 1-7).

115V ac line adapters

AC adapters are small black modules that plug into an ac outlet, providing dc for running a portable radio or other battery powered appliance. An ac adapter (eg #56-281 $7.95 *Hosfelt* 12V @ 1.5A), followed by a three terminal regulator makes an excellent dc power supply. Two adapters are required to make a ±12V bipolar supply.

Most ac adapters consist of a transformer, full wave bridge rectifier, filter capacitor and a cord with jack for the dc output. AC adapters are rugged, reliable and can be used instead of the transformer and bridge in Figure 4-2.

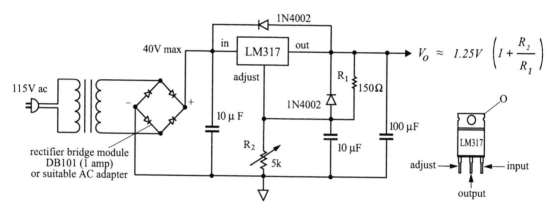

Figure 4-2. This variable dc power supply (1.5V to 36V at one amp), can be constructed from an ac adapter and a variable voltage regulator. Adapters can be used in series or parallel (just like batteries), to provide higher voltage or current outputs. Bipolar supplies need two adapters and a common ground between the two outputs.

Unlike regular power supplies, ac adapters of the same type can be used in series or parallel combinations, to give higher voltages or more current. They also make fine bipolar supplies, with low ripple and noise when properly filtered. Transformers can be used in this manner too, just be sure to phase outputs correctly or two transformers will provide less power than one!

Reducing power supply ripple

All dc power supplies have output ripple, and for simple supplies this usually increases with current drain. On a well designed linear supply, ripple may increase less that 10% from no load to full load, on an unregulated unfiltered ac adapter it may increase 1000% or more.

Most ac line powered linear supplies use full wave rectification, then output regulation, followed by a filter capacitor to smooth the final output. An appropriate filter capacitor can be selected if ripple frequency, working current and desired ripple amplitude are known. A numerical example in Figure 4-3 shows the procedure for calculating the size of a filter capacitor from these parameters.

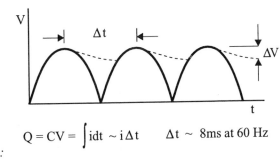

$$Q = CV = \int idt \sim i\Delta t \qquad \Delta t \sim 8\text{ms at } 60 \text{ Hz}$$

example:

If 0.5V ripple is required at 1A the required smoothing capacitor will be :

$$C \sim \frac{i\Delta t}{\Delta V} = \frac{1 \times .008}{0.5} = 16000 \ \mu F$$

Figure 4-3. Output supply ripple always increases with load on simple linear power supplies. As indicated by the calculation in this figure, very large output capacitors are required to provide adequate filtering on high current supplies

Switching noise on a commercial switching power supply is only marginally improved by adding filter capacitors. This is mainly due to the high inductance of large capacitors and because high frequency hash crosses many boundaries by capacitive coupling.

An inexpensive very low noise power supply

Batteries are the lowest noise reasonably priced power sources, with noise being undetectable except for the most discriminating task. A rechargeable battery continuously charged at slightly more than the load rate acts as an excellent filter, turning a good power supply into a first rate low noise, high stability source. This only works for linear supplies, switching supply noise is not substantially reduced by this technique.

A brute force variable voltage linear supply with an isolation transformer

A variable autotransformer, followed by a second fixed turns ratio transformer, bridge rectifier and filter as shown in Figure 4-4, make a simple rugged unregulated linear supply useful for many applications. A second transformer is necessary for user protection. The outputs from this second transformer are floating (ie. they are not referenced to ac line power). A step-up, step-down, or a 1:1 turns ratio isolation transformer can be selected depending on current/voltage requirements. A 115V to 115V one amp isolation transformer is available from *Hosfelt* for $14.95 #56-278.

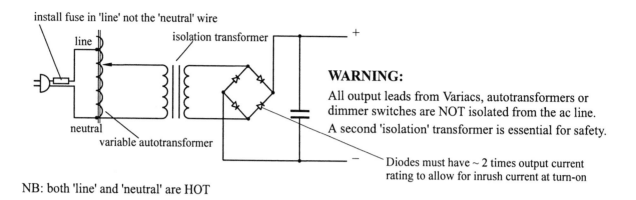

install fuse in 'line' not the 'neutral' wire

line

isolation transformer

WARNING:

All output leads from Variacs, autotransformers or dimmer switches are NOT isolated from the ac line. A second 'isolation' transformer is essential for safety.

neutral

variable autotransformer

Diodes must have ~ 2 times output current rating to allow for inrush current at turn-on

NB: both 'line' and 'neutral' are HOT

Figure 4-4. This rugged but simple variable voltage linear power supply can be tailored to give high voltage or high current outputs. An isolation transformer is essential for safety reasons. Solid state dimmer switches sold for controlling incandescent lamp intensity are inexpensive and can be used, but a variable autotransformer is better. An isolation transformer is also required when using a dimmer switch.

Costs can be reduced by using a solid state dimmer instead of a variable autotransformer, and the soldering iron controller depicted in Figure 7-8 is convenient for this purpose. Nevertheless, this cost saving comes at a price. Unlike a variable autotransformer which provides an easily filtered output over its entire range, dimmer switching noise scales with loading. It is also wise to check any dimmer switch to ensure its output tracks the knob setting before using it with valuable equipment, because some solid state dimmer switches will not give very low voltage outputs unless they are heavily loaded.

Some Variacs™ provide voltages from zero to 140V or more from a 115V ac input, an added bonus. Output from a Variac varies linearly with control knob setting unlike dimmer switches. New Variacs are expensive but well worth the money. Small surplus variable autotransformers are available from *C & S Sales* with prices starting at $22.50.

For obvious safety reasons all parts at line potential in Figure 4-4 must be enclosed, and the hot ac input line fused appropriately. A fuse incorrectly installed in the neutral line will leave transformer primary leads 'hot', even after the fuse has blown - a dangerous situation.

Curing noise problems by decoupling power lines

When individual electronic portions of a project test ok but misbehave as a composite unit, use an ac coupled oscilloscope to check for noise on the power rails. It is surprising how much noise can be transferred to power lines by motors, relays and other high current elements.

Noise generators can be traced by removing suspected elements sequentially while monitoring the power rails. Culprits should be decoupled with 0.1μF ceramic capacitors placed across power lines where they enter the noise source, make these leads short. Electrolytic or tantalum capacitors can be installed on the power bus and 0.1μF ceramics added as described in Figure 3-14.

If interference persists despite these efforts, separate supplies should be used for noisy equipment. For example, ac powered projects may employ switching power supplies for motors etc., but use ac adapters exclusively for logic circuitry. A common ground must usually be provided between all supplies but connect all common leads at **ONE** point only. Noise is often less problematic with battery powered projects, especially when using low internal impedance cells such as lead acid and nickel cadmium batteries.

"What's the difference between signal and power leads?"

Beginners sometimes use signal and power leads interchangeably but this can cause problems. For example, in Figure 4-5 part of a project (apparatus in the figure), draws 10 amps and is connected to a 12V battery by two 4ft lengths of #24 AWG wire.

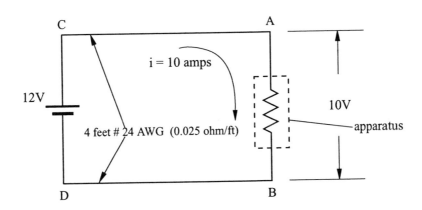

Figure 4-5. Voltage drops across small wires used to carry large currents can be quite large as shown in this figure. Separate 'signal' lines are sometimes necessary when monitoring potentials on high current lines as explained in the text.

To save wire and time students sometimes monitor the apparatus voltage at C and D, but this measurement will be in error by 2 volts because thin wire produces a considerable voltage drop at high currents.

In this example each 4ft length of wire has a resistance of 0.1Ω (0.025Ω/foot) and drops 1 volt. Even #16 wire gives an error of 300mV, which is unacceptable for many feedback applications. The solution is to run a separate pair of monitoring wires directly to A and B. These may be very thin if necessary because they carry only small signal currents.

Hindsight is wonderful! By looking at this simple circuit one may wonder how anyone can make this mistake. The beguiling scenario often plays like this: A high current device runs at a low duty cycle, therefore thin wires do not overheat to alert the experimenter of impending problems. Cabling has been neatly bundled and adding extra wires is difficult or untidy. Feedback sense signals are taken from C and D and then *"things just don't seem to work as they should."*

If you think only neophytes make this error you are mistaken. Many professionally designed printed circuit boards are discarded because traces are too narrow or copper cladding too thin. Unfortunately CAD programs seldom detect this subtle but well known problem, when a design is taken from breadboard to production.

Voltage converters

Autonomous robotic projects often need bipolar voltages (eg. ±12V or ±5V) and these can be derived using a separate battery for each voltage required. Another approach is to use a single battery followed by a converter generating both plus and minus voltages, referenced to a common 'ground'. For low power applications this is often more convenient, less expensive, saves weight and reduces battery maintenance.

It is easy to change ac to dc with rectifying diodes but more complicated to convert dc to ac, especially if appreciable power is required. Price is a rough 'complication indicator' and that is why a 12V, 20 amp battery charger operated from 115Vac, costs far less than a wimpy 12V inverter providing 115Vac at 1 amp.

Changing a positive voltage to a negative voltage or vice versa is similarly involved and also expensive at high powers. Usually dc current is chopped, providing alternating current that is then manipulated using transformers, voltage multipliers, dc restorers etc. - all well established techniques.

Three conventional low current converters listed below are handy for powering a few ICs or voltage references, (some external components must be added by the user).

DIP voltage converters: LTC1044CN8 $2 LT1054CN8 $3 (Linear Technology), *Electrosonic* ICL7660 $1.75, *Active Components*

LTC1044/ICL7660 : converts +5V to ±5V at 10mA
LT1054   : converts +5V to ±12V at 25mA or converts +12V to -5V at 200mA

One of the best options for generating bipolar or any additional voltages is to use extra batteries, as stated previously. Separate batteries not only solve interference problems, they also remove power restraints that sometimes inhibit design choices when low current 7660 type converters are employed.

DC to dc converters

DC to dc converters are usually the best method for producing bipolar voltages from a single battery, they also have higher output capability and better regulation than 7660 type converters.

DC to dc converters have been used for decades in the space program and are now available from surplus outlets for amateur use. Typical dc to dc converters are depicted in Figure 4-6. Table 4-5 lists some devices available on the surplus market, (these are brand new units, ideal for hobby use).

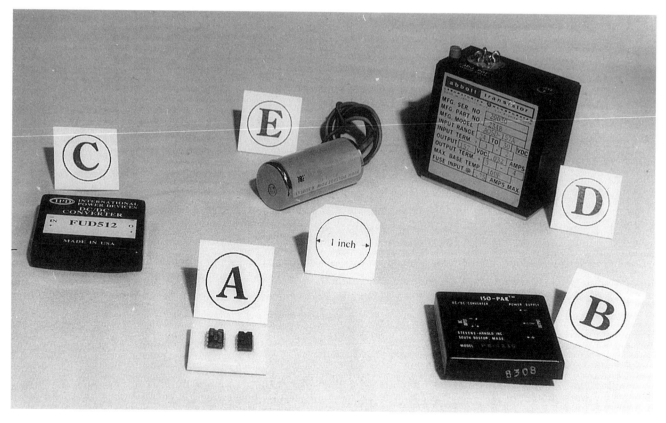

Figure 4-6. Some dc to dc converters are available from surplus outlets at attractive prices - new costs are usually prohibitive for hobbyists. Items [B] and [C] in the following list are available from *Hosfelt*:   [A] 7660 +5 to ±5V/10mA, LT1054 +5V to ±12V/25mA.   [B] +12V to ±15V/170mA. [C] +5V to ±12V/415mV.   [D] +28V to 150V/32mA. [E] Up to 1500V dc at 1mA with 0.1% ripple at full load. Output potential is proportional to the input supply voltage. [B] and [C] from *Hosfelt*.

70

Table 4-5. Surplus priced dc-dc converters

stock #	price	input	output	supplier
56-226	$6	+12V	±15V/33mA	*Hosfelt*
56-304	$9	+5V	±12V/415mA	*Hosfelt*
56-305	$6	+5V	±12V/40mA	*Hosfelt*

New prices on dc to dc converters are steep, placing them beyond the means of most amateurs. Surplus units listed in this table are bargains. They have high efficiency, good input/output regulation and low noise ripple.

Voltage regulation

Do you know how to run a 6V portable CD player from a 12V automobile lighter socket? It can be done for less than $1 as shown in Figure 4-7.

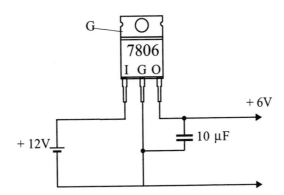

Figure 4-7. Portable radios, tape and CD players can be operated from an auto lighter socket using this 12V to 6V converter. An LM317 variable regulator can also be used to provide any voltage from 3 to 10V at 1 amp. Protection diodes are not required for fixed voltage regulators but must be used with an LM317 to protect the regulator in case of a short circuit.

Three terminal regulators such as the LM7806 (59¢ *Hosfelt*), are inexpensive versatile devices capable of regulating dc voltages with a minimum of added components. They are widely available in positive and negative versions at fixed voltages (typically 5,6,8,12,15,24 volts), or as variable regulators providing outputs from ±1.2V to ±37V. Pinouts and essential features for fixed and variable voltage regulators are given in Figure 4-8.

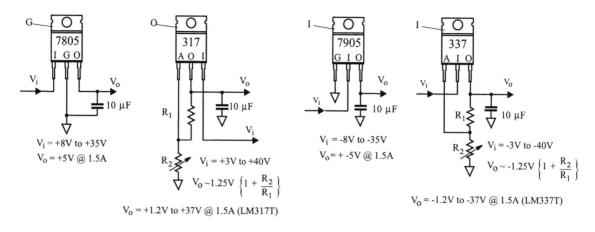

Positive voltage regulators   Negative voltage regulators

Figure 4-8. Three terminal voltage regulators may look the same but sometimes have different input/output pin configurations. Always check pinouts before wiring, because a regulator can be destroyed if installed incorrectly.

A voltage drop occurs across any regulator (for the same reasons covered in Question 3 of Chapter 2), and devices in Figure 4-8 are no exception. Allowance must be made for these drops by providing extra voltage at the regulator's input terminals. For example an extra 1.5V is required at 100mA, and 2.5V headroom should be provided when current consumption increases to 1amp.

Constant current regulation

Current regulation is required less often than voltage regulation in projects, but if you need a current regulator try the circuit in Figure 4-9, it is one of the simplest and very effective. This constant current regulator can also be used for charging nickel cadmium batteries or anywhere an invariant current supply is required. Input voltage in the figure can be increased if a larger heat sink is installed. Be sure to install an appropriate heat sink for large currents or the regulator will shut down.

    With an appropriate pass transistor, the circuit in Figure 4-10 can be used in projects to control even very large currents. This circuit is useful when a load is not required to be exactly at ground or power supply potential. Usually R is small and this presents no serious difficulty. The principle used in this circuit is fundamental to precision regulation in many applications.

    Field effect current regulator diodes are also available from ~0.22mA to ~4.7mA but hobbyists will find the circuits of Figures 4-9 and 4-10 more generally applicable. Both circuits can handle large currents and are constructed from readily available parts.

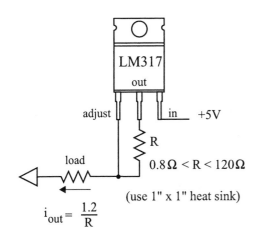

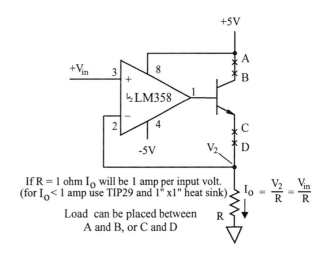

Figure 4-9. constant current regulator.
redrawn with permission from National Semiconductor's Linear Databook (1982)

Figure 4-10. constant current source

Batteries

Terms such as energy, power, work, weight, mass, speed and velocity have specific meanings but are so frequently misused in casual conversation it is small wonder confusion exists. Roboticists should appreciate the distinction between energy, work and power, and examples given in this chapter show how these terms are applied in practical problems.

Energy can be stored in many forms. For example, rotational energy in a spinning flywheel can be downloaded through a generator to provide electrical current, or spun up to speed to restore its initial energy. This might make a good project, but from a practical viewpoint a roboticist needs a convenient portable energy source, and for most autonomous project applications it's hard to beat a battery.

Never discount any energy sources at the planning stage, because they may be more efficient than batteries for certain applications. For example, butane or camping fuel can provide far more heat than a battery of the same weight, and might be a solution if a heater is required. If your uncle is a NASA bigwig perhaps you can borrow a high-tech fuel cell - the most efficient regenerative energy source ever devised.

Unlike capacitors, batteries do not store electrical charge, they generate electric current through chemical reactions. Once these chemicals are depleted, the battery is flat. Rechargeable, or 'secondary' batteries can be regenerated by putting back more energy than was taken out, (typically 50% more). Any battery can be recharged to some extent from a half-wave rectified source. However, some 'primary' or one-time use batteries may explode if they are over charged. Lithium and mercury cells are especially hazardous.

Most chemical reactions go more slowly as the temperature drops, and at low temperatures a battery's shelf life due to self discharge can be extended considerably. Weather balloon telemetry systems use a dry battery that can be stored 'indefinitely'. It is activated with water just before launch. Tiny mercury, silver-oxide and lithium batteries[1] are used in watches and hearing aids. Polaroid makes thin sheet Pola Pulse ® 6V batteries that can put out 10 amps for a second or two. Researchers have examined all the promising elemental combinations to make the best batteries, so it is unlikely you can beat the big boys by using kitchen chemicals.

---

[1] Literally speaking a battery is a series or parallel combination of 'cells', (the fundamental electrochemical units that produce electric current). Try asking a salesperson for a 'cell' to power your hearing aid!

A useful book from Radio Shack contains discharge curves, and specifications on dozens of batteries. Basic data on a much wider range of cells is given in a CRC handbook listed below.

> Enercell Battery Guidebook 62-1304, 232 pages $5.99 *Radio Shack*
> Handbook of Tables for Applied Engineering Science   *CRC Press* (in technical libraries)

Home made electrical cells

Almost any two dissimilar metals dipped in an acidic solution will generate an electrical current. Pieces of aluminium and copper foil (0.5" x 0.5"), in lemon juice develop about 0.5V. If we put a 1kΩ resistor across this cell, its voltage drops to 0.2V, showing it produces about 200μA. This is pathetic performance, when even a tiny watch battery can deliver one thousand times this current at 3V.

So if voltage alone is not the key to battery performance, how many parameters are important when choosing batteries for a project? A short list is given below with some dependent parameters.

Initial cell voltage	(open circuit voltage for a new or freshly charged cell)
Plateau voltage	(A constant potential region during discharge)
End point voltage	(minimum operating voltage)
Internal resistance	(determines short circuit current capability)
Shelf life	(governed by self discharge characteristics)
Charge procedures	(fast, normal, float charge rates, fixed current or fixed voltage)
Float life	(cell life under a minimum charge rate that provides good service)
Temperature characteristics	(cell life, current and voltage performance etc.)

Note this list does not mention electrical noise or ripple, that is because batteries have the lowest noise and ripple of any readily available source. If your project is plagued by noise, you can do no better than use battery power for your circuitry.

Some batteries commonly used for robotics are shown in Figure 4-11.

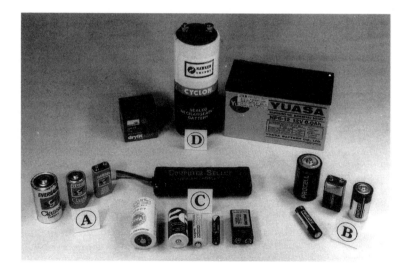

Figure 4-11.   Although many types of batteries are sold, only a few are suitable for robotics because of cost, availability, voltage, and current capability. Overall, hobbyists favor SLA and NiCd batteries for medium and large scale projects, and alkaline or heavy duty cells for low power devices. Very heavy power needs are least expensively handled with lead acid auto batteries.   [A] heavy duty cells.   [B] alkaline batteries give better performance than heavy duty cells.   [C] nickel cadmium batteries are most cost effective for projects. [D] SLA (sealed lead acid) batteries are excellent performers with stable voltage output.

Battery characteristics

Microprocessors, op amps, optical detectors and other circuitry require efficient regulated power sources, and these can be derived from almost any type of battery. However, batteries that work well in smoke detectors and small toys where voltage stability is a secondary consideration, are not always the best choices. Data on cell performance given in this section will enable you to select an appropriate battery for your specific application.

Discharge performance curves

Alkaline and ZnCl (zinc chloride or heavy duty) are inexpensive one-time use primary cells. They provide a continuously decaying voltage output under a constant current load, as shown in Figure 4-12. Although these cells continue to provide power below one volt, this low voltage energy cannot be used efficiently if regulated outputs are necessary.

Rechargeable alkalines initially have the same performance as regular alkalines, but because their total stored energy decreases with each recharge (for a maximum of about 25 charges), they are tricky to use if reliable and repeatable performance is a requirement.

Both NiCd (nickel cadmium or 'Nicads') and SLA (Sealed Lead Acid or 'gel cells') have a useful flat voltage region or plateau, of predictable duration if batteries are given proper care. This chapter shows that 'caring' is an important factor. If you look after batteries they will reward you with fine performance, neglect or abuse may require a new set of batteries.

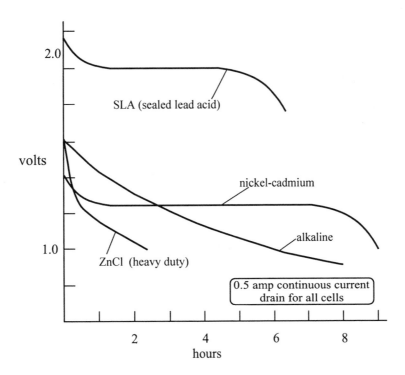

Figure 4-12. These discharge characteristics for D size units, show both NiCd and SLA cells have steady plateau voltages that are easily regulated with little energy loss. Regulating alkaline and heavy duty cell voltage is more difficult, because considerable energy must be wasted during initial discharge (0 to 2 hours) if a stable regulated voltage output is required from 0 to 6h.

Battery ratings

Manufacturers assign their batteries a 'C' rating equal to the battery's energy capacity measured in ampere hours (Ah), when discharged at a C/10 rate.

Example: C ≈ 4Ah for a D size nickel cadmium cell.

Therefore, if a NiCd 'D' cell is discharged at 0.4A (the C/10 rate), it will take 10 hours for cell voltage to drop to 1.0 volt. This result can also be inferred from Figure 4-12. Additional data for other types of cells are given in Table 4-6.

Table 4-6.   Ampere-hour capacities for D cells, (bench test results on new cells).

discharge rate (A)	NiCd	alkaline	ZnCl[a]	SLA[b]
0.1	5.9	8.7	3.4	2.8
0.5	4.3	3.3	1.4	2.5
1.0	3.8	1.1	0.75	1.8
initial voltage (V)	1.25	1.58	1.60	2.15
final voltage (V)	1.00	1.00	1.00	2.00

[a] 'heavy duty' cell   [b] Sealed Lead acid or 'gel' cell

D cell capacities in ampere-hours (Ah) to 1 volt/cell at 20°C are listed here, but the capacity of any sized cell or battery is closely proportional to its volume. Procedures for assessing a battery's capacity from its size are explained in the text, and volume ratios for different cell sizes are given in Table 4-9.

Battery energy and power - estimating battery life

If you were given a flashlight, two alkaline D cells and a bulb rated for 2.4 volts at 0.5 amp, could you calculate how long the bulb will stay alight?

Two cells must be connected in series to give 2.4 volts, with 0.5A passing through each cell. Table 4-6 lists the capacity of an alkaline D cell at 3.3 Ah, and this energy is sufficient to keep the lamp running for 6.6 hours at 0.5 amps.

As a check we can compare total energy used by the lamp, with that provided by the battery:

Lamp power = 2.4V x 0.5A = 1.2W

Energy used by lamp = power x time = 1.2W x 6.6h x 3600   = 2.85 x 10^4 Ws   = 2.85 x 10^4 J

Energy used by two D cells = 2 x 3.3Ah x 3600 x 1.2V   = 2.85 x 10^4 J

(1hour = 3600seconds   1watt-second (Ws) = 1joule = 1J)

These calculations are only approximate, because lamp filament resistance changes as the battery gets weaker, as shown in Table 4-7.

Flashlight bulbs last much longer when run from NiCd batteries, because they are not subjected to the high turn-on voltages (~3.2V) that occur when alkaline or ZnCl cells are used.

Table 4-7. Flashlight bulb filament resistance

V	i (A)	filament resistance (Ω)	
0.0	0.00	0.67	cold resistance
1.0	0.33	3.03	
1.5	0.38	3.95	
2.0	0.44	4.50	
2.4	0.475	5.05	**nominal operation**
2.5	0.49	5.10	
3.0	0.53	5.66	
3.25	0.56	5.80	

Flashlight bulb operating characteristics given here show filament resistance increases with temperature. All tungsten incandescent lamps exhibit similar behavior, and this phenomenon provides a self current limiting feature, or ballast effect.

Energy and volume density

Battery manufacturers try to pack as much energy into each cell as possible, while keeping cell weight and volume low. They have to juggle other parameters too, such as discharge characteristics, storage life, high and low temperature performance, so their task is not an easy one.

Energy and volume densities are important for electrically powered aircraft, blimps and many other robotic projects and typical values are given in Table 4-8. Fuel cell and gasoline equivalents are also included for comparison.

Ratings in Table 4-8 can vary considerably with temperature and test conditions. Nickel cadmium and lead acid batteries outperform other cells at low temperatures. Unlike NiCd and SLAs, alkaline and ZnCl performance increases significantly for intermittent operation. This is because rest periods permit inhibiting chemical by-products to dissipate by diffusion.

Table 4-8. Energy densities for batteries, fuel cells and gasoline.

	NiCd	Alkaline	ZnCl	SLA
Wh/kg	34	27	16	29
Wh/cc	0.11	0.08	0.03	0.11

Fuel cell	Gasoline
10 - 3500[a]	13250[b]
0.03 - 11	9.5

(For D size cells, 0.5A continuous operation at 20C)
1cc = 0.061cu.in.   (0.11Wh/cc = 1.8Wh/cu.in).
1kg = 2.2lbs   (34Wh/kg = 15.5Wh/lb)

[a] thermal efficiency to 75% is possible
[b] for 100% fuel energy conversion

Weight and volume energy densities for batteries, fuel cells and gasoline show why gasoline has a distinct advantage when used in automobile engines. Fuel cells show the greatest promise as an alternate power source for automobiles.

It may be concluded from Table 4-8 that fuel cells and gasoline have a clear advantage over batteries for automobile propulsion, because of their higher energy densities. Recent trials of fuel cells in buses show these also meet DOE and US car makers' requirements for passenger automobiles. Ballard (Vancouver, Canada) fuel cells give buses a 250-mile operating range without refuelling, and have an energy density of 2500Wh/kg.

Battery dimensions

Dimensions vary even among cells of the same size mainly due to packaging, but typical outside dimensions for readily available consumer products are listed in Table 4-9, terminal projections are not included.

Table 4-9. Dimensions of popular cells and batteries

Size	outside dimensions	volume (cc)	Volume ratio (D cell =1)
AAA	1cm dia. x 4.2cm	3.3	0.067
AA	1.4cm dia. x 4.7cm	7.2	0.15
C	2.5cm dia. x 4.6cm	23	0.47
D	3.3cm dia. x 5.7cm	49	1.00
9 volt	1.7 x 2.5 x 4.5cm		~0.06

Battery and cell dimensions given in this table can be used to calculate the energy capacity of any size NiCd, SLA, alkaline or other cell. This is because energy content scales closely with battery volume. (1 cu.in. = 16.4 cc)

Calculating energy capacity for any battery size

A large battery will evidently last longer than a small battery of the same type for the same current drain. But how much longer?

Well, a battery's stored energy is proportional to its electrolyte volume. This means the energy capacity of a cell or battery can be estimated, as shown in the following example:

Example: How long will an alkaline AAA cell provide a continuous current of 0.1A before its voltage drops to 1 volt?

From Table 4-6:	D cell capacity	= 8.7Ah at 0.1A
From Table 4-9:	Volume ratio (AAA/D)	= 0.067
	AAA capacity	= 8.7 x 0.067 = 0.58Ah (cell lasts for 35 minutes)

This method can be used to estimate the approximate capacity of larger or smaller cells.

Battery cell weights

Even for cells of the same type and size, cell weight may differ by 10% or more, so weigh each cell if weight is critical for your application. Some manufacturers package C size NiCd cells into a D size case, and these are often sold at D cell prices. If you suspect a D cell is a lightweight, it is probably a masquerading C cell. Battery and cell weights of the most commonly used items for robotics are listed in Table 4-10.

Table 4-10. Battery weights

size	NiCd	alkaline	ZnCl	SLA
AAA	10g	12g	10g	
AA	22g	22g	18g	
C	52g	66g	48g	
D	158g	142g	102g	180g
9 volt	36g	46g	38g	

Battery and cell weights are listed in grams, but listed weights vary slightly depending on a manufacturer's packaging methods. (28.4g = 1 oz)

Internal resistance of batteries

Every battery has the equivalent of a resistor inside each cell, limiting the maximum current a battery can provide. This 'internal resistance' is an important battery parameter, because it governs voltage available at a battery's terminals when it is under load.

A cell's electrochemical potential V_o is not altered significantly when the cell is loaded by an external resistor R. Therefore, V_o of Figure 4-13, can be measured at A, B with a DVM, when R is removed. This is an accurate measurement of V_o because no appreciable current flows through r, (see Question 2 of Chapter 2).

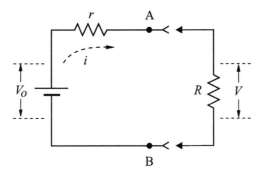

Figure 4-13. All cells have some internal resistance, (r in the figure). The value of this resistance can be determined by using an external load R, a voltmeter, and procedures given in the text.

Referring to Figure 4-13, internal resistance r is determined as follows:

[a]   With R removed, V_0 is measured at battery terminals A, B.
[b]   With R inserted, V is measured at A, B. Also $V = i R$.
[c]   By Kirchoff's second rule: The fundamental cell potential V_o must be equal to the sum of all voltage drops in the circuit, ie. $V_o = i(r+R)$. Internal resistance r can now be calculated using this expression.

Measuring internal resistance

Warning: Internal resistance r could be determined from measurements of open circuit voltage and short circuit current i_s using the following relation. However, this can be a dangerous procedure for amateurs.

$$r = \frac{V_0}{i_s}$$

Nickel cadmium and lead acid cells can deliver very high currents that may cause a fire or severe burns. Industrial accidents have been reported where metal watch bands have fused across the terminals of high capacity batteries. In addition, cells or batteries can be destroyed or have their performance permanently impaired by a short circuit test.

Internal resistance can be measured safely and accurately as follows:

[1]   Because r is usually less than 1Ω, short lengths of flexible heavy gage wire must be used for these tests. Three separate leads are more flexible than one heavy lead, and soldered ends make good contact pads as shown in Figure 4-14.

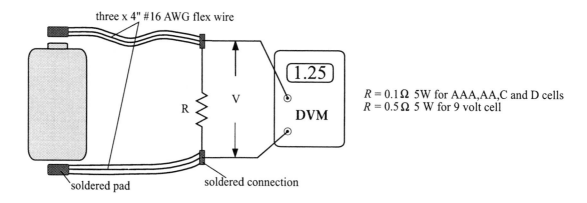

Figure 4-14.   Heavy gage flexible wires making good contact with the battery electrodes, are essential for measuring internal resistance accurately. Three thin stranded wires shown have the same current carrying capacity, but are more flexible than a single heavier gage wire. 'Monster' #12 gage hi-fi speaker wire is ideal for these tests.

[2]   Touch soldered wire ends to the cell or battery firmly but *QUICKLY* and record V.
Do not wait for the DVM reading to settle to a stable value - by that time the battery will be flat.

[3]   Measure V_0 with R removed from the battery.

The following results were obtained for a single nickel cadmium D cell:

$$V = 0.97\ volts \qquad R = 0.1\Omega \qquad i = \frac{V}{R} = 9.7\ amps$$

$$V_0 = 1.36\ volts \qquad r = \frac{V_0}{i} - R = \frac{1.36}{9.7} - 0.1 = 0.04\Omega$$

Even this small cell packs a wallop, and in a short circuit test can deliver a short circuit current:

$$i_s = \frac{V_0}{r} = \frac{1.36}{0.04} = 34 \; amps$$

Internal resistances for some common cells and a 9V battery have been measured using this method and are listed in Table 4-11. These results show why nickel cadmium or lead acid batteries should be selected whenever high current performance is required.

Table 4-11. Internal resistance of common cells

size	NiCd	Alkaline	ZnCl	SLA
AAA	0.06Ω	0.3Ω	0.46Ω	
AA	0.04Ω	0.24Ω	0.41Ω	
C	0.05Ω	0.18Ω	0.26Ω	
D	0.04Ω	0.12	0.20Ω	0.03Ω
9 volt	1.5Ω	2.6Ω	8.0Ω	

Internal resistances for common batteries listed, show why SLA and NiCd batteries are favored by hobbyists for operating motors and other high current equipment.

Testing batteries for their remaining capacity

"No quick and simple test will determine a battery's available energy capacity"

"What about battery voltage?"

Open circuit (no load) battery voltage is an unreliable measure of remaining capacity. A well-drained battery may indicate almost 95% of its full charge voltage if it has not been used for a week or so. Similarly an almost flat battery will have the same voltage as a new battery after only a short charge, though it may be incapable of doing much useful work.

"How about battery testers?"

A battery tester will show whether a cell is exhausted but can give no reliable measure of remaining capacity. A good checker must apply a load of at least C/5, but most testers employ much lighter loads.

Although no quick and simple test exists for testing primary battery capacity, a methodical experimenter can foretell almost exactly how lead acid and nickel cadmium batteries will perform using procedures outlined below.

SLA (Sealed Lead Acid) Batteries

SLAs are manufactured in cylindrical and rectangular packs, and available as 2V cells or 6V and 12V batteries, in a variety of capacities. Because they have no 'memory' effects, are easily mounted, and can deliver high currents at steady voltage levels, SLAs are usually the #1 choice for heavy duty robotic applications.

A simple method to determine available capacity of sealed lead acid cells

Remaining capacity at a C/10 rate for SLA cells can be estimated from Figure 4-15, which is accurate to about 20% if the cell has not been used for 24 hours. Accuracy improves to 5% with no activity for 5 days. This test is simple but not quick, because of a mandatory 24 hour hiatus.

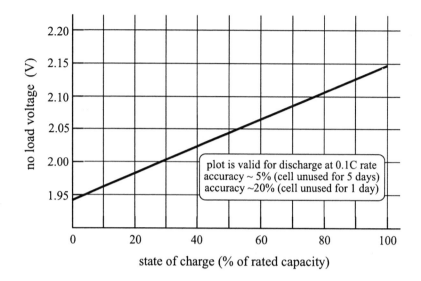

Figure 4-15. This useful plot shows the remaining capacity of an SLA cell as a percentage of its nominal maximum value. If a cell has not been used for 24 hours, cell voltage can be used to predict remaining capacity to within 20%. There is no similar test for NiCd, alkaline or heavy duty regular batteries. *Figure redrawn by permission of Hawker Energy Products Inc. 617, North Ridgeview Drive, Warrensburg, Missouri. 64093-9301. (816) 429-2165.*

Be careful when buying used SLA batteries. Take your voltmeter to the store and measure battery voltage before buying. Some surplus items are sold 'AS IS' and cannot be returned for a refund. If SLA cell voltage drops below 1.8V (10.8V for a 12V battery), it will not accept a full charge.

SLA float, service and storage life

SLA batteries under a float or trickle charge at C/500, will last for more than 8 years at 20°C. Service life depends on the type of loads and their frequency. Continuous deep discharge results in a shortened life of about 200 cycles, but this rises to ~2000 cycles for light work loads.

A battery's performance deteriorates if it is left uncharged for more than two years. Cold temperatures extend shelf life and stored capacity, but high temperatures accelerate energy loss dramatically (ie. 0% of a full charge remains after four months at 40°C/124°F).

SLA gassing, electrolyte leakage and mounting orientation

Hydrogen permeates through all materials to some extent, and SLA casings are no exception. Under normal circumstances the minute amounts of hydrogen liberated during charging are harmless, and readily dissipate with normal ventilation. However, gas tight containers should never be used to encase SLAs, since gas accumulation could reach explosive critical mixture concentrations (4 to 74% H_2 in air).

Electrolyte leakage does not occur, unless structural integrity of the casing has been compromised, so SLAs can be mounted, used and charged in any position.

Charging SLA batteries

Constant voltage charging is recommended since this is the simplest, quickest and most efficient method for SLA cell regeneration. Sealed lead acid battery chargers should have a current capability of at least 2C and be set at 2.27V per cell (ie 13.6V for a 12V battery). This voltage is also safe for trickle or float charging SLAs.

SLA open circuit and endpoint discharge potentials

Open circuit voltage after cell inactivity for 24h is a reliable state of charge indicator, as shown in the graph of Figure 4-15. For maximum service life a cell should not be discharged below 1.6V even for short durations.

Suppliers of SLAs and chargers

Hawker Energy Products Inc., Newark, Radio Shack ,Tower Hobbies, and an increasing number of hobby and electronic outlets stock SLAs. *Tower Hobbies* prices start at $9.99 for 6V 1.3Ah and $19.99 for 12V 7Ah batteries, chargers are $15 each for either voltage.

Nickel cadmium rechargeable batteries

Testing NiCd batteries

No quick or simple capacity test exists for NiCds, but thousands of model plane and model race car enthusiasts use similar procedures to those outlined below, to get predictable high performance. They are the only known techniques for reliably assessing rechargeable nickel cadmium cell capacity.

Charging and rejuvenating NiCd batteries

Nickel cadmium cells benefit from exercise. Periodical depletion until they are almost exhausted (to 1.0V/cell), followed by charging at a C/10 rate, gives cells a new memory and restores their muscle power.

The following procedures are recommended for new cells, or in cases where individual cell performance history is unknown.

[1]   Discharge each cell to 1 volt by running batteries to depletion in a portable radio, flashlight, or by connecting a discharge resistor across each cell. Resistors listed in Table 4-12 limit discharge current to a C/10 rate and no cell damage will occur even if cells go below 1.0 volt for a few hours. Installing these resistors on single cell battery holders, makes discharging and recharging more convenient.

[2]   NiCd cells should be charged at C/10 constant current for 1.5 times their C capacity (Table 4-12).

[3]   Discharge cell at a C/10 rate and record the time to reach 1V/cell.

Repeating [2] and [3] a few times restores cells to their maximum capacity. Future performance can then be predicted from discharge records, because cell behavior is repeatable after rejuvenation.

Table 4-12. Discharge parameters for NiCd cells

size	C rating (Ah)	discharge resistor (~C/10)	discharge time[b] to 1.0 volt/cell	charge time (h)	C/10 charge current (A)
AAA	0.2	47Ω (0.5W)	8h	15	0.02
AA	0.5	27Ω (0.5W)	11h	15	0.05
C	2	4.7Ω (0.5W)	8h	15	0.20
D	4	2.7Ω (1W)	9h	15	0.40
9 volt[a]	0.1	1kΩ (0.5W)	12h	15	0.01

[a] battery contains 7 cells   [b] for a fully charged cell

Discharge resistors and C/10 charging times for various sizes of nickel cadmium cells are listed in this table. Fully discharged cells can be safely charged for rates and times given in the last two columns. Fast charging is ok too, if cell temperature and open circuit voltage are monitored to prevent overcharging.

Fast Charging nickel cadmium cells

Constant current charging up to a 0.4C rate is ok for most NiCd cells. Higher charging rates are possible, but cell temperature and open circuit voltage must be continuously monitored, to prevent overcharging.

Eliminating shorting whiskers

Dendrite formations sometimes internally short a nickel cadmium cell. Open circuit voltage drops to zero, and the cell will not accept a charge. Connecting a 12V/1A power supply across the cell for one second zaps whiskers, and recharging is then possible. This is not a complete cure, because cell capacity decreases slightly with each new growth of whiskers, but it does extend cell life considerably.

Caution: Connecting a 12V supply across a single NiCd cell for more than a few seconds may permanently damage a cell or even cause it to explode from overheating.

Erasing memory effects

Partial charge/discharge cycles reduce the working capacity of a nickel cadmium battery, but application of steps [1], [2] and [3] will restore it to good health. This phenomenon is termed 'memory' because the cell keeps an internal chemical record of its last maximum performance, and does not try any harder subsequently!

Cell voltage reversal

If a NiCd cell is discharged at a very high current to below 0.5V, its output polarity may reverse. This polarity reversal can sometimes be corrected by discharging the cell to less than 0.1 volt, using a C/10 discharge resistor from Table 4-12. A shorting wire is then be used to bring the cell voltage to zero, followed by a regular charge at a C/10 rate.

Storage life and float charging

Nickel cadmium cells lose about 25% of their full charge in about 30 days by self discharge, and are down about 90% in six months. A float charge at C/100 will keep cells topped up without damage. With care NiCds will last 10 years or more and cells can be used for more than 1000 charge/discharge cycles.

Nickel cadmium battery packs

Commercial battery packs are series strings of individual cells, matched for equal performance by the manufacturer. After prolonged use (or abuse), a weak cell may not accept a charge and other cells are then subjected to above normal charging rates, hastening their demise too. Checking individual cell voltages of battery packs periodically forestalls such problems, since frequently only one cell need be discarded. If a battery pack is sealed in shrink tubing, it may be possible to poke a pin through this covering to reach each cell's electrodes for measurement.

Replacement cells should definitely be of the same capacity and preferably from the same manufacturer. For example 800mAh NiCd AA cells are available and these should not be mixed with 500mAh AA types. Incidentally, 800mAh cells have higher internal resistance than 500mAh cells so the latter have superior high current capabilities.

Temperature performance of NiCds

NiCds retain about 50% of their room temperature capacity even at -5°F (-21°C), and easily outperform alkaline and ZnCl cells at low temperatures. Of the four types of cells discussed in this chapter SLA batteries are the best choice for very low temperature operation, retaining more than 50% of their capacity at -18°C (0°F).

Soldering nickel cadmium batteries

Soldering to regular NiCd cells can be tricky even for a pro, because excessive heat can cause internal cell damage. Cells with stainless steel tabs suitable for soldering, are available from most large hobby outlets.

NiCd battery cyclers and chargers

Nickel cadmium battery cyclers are currently available from $80, but prices are dropping with increasing sales. Advertisements in *'Radio Control Car Actions'* and *'Model Airplane News'*, follow the latest technologies.

Chargers for NiCds are widely available from inexpensive to exotic, and building a charger is uneconomical unless one has a bin full of spare parts.

Where to buy nickel cadmium batteries

Tower Hobbies sells 7.2V and 8.4V 1.4Ah C cell battery packs for $16.99, and individually tabbed or regular 600mAh AA cells from $1.79. *Radio Shack* also stocks cells and battery packs. Hardware stores now sell 7.2, 8.4, 9.6V and 12V NiCd battery packs for cordless power tools, and some drugstores keep rechargeables on their shelves.

Choosing a battery for your project

Small projects

Battery current drain and run times determine which type of battery is best for a project. Cash is also an important element in the selection process. For small projects, operated infrequently at current drains of less than 500mA, alkaline or heavy duty ZnCl batteries are a cost effective and convenient choice.

An investment in good rechargeable batteries is never wasted. They can be used for many different projects, and can always be sold to other roboticists or surplus outlets.

Batteries for large motors etc.

When operating large motors, powerful electromagnets, high powered lamps etc., SLA or NiCds should be given first consideration. Initial costs are higher but steady plateau voltages, low internal resistance and repeated use at new battery performance after charging, are major benefits.

Because of their unfavorable voltage/time discharge characteristics, it is difficult to obtain regulated power from alkaline or ZnCl cells without large energy losses.

Batteries for high energy projects

A large footprint hovercraft can lift a 150lb person using electrically powered propellers, but it will not run very long on small alkalines. Automobile batteries are the best choice when brute force is required.

Congenial managers of auto service centers will sometimes give good 'trade-in' batteries away, in exchange for a smile and a few kind words - it's hard to beat that price. Always explain why you need the battery, how hard you've worked to pay for other project items, and why you came to the best managed auto service center in town.

"Can I run everything from one or two batteries?"

In principle one or two large batteries and a few dc to dc converters will provide all the bipolar capability needed for any project. In practice, fewer problems arise when logic and control circuitry are run from separate batteries, using appropriate regulators or dc to dc converters. It is always best to keep heavy duty noisy equipment separated, unless a test shows full compatibility with all other elements.

Deriving special voltages from battery taps is seldom a good policy with rechargeable batteries. Depending on current drain, this procedure sometimes results in unbalanced cell use and subsequent overcharging, shortening battery life. If cells are installed in battery holders, battery taps are ok, but cells should then be recharged individually.

"What about nickel-metal-hydride batteries?"

These cells deliver about twice the energy density of older NiCds, but at three times the cost. NiMH cells work well for low current drain applications, and are widely used in laptop computers. Fliers and model race car enthusiasts still prefer NiCds, and they know what is best for high current performance

Improved 950mAh nickel cadmium AA sized cells have almost 80% of NiMH energy density, while costing only half as much. A down side to NiMH and high energy NiCd batteries, is increased internal resistance, which translates into lower peak current capability.

Comparing humans and machines

In our hi-tech society, many wonderful fruits of science and technology are taken for granted. On the other hand it can be chastening to see that although great strides have been made in technology, we are still a very long way from duplicating animal performance and efficiency. Comparing human performance with that of machines is informative for hobbyists, since it places their own mechanical robotic efforts in perspective.

As a practical case example, we will examine the problem of moving a 130lb (59kg) person over a distance of 50 miles (80.5km) by automobile, powered bicycle, and using muscle power alone.

Real automobile efficiency

Although most people know fuel is not used with complete efficiency in cars, few realize about 85% of gas put into the tank is wasted as heat or chemical by-products.

A rough budget in Table 4-13 details how energy is used in an internal combustion automobile engine. This is not the whole story of course, since stop and go traffic or high speed driving increases gas consumption, and older heavier vehicles are less fuel efficient.

Table 4-13. Energy budget for a six passenger automobile

	energy usage *
exhaust and cooling system	60%
friction, transmission and drag force	20%
air conditioning, lights and accessories	5%
drive power at wheels	15%

* Vehicle performance = 25 mpg (9.4 liters/100km) at 50 mph

Automobile energy usage itemized in this table shows only 15% of gas tank fuel is used to propel a standard automobile. Even with a perfect internal combustion engine and ideal zero friction components, 70% of fuel energy is still wasted.

Automobile energy and power requirements for a 50-mile trip

An automobile with a 25 mpg rating uses 2 gallons (7.57 liters) of fuel to travel 50 miles:

$$Energy \ used \ by \ automobile \ for \ 50 \ mile \ trip \ = 2 \ x \ 1.3 \ x \ 10^8 = 2.6 \ x \ 10^8 J \ \ ^2$$

[2] 1 gal (3.785 liters) of gasoline = 1.3×10^8 J (joules). This is called the 'fuel energy equivalent' and is determined by measuring heat energy liberated when fuel is burned in a calorimeter. The same technique is used to measure food energy, although some dieters claim it takes more calories to chew and swallow celery than are released in the gut.
  A joule (J) is a unit of work or energy = force x distance   [1 joule = 1Newton-meter (Nm)]

"What is a joule?"

You can do one joule of work very simply by performing steps [1] and [2] below:

[1]   Place a 1kg [3] (2.2 lb) weight on the floor.

[3] A kg is a unit of mass but 'weight' is a force. On earth, one kilogram has a weight equal to *mg* or 9.81N.

[2]   Lift the 1kg mass to a height of 1 meter (39.4") in one second. Energy used is: *mgh* = 9.81 x 1 = 9.81N

You have just done 9.81 joules of work, and the weight has acquired 9.81 joules of potential energy. You also applied power at the rate of 9.81 watts (W), because:

$$energy \ = \ power \ x \ time \qquad (for \ example: \ 1 \ kilowatt \ hour \ = 1000 \ W \ x \ 3600 \ s \ = 3.6 \ x \ 10^6 J)$$

[3]   Repeat [1] and [2] **13.3 million times**. This takes about 10 months if the weight goes up and down in two
       seconds.

You have now used the same amount of energy available from 1 gallon (3.79 liters) of gasoline!

Useful automobile power

If our auto is driven on level ground, at constant speed its average total power expenditure is:

$$Average\ total\ power = \frac{energy\,(J)}{time\,(s)} = \frac{2 \times 1.3 \times 10^8}{3600} = 72.2\ kW$$

Referring to Table 4-13, we note only 15% of this power is used for moving the vehicle:

$$Propulsion\ power\ at\ automobile\ wheels = 72.2 \times 0.15 = 10.8\ kW\ ^4$$

---

[4] An electric clothes dryer heater uses approximately 4kW, and total top burner and oven power on a full size (30") electric stove is about 10kW. By comparison a relaxed adult has a heat output of 80 kcal/h or 93 watts.

A 'perfect' gasoline auto engine

An equation developed by French engineer Sadi Carnot in 1824 enables us to calculate the maximum possible efficiency for an ideal or perfect heat engine. Even Mercedes or Lexus cannot reach this efficiency with a practical engine.

$$Efficiency = 1 - \frac{T_c}{T_h}$$

T_c = cold reservoir temperature
T_h = hot reservoir temperature

For a water cooled auto engine we can use a cylinder temperature T_h = 150°C (423K), and on a warm day T_c will probably go no lower than about 25°C (298K). The thermodynamic efficiency is:

$$Efficiency = 1 - \frac{298}{423} = 0.296$$

So conventional gasoline engines have an upper efficiency limit of about 29.6% and this cannot be attained in practice. Diesel engines can have higher thermal efficiency (30 to 40%), but are subject to most other losses listed in Table 4-13.

Energy used during a 50 mile run

A good marathoner weighing 130lbs, can run 50 miles comfortably in 6 hours and 40 minutes (eight minutes per mile), using about 60 Calories [5] of energy for each mile*. Top performers cover 50 miles in less than 5 hours so this is an easy jog for a trained athlete. The 'gross' or total food energy used by a runner will be:

$$Gross\ food\ energy\ intake\ required\ to\ run\ 50\ miles = 50 \times 60 \times 10^3 \times 4.2 = 1.26 \times 10^7\ J$$

---

[5] Calories (uppercase 'C'), quoted by physicians, dieters and food labels are really kilocalories. i.e. 60 Calories = 6 x 10⁴ calories where 1 calorie ~ 4.2 J

* Reference data on energy expenditure varies widely and untrained runners may use 90 Calories/mile.

Food to energy conversion for humans

Experimental measurements on humans, show a body has an efficiency of about 24%. In other words about one quarter of food eaten is converted into useful work or energy.

Using Carnot's equation again, we can calculate how hot our muscles must be to give this efficiency. Assuming it is a warm day ($T_c = 20°C$):

$$Efficiency \ = \ 1 \ - \ \frac{T_c}{T_h} \qquad 0.24 \ = \ 1 \ - \ \frac{293}{T_h} \qquad T_h \ = \ 386K \ = \ 112.5°C$$

WOW! blood temperature must be a little **above the boiling point of water!**

This calculation illustrates how careful one must be in using equations. Humans are not heat engines and do not convert energy according to the Carnot cycle. Animal energy is produced through complex chemical interactions at the cellular level, similar to a fuel cell.

Food requirements for a 50 mile run

Food can be classified into two main groups for energy purposes.

[1]   Fats: These include alcohols, oils, and greases such as butter, animal fat etc.
         Fat fuel energy equivalent   ~ 9 kcal/g   = 3.78×10^4 J/g

[2]   Carbohydrates: Starches, sugars and proteins
         Carbohydrate fuel energy equivalent [6]   ~ 4 kcal/g   = 1.68×10^4 J/g

[6] starch ~ 4kcal/g,   protein ~4.3kcal/g,   milk sugars ~4kcal/g,   fructose or glucose ~3.8kcal/g

A gram of chocolate contains about 4 kilocalories, or has an energy equivalent of 1.68×10^4 J/g. For a 50 mile trip a runner needs to eat about 1.5 pounds [7] of chocolate or its equivalent:

$$Chocolate \ intake \ = \ \frac{1.26 \times 10^7}{1.68 \times 10^4} \ = \ 750g \ \ (1.65 \ lbs)$$

Vodka has roughly the same energy density as chocolate, because 80° proof spirits are only 40% pure alcohol (ethanol). If imbibing only, a runner needs about 30 oz of vodka to stagger 50 miles.

[7] Dieters are often surprised how little weight they lose through exercise, and these calculations substantiate their worst fears. To lose one pound of body fat, one must walk or run about 35 miles! Trained runners lose less weight per mile, because their bodies are more efficient energy converters when doing their specialty.

Runner's average power output during a 50-mile 400 minute run

Recalling only 24% of food energy intake converts to useful work, average power is:

$$Runner's \ net \ average \ power \ output \ = \ \frac{energy}{time} \ = \ \frac{1.26 \times 10^7 \times 0.24}{400 \times 60} \ = \ 126 \ watts$$

A computer interfaced bicycle ergometer makes a useful project, but even fit students are surprised at their low performance levels when testing their handiwork. Generating 300 watts for 5 minutes is extremely difficult for the average person, (current world record by Eddy Merckx in 1972 is 455 watts for one hour).

Energy and power required for riding a bicycle

A bicycle is one of the most efficient means of transportation ever devised, requiring about 28% of a runner's energy over the same flat course [8].

Energy and power requirements for a rider can be obtained by scaling the runner's data:

$$Bicyclist's\ gross\ energy\ consumption\ =\ 1.26 \times 10^7 \times 0.28\ =\ 3.53 \times 10^6\ J$$

$$Bicyclist's\ net\ average\ power\ output\ =\ 126 \times 0.28\ \approx\ 35\ watts$$

[8] Top bicyclists have ridden at over 20mph continuously for 1000 miles and can ride 100 miles in under 4 hours.

Gasoline powered bicycles

Gasoline powered bicycles are manufactured, and add-on power units sold for converting standard touring machines. These usually develop ~1.5 horsepower (1.1kW) from a 50cc two stroke engine [9]. Gasoline bicycle engines propel bike and rider at up to 35mph on level ground, but rider assistance is usually required on steep hills to prevent motor stalling.

A gasoline engine powered bicycle has power train losses of about 5%, and its overall efficiency is roughly 20%. Fuel consumption on a flat course with no wind is:

$$Fuel\ for\ gasoline\ bicycle\ =\ \frac{3.53 \times 10^6}{1.3 \times 10^8 \times 0.2}\ =\ 0.136\,gal\ (514\,ml)\ =\ 370g\quad (\rho_{gas} = 0.72\ g/cc)$$

[9] Model aircraft engines with 50cc capacity develop twice the power of a bicycle engine. However, model aircraft engines are not designed for continuous running, and operate at higher revs (>7000rpm), requiring a gear reduction unit.

Electrically powered bicycles

Acceptance of electrically powered bicycles has been equivocal because their motors are less rugged, and trip lengths only a fraction of those possible using gasoline engines. DC brush motors and either wet or sealed lead acid (SLA) batteries are usually employed with a current regulator. Recharging batteries when coasting downhill has been tried with limited success, but seems to undergo a periodic renaissance.

Short trip capability, increased vehicle weight, and inconvenience of recharging have kept electric bicycle popularity at a low level. Nobody really enjoys pedalling home with an extra load of dead batteries and a heavy electric motor.

Large dc brush motor efficiency is about 60%. However, battery efficiency is reduced by a factor of 0.67 because 150% of energy used must be provided for each battery recharge. Overall battery powered bicycle efficiency is consequently about 35%, after power train losses are included.

Battery requirements for an electric bicycle

Battery weight and volume for an electrically powered bicycle can now be calculated using SLA data from Table 4-8, and applying an overall efficiency factor of 0.35:

SLA energy density = 29Wh/kg = 29 x 3600 = 1.04×10^5 J/kg
SLA volume density = 0.11 Wh/cc = 0.11 x 3600 = 396 J/cc

$$Battery\ weight\ = \frac{3.53 \times 10^6}{1.04 \times 10^5 \times 0.35} = 97kg\ (214\ lbs)$$

$$Battery\ volume\ = \frac{3.53 \times 10^6}{396 \times 0.35} = 2.55 \times 10^4 cc\ (29.4cm)^{3\ [10]}$$

[10] This is equivalent to the volume of six (6" x 6" x 7.5") auto batteries.

Total weight of bicycle, rider and battery = 404 lb   (i.e. 60+130+214)

Iterative design is a fact of life in engineering whether one is planning to build a bridge, skyscraper, jetliner or a battery powered bicycle. First estimates often fall short of initial expectations forcing reappraisals that lead to compromises, or alternate solutions.

Batteries now account for 53% of total weight. The bicycle frame, wheels and gear train must consequently be heavier, and extra batteries will be necessary just to carry the batteries. A battery powered bicycle using current technologies for a 50-mile trip seems inappropriate, and extra computational efforts will bring diminishing returns.

Critique

Admittedly this study has compared apples and oranges to some extent, but some salient points are apparent from results summarized in Table 4-14.

Although wheeled transportation is clearly superior, our runner does reasonably well, and if equipped with in-line skates will approach the performance of a bicycle rider on level ground. The automobile has been treated unfairly because it could easily carry six or more people for the same energy, but is very inefficient, even with this allowance.

Battery powered bicycles are clearly inferior for long distance travel without frequent recharging, and electrically powered automobiles are similarly disadvantaged. Our calculations explain why mileage figures for electric cars seldom include energy for air conditioning, or adequate cold weather heating. Poor performance of electric cars in cold weather or over hilly terrain, are also never mentioned by their proponents. Energy requirements for all articulated machines escalate rapidly with complexity, and if a machine must also be self-propelled, battery weight can be a very serious impediment.

Table 4-14. Energy usage and power requirements for various transportation modes on a 50mile trip

	automobile (gasoline)	runner	bicycle (human power)	bicycle (gasoline)	bicycle (battery)
energy (gross)	2.6×10^8 J	1.26×10^7 J	3.5×10^6 J	3.5×10^6 J	3.5×10^6 J
fuel quantity	gasoline 2 gal / 12 lb	chocolate 1.5 lb	chocolate 0.4 lb	gasoline 0.13 gal / 1.1lb	battery[†] 214 lb
power (net)	10 kW	0.126 kW	0.035 kW	0.035 kW	0.035 kW

[†] equivalent to six fully charged 12V automobile batteries

Energy, fuel and power requirements for moving a 130-lb person 50 miles, by several transportation modes are given in Table 4-14. A car uses about seventy-five times the energy required by a bicycle rider, and its fuel is energy intensive, because oil reserves take millions of years for restoration. Even an unaided runner gets by with about one twentieth of an automobile's energy needs. Battery powered cars and bicycles still await a quantum leap in electrical cell performance.

Perhaps the most important lesson to be learned from this exercise is how impractical and inefficient a perambulating articulated robot, resembling a human runner would be for a 50-mile trip.

Nevertheless, this situation can change quickly because contemporary fuel cell technology research is revealing some tantalizing possibilities. Polymer fuel cells now operating at about 60% thermal efficiency have an energy density of 2500Wh/kg, which is about one hundred times better than batteries presently used for electric vehicles. Buses powered with these new fuel cells are being road tested, and hopefully fuel cells will eventually appear on surplus markets for hobby use.

Sensors

Transducers are some of the most useful devices in robotics, because of their ability to transform energy in one form, into energy of another form. For example, pressure fluctuations in air can be detected by our ear transducers, then converted into electrical impulses for the brain to recognize as sound. Microphones also detect and convert sound to electrical current, and speakers can transduce this current to duplicate the original sound waves. In this book, sensors are loosely defined as those transducers such as accelerometers, microphones, photo detectors, thermometers etc., that do not produce significant force.

Approximate measurement capabilities of commonly used sensing methods are rated on a relative basis in Table 5-1. Compound qualities such as torque, spin rate, weight, force, pressure, sound levels etc. can be determined from measurements of displacement, velocity, acceleration . . . which are in turn based on fundamental units of mass, length, time, temperature etc. This means a hobbyist can measure just about anything, by choosing appropriate sensors and applying them in a suitable manner.

Optical sensors now account for many precision measurement and sensing techniques. They are not discussed in this chapter, but in-depth treatment of these devices is given in a future book in this series.

Table 5-1. Relative sensing capabilities of various transduction methods

**** = best	displacement	velocity	acceleration	time	current	temperature	flow
resistive	***	***	***	**	****	***	**
capacitive	**	***	***	**	*	*	**
inductive / magnetic	**	***	***	**	**	*	**
optical	***	***	***	***	*	**	***
piezo electric	*	***	***	****			**
thermal					**	****	***
thermoelectric					**	***	
mechanical	****		*	***		*	***

Hundreds of ingenious devices have been invented for measuring specific physical quantities, and a few are listed below

bimetallic thermal switch
bellows pressure switch
electrolyte or mercury switch
tilt sensors (capacitor or pendulum)
Pitot tubes
ionization pressure gage
ultrasonic pressure gage
vibrating wire pressure gage
Geiger counter

optical pyrometer (high temp measurements)
eddy current proximity switch
LVDT displacement sensor
ultrasonic distance gage
piezo electric microphone
strain gage accelerometer
strain gage torque meter
laser and LED range meters
ultrasonic flowmeter

These instruments are engineering gems, and a great deal can be learned by understanding their operating principles. Roboticists at all levels will benefit by studying such devices, and information is available at good technical libraries. Two useful books on transducers and techniques are listed below.

Transducer handbook H.B.Boyle, Newnes Linacre House, Jordan Hill, Oxford, UK OX28DP $43.
Sensors and Transducers Ian R. Sinclair, Newnes Linacre House, Jordan Hill, Oxford, UK OX28DP ~$50.

Understanding transducer specifications

Accuracy and precision

" Isn't precision the same as accuracy?"

    No. If a five-digit DVM displays a lithium battery's voltage as 3.1001volts this is a precise reading because of the number of digits. However, it may be inaccurate if the true battery voltage is 3.2001V. In this case a value of 3.2V would be more accurate, because it differs by only 0.0001V from the true voltage, *(accuracy being the difference between measured and true value)*
    Our faulty five digit meter accuracy is: (voltage difference)/true voltage = 0.1/3.2001 = 0.031 or 3%. This is poor performance considering a 3-digit voltmeter has an accuracy of ~ 0.5%.

Repeatability

If our five-digit DVM reads low by the same amount for all voltages this is called a systematic error, and is easily corrected by adding the error difference to every reading. Measurement repeatability gives us confidence to make such corrections, and the meter can also be calibrated with a reference source, extending its useful life.

Linearity

Manufacturers strive to achieve good linearity in their sensors. This makes them easier to use without added complications of linearization, look-up or correction tables. For most robotic projects 1% linearity is fine, but in some military applications a non-linearity of 0.0001% may be barely acceptable. Overheating or other stressing can permanently affect a transducer's linearity as shown in Figure 5-1, indicating the effect of overloading a spring balance.

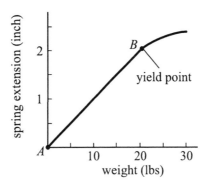

Figure 5-1. A spring balance responds linearly up to point *B,* but if loaded beyond 20lbs the spring will be permanently deformed changing the scale's calibration. Pressure gages and many other transducers can also be damaged by overstressing.

Examine Figure 5-2 carefully and see if you can tell why the lower curve is nonlinear.

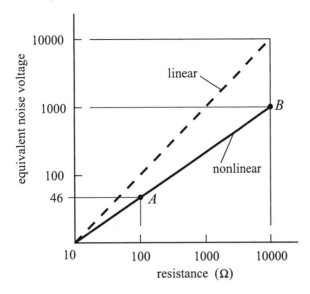

Figure 5-2. Linear and nonlinear logarithmic plots

The lower line in Figure 5-2 has a 100-fold increase in resistance from 100 Ω to 10,000 Ω (A to B), but equivalent noise voltage increases only 1000/46 or 21.7 times over the same interval. This nonlinear curve appears linear only because it is plotted on logarithmic scales (both axes increase logarithmically). Electronic data are often compressed by logarithmic plotting, and a linear scale for data in this figure at 0.1"/Ω would be 83 feet long.

It is important to remember that straight lines on logarithmic plots represent linear relationships only if both axes have the same scale size, and a straight line is at 45 degrees to each axis.

Resolution

From a practical viewpoint, resolution is the smallest incremental change that can be measured when monitoring a transducer's output. Table 5-2 lists approximate resolutions for some typical 'rulers'.

Table 5-2. Comparison of several 'ruler' resolutions

measuring technique	resolution (m)	resolution (inch)
steel rule	2×10^{-4}	10^{-2}
slide vernier	2×10^{-5}	10^{-3}
screw micrometer	2×10^{-6}	10^{-4}
differential screw micrometer	7×10^{-8}	2.8×10^{-6}
piezo micrometer	5×10^{-8}	2×10^{-6}
optical interferometer	1×10^{-9}	4×10^{-8}

Measuring very small differences in length, voltage, current or any other parameter takes sophisticated and expensive equipment. In acoustic emission studies broadband piezoelectric transducers have been used to detect displacements of 10^{-14}m.

Digital measurement resolution

If a perfect sensor is monitored using an analog to digital converter, measurement resolution cannot be better than the least significant bit ($\pm$ LSB). For 16 bit conversion of a 5 volt input the smallest resolvable voltage step will be about $80\mu V$ ($5V/2^{16}$), corresponding to a measurement resolution of 0.0015% ($100/2^{16}$). Working with such small voltages is beyond the capabilities of most hobbyists, and is also unnecessary because projects seldom need resolutions better than a few percent. Table 5-3 shows the percent resolution, and smallest resolvable voltage step for various analog to digital conversions.

Table 5-3. A to D converter resolution

conversion bits	1 bit voltage step	resolution
4	0.31 V	6%
8	0.02 V	0.4%
12	0.003 V	0.024%
16	76 μV	0.0015%

Resolution is excellent for 12 and 16 bit A/D converters, but these units are more expensive and tricky to use for most hobbyists, because of the care needed for small signal handling. An 8 bit converter is fine for most projects, it is less expensive, fast, and easy to use.

Analog measurement resolution

Analog measurement resolution is usually better than digital sampling, because most transducers have analog outputs. An all analog system has better accuracy, precision, frequency response and dynamic range, because fewer losses or distortions occur when a transducer's output is monitored directly without conversion.

However, benefits of digital signal handling often outweigh these analog advantages, and digitization is used except for those cases where maximum speed or basic simplicity is a requirement.

"How can a potentiometer have infinite resolution?"

'Infinite' resolution really means the resistive element is a continuous strip with no discontinuities, (ie. it is not a wire-wound pot). Measurement resolution with an infinite resolution potentiometer ultimately depends on factors such as power supply ripple and noise, A/D resolution, system bandwidth or frequency response.

Sensor frequency response

Many sensors rely on physical or chemical changes, often at a molecular level, to produce an electrical current, a change in resistance, capacitance, or some other parameter. Calculating the maximum frequency response for transducers of this type is straightforward. It is nearly always limited by the sensor's output capacity, which is never less than about 5pF.

Measuring very high speed signals requires a terminated coaxial 50Ω cable to feed the transducer's output to an oscilloscope or other instrument. Rise time for a 5pF sensor with a 50Ω load is about 1/RC or ~ 0.25 ns, this corresponds to a frequency bandwidth of 1.4 GHz, where bandwidth is ~ 0.35/{rise time}.

This performance is seldom achieved in practice, because many sensors have much larger output capacities, or resistances >50Ω. Sometimes transducer size may be a limiting factor too. For example, diaphragm pressure gages cannot change their shape rapidly in response to a rapid change in pressure, because of inertia or other restraining effects. Consequently, such gages have poor high frequency response.

Hysteresis

Physics texts normally show hysteresis curves for ferromagnetic materials to explain the results of magnetization and demagnetization. Hysteresis literally means 'to be behind' or 'lag'. This effect occurs to some extent in most transducers, because no material or process is perfectly elastic. When a transducer returns exactly to its initial shape or condition after it has been deformed or excited, it is said to be elastic.

It surprises some people when they are told that a glasslike substance such as quartz is one of the most elastic materials known, and rubber bands are less elastic than steel. A simple rubber band balance in Figure 5-3 can be used to demonstrate hysteresis, and show that rubber has only fair elasticity.

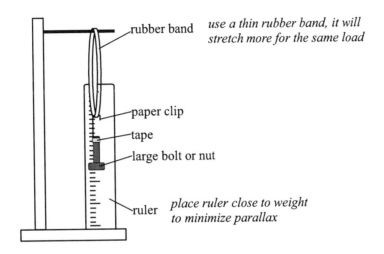

use a thin rubber band, it will stretch more for the same load

place ruler close to weight to minimize parallax

Figure 5-3. This rubber band weigh-balance can be used to measure the hysteresis of rubber, linearity of extension with load, and show that rubber has only fair elasticity except for small extensions.

[1]   Place a weight on the rubber band
[2]   Record the band's extension by observing weight position on the ruler
[3]   Pull the band until it extends to twice its unloaded length, then release gently.
[4]   Note the weight does not return to its position in [2].
[5]   Remove the weight and allow the rubber band to sit unloaded for a few minutes, then repeat [1] and [2]. The rubber band will have recovered to its original length.

Delay or lag in recovering its original tensile properties, denotes hysteresis and inelasticity in the rubber band. A quartz fiber will recover to its initial length almost exactly after stretching. You will need either a very long fiber or a microscope to measure the minute extensions of a loaded quartz fiber.

Most mechanical transducers suffer from hysteresis to some extent, because they behave in an analogous manner to this rubber band example.

Sensitivity

"What is the sensitivity of an electrical transducer?"

The sensitivity for all transducers is simply (output)/(input). This is true for an operational amplifier as shown in Equation 5.1 and for a loaded cantilever beam in Figure 5-4.

$$(5\text{-}1) \quad op\ amp\ sensitivity = \frac{output}{input} = \frac{V_{out}}{V_{in}} = gain$$

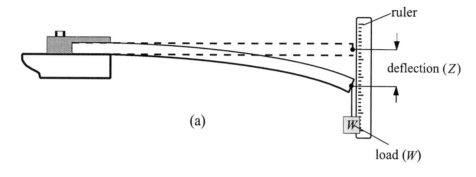

(a)

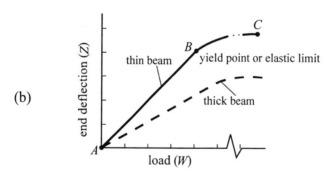

(b)

Figure 5-4. A cantilever beam with a weight W at its free end, will bend with an end deflection proportional to small loads as shown in (b). Overloading produces permanent deformation after the yield point, and the beam eventually breaks close to the clamp, where strain is a maximum. Thin beams are more sensitive to loads than thicker beams. Consequently, a thick beam has a smaller slope as suggested by the dotted curve in (b)

If a weight is attached to the free end of the beam in Figure 5-4 (a), its end will deflect a distance Z from its unloaded position. Deflection increases linearly with load from A to B as shown in Figure 5-4 (b), and the beam breaks at C with gross overloading.

The cantilever acts as a transducer, converting force into displacement and its sensitivity is Z/W as shown in eqn.5.2. A thick beam bends less for the same load, so it will have a lower sensitivity as indicated by the dotted line in Figure 5-4 (b).

$$(5.2) \quad cantilever\ beam\ sensitivity = \frac{output}{input} = \frac{Z}{W}$$

98

Transverse sensitivity

If the beam in Figure 5-4 is much wider than its thickness it will deflect less when loaded across its width, and therefore has low transverse sensitivity.

Ideal transducers only respond to excitation along their principal axis, and have zero transverse sensitivity. Unfortunately this is not so with practical transducers. Accelerometers, gyroscopes, paddle flow meters, torque wrenches, weigh scales, tilt meters and many other instruments, have measurable transverse sensitivity that may be troublesome at times.

Transverse sensitivity is usually specified as a percentage of the longitudinal, or principal axis sensitivity, for example: < 1% of full range for 0° to 90° tilt.

Temperature sensitivity and other effects

Output signals from every transducer are affected by temperature and in some cases by vibration, pressure, nuclear radiation, humidity and other phenomena. Manufacturers test their products for any effect that is likely to influence performance, but usually TC (or temperature coefficient) is the most important one for a hobbyist. TC's are usually quoted as a fraction of full range output eg. TC = 0.1% of full range/°F.

Resistive transducers

Some of the simplest and most reliable sensors are based on resistance techniques. Potentiometers, thermistors, hot wires, strain gages and conducting rubber, are all useful for robotic monitoring. Beginners may have some apprehension when contemplating the use of an unfamiliar method. Hopefully these concerns can be allayed, by showing how easily these devices can be incorporated into projects.

Potentiometers

Because of their simplicity, slide and rotary potentiometers are favored by many roboticists. They are inexpensive, reliable, accurate, widely available, and can be incorporated into designs with a minimum of mechanical and electrical effort.

When a dc voltage is applied across the ends of a potentiometer, wiper voltage is directly proportional to its position along the resistive element. This means both rotary and linear motions can be continuously monitored by sampling a potentiometer's wiper voltage.

Monitoring rotary motion

Multi-turn rotary potentiometers shown in Figure 2-6, are best for monitoring motion. They turn freely, and are easily coupled to the shaft being monitored with a piece of heat shrink tubing. Pots with reasonable friction can be connected with an inexpensive flexible coupler, made from a stiff coil spring or heavy rubber/plastic tubing (see Chapter 8). Some pots are so hard to turn, they must be rigidly coupled to a motor etc.

Ten turn precision potentiometers are sold by *Hosfelt* 100Ω to 100kΩ (#38-166 to #38-175) $9.99, with linearity < 0.5%, and these are useful for monitoring shaft rotation.

Continuous rotation precision potentiometers are usually expensive, but because of their simplicity and accuracy are sometimes preferred to optical encoders. All continuous rotation pots have a dead band of about 10 to 40 degrees, depending on wiper width. Servo Systems is one of the best surplus sources for special potentiometers, they carry a wide selection of rotary and longitudinal pots with linear, sine, cosine and other special tapers. Continuous rotation pots from *Servo Systems* start at $7.50.

Monitoring longitudinal motion

Monitoring large longitudinal motions with high accuracy is difficult and usually expensive, so hobbyists always welcome inexpensive solutions. Home made linear optical encoders can be made in any length and cost very little. Some other techniques described below and in Figure 5-59, are suitable over short distances.

Rectilinear potentiometers

Slide and long stroke potentiometers in Figure 5-5 (a) are suitable for monitoring small longitudinal displacements. Slide pots can be salvaged from audio equipment and are made with strokes up to a few inches. An open frame 1.8" slide pot is sold by *Newark* 87F2106, $4.19. Most slide pots need a force of about 5 oz (1.4N) to move the wiper, because of stiction (static friction).

Long stroke potentiometers are sold by *Servo Systems* but generally these are too expensive for hobby use. Servo System's catalog lists two potentiometers with 0.04" resolution and 1% linearity. An 8" stroke PR-259 at $211, and a 12" stroke PR260 for $309.

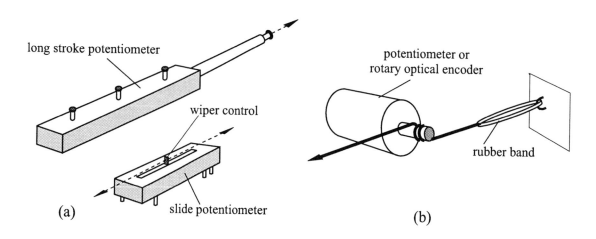

Figure 5-5. (a) Because resistance is proportional to wiper position, small longitudinal displacements may be measured inexpensively with a slide potentiometer. (b) Rotary potentiometers can be used too, with rubber bands providing a restoring force. Linear or rotary optical encoders can be fabricated from photocopied transparencies.

A restoring force for a rotary pot in Figure 5-5 (b) can be provided by an extension or flat spiral spring, but rubber bands give greater flexibility. Like springs, rubber bands can be used in series or parallel combinations.

An excellent commercial 10" stroke linear extension transducer, employing the rotary potentiometer principle shown in Figure 5-5 (b), is sold by *Servo Systems* PR-264 $199.

A simple force multiplier (screw driven carriage)

A machine nut driven by a rotating threaded rod in Figure 5-6, develops impressive force, and its rider's position can be determined from screw pitch and rotation information. If the screw is driven by a dc brush motor, a rotary encoder or revolution counter is often employed to keep track of angular rotation. Sometimes the lead screw is driven by a stepper motor, and rotation data are obtained from step control information sent to the motor.

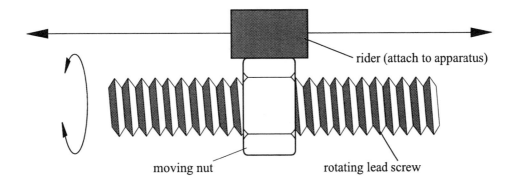

Figure 5-6. If a nut is restrained from turning it will move along a rotating screw thread with considerable force. Even small motors can move heavy objects using this principle. A rider's position along the screw can be determined from screw pitch and lead screw rotation. Long lengths of threaded rod in various diameters are available from hardware and hobby stores.

Strain gages

A strain gage can be used to weigh a 600,000 lb (273t) jumbo jet before take off, or determine its center of gravity to within an inch or so. It can easily tell if a bridge overpass has expanded by only one millionth of its original length (1 microstrain), or detect the force produced by someone blowing on a small plate. One might imagine that sophisticated and expensive equipment is required for such applications, but hobbyists can duplicate these measurement capabilities for less than $10.

The fundamental operating principle of a strain gage is very simple, relying on the fact that a wire's resistance changes when it is stretched or compressed longitudinally. This is illustrated in Figure 5-7 where a wire (1) is stretched to form a longer but thinner wire (2), with reduced cross sectional area and higher resistance R.

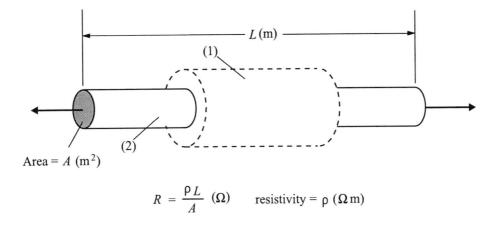

$$R = \frac{\rho L}{A} \ (\Omega) \qquad \text{resistivity} = \rho \ (\Omega\,m)$$

Figure 5-7. A wire stretched from its original shape (1) to a longer, thinner shape (2), has increased its resistance to R. Using this principle, electric foil strain gages can measure tension or compression simply and accurately.

"What is strain?"

Strain is the fractional change in length when a body is stretched or compressed. In Figure 5-8 a solid bar has been stretched by ¼ of its original length so the strain is 0.25 or 25%. This is a gigantic strain for a bar of steel, aluminum, Plexiglas™ or wood. All these materials will be permanently deformed or break before they can be stretched to such an extent. Although cast iron, brittle steels and concrete are strong in compression they cannot be stretched more than 0.5% before breaking.

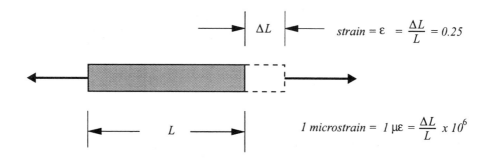

$$strain = \varepsilon = \frac{\Delta L}{L} = 0.25$$

$$1\ microstrain = 1\ \mu\varepsilon = \frac{\Delta L}{L}\ x\ 10^6$$

Figure 5-8. Stretching a solid produces a positive strain equal to + ($\Delta L/L$). If the bar is longitudinally compressed by the same amount, the compressive strain will be - ($\Delta L/L$). The symbol ϵ is used to denote strain.

An enjoyably readable layman's book on stress, strain and much more is:

The New Science of Strong Materials or Why Things Don't Fall Through the Floor. by J.E.Gordon
Penguin books ISBN 0 14 02.0920 4 ~$8 *Viking Penguin Inc*. NY 10010.

Commercial strain gages

Line drawings of some modern electrical foil strain gages manufactured by Vishay's Measurement Group Inc. are shown in Figure 5-9.

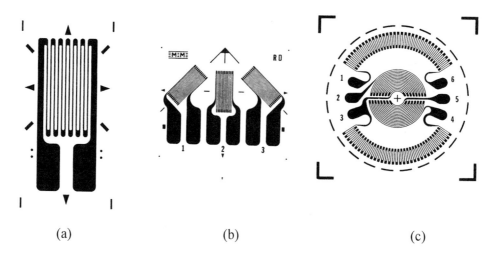

(a)   (b)   (c)

Figure 5-9. (a) Strain gages employ a folded elongated path to enhance resistance and gage sensitivity in a preferred direction. Heavy areas are solder pads.   (b) Three element rosette allows surface strain monitoring in three directions simultaneously.   (c) A diaphragm gage for measuring plate deflections.
Manufacturer's drawings courtesy of Measurement Group Inc. Vishay.

Foil strain gages are made by photo-etching techniques that leave a thin (~ 0.0002") metal layer attached to an insulating backing (~ 0.002"). The conductive grid is a careful mix of low thermal expansion metals such as nichrome™, constantan and platinum tungsten alloys. These special metals reduce unwanted expansion due to temperature changes and maximize output signals due to strain.

Gages are also made from materials chosen to minimize strain signals and enhance temperature sensitivity. These devices are called thermal gages and are used to measure temperature. They have very fast response times and other desirable characteristics.

Some typical operating parameters for electrical foil strain gages are given below.

operating temperature range	-270 °C to +400 °C
strain limits	±1% to ±10% (ie. 100,000 microstrain max)
fatigue life	10^7 cycles max at ±2200 microstrain
active gage area	0.4mm x 0.4mm to 150mm x 8mm
gage factor	2 to 4.5
non-linearity (ΔR/strain)	approximately 0.02%
maximum dc gage current	approx 50 mA
resistance	discrete values from 60 to 3000 ohms ~ ±5%
dynamic response	depends on gage length (~ 40kHz for a 6 mm gage)

Considerable research and development effort has brought strain gage technology to its present level of sophistication. Modern strain gages are easy to use, and all their important characteristics are well understood.

Some semiconductor strain gages have more than 200 times the sensitivity of a foil gage. These are difficult for amateur use, because of their very high temperature coefficient. Some inexpensive transducers employing semiconductor technology are available for hobby use, and a pressure gage using ion implanted piezo-resistors is described in a future book in this series. Magnetoresistive devices can be used to make angle and position sensors, using techniques described later in this chapter.

Strain-gaged cantilever beams for robotic applications

A strain-gaged cantilever beam made from metal or plastic, can be used for a variety of robotics measurements as shown in Figure 5-10. Gages are bonded to beams with crazy glue (cyanoacrylate adhesive), and fine wires are soldered to the pads. It is very important to provide stress relief for signal wires, which must also be thin, flexible, and supported with transparent tape. If you follow the guidelines given in this chapter, your sensing beams will be simple rugged devices, with good measurement capabilities.

Principles employed for devices illustrated in Figure 5-10 are embodied in many commercial instruments, often with an added wrinkle to improve performance. For example, many accelerometers use a central load on a circular diaphragm. A strain gage signal is then used with feedback control to apply a restoring force, maintaining the weight at its rest position. This technique enables an accelerometer to operate linearly to one part in a million or better, by reducing non-linearities due to diaphragm displacement. Speaker coil transducers are often used to apply this restoring force, and coil current is monitored as a measure of acceleration. Some other uses for speaker/voice coil transducers are described in Chapter 6.

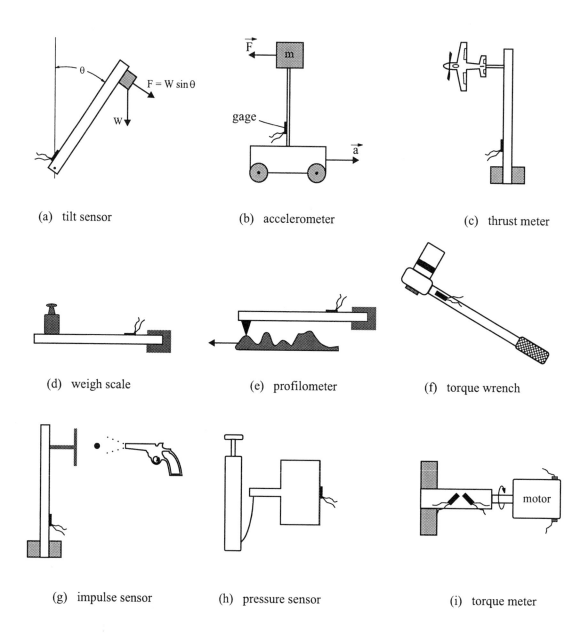

(a) tilt sensor

(b) accelerometer

(c) thrust meter

(d) weigh scale

(e) profilometer

(f) torque wrench

(g) impulse sensor

(h) pressure sensor

(i) torque meter

Figure 5-10. Strain gages can be used for sensing in many different applications, as shown in these diagrams. (a) This sensor can measure tilt accurately over $\pm 90°$ if readings are corrected for the sine factor. It also measures to ~ 2 degree accuracy over $\pm 60°$ without sine correction. (b) Weight lags when the cart moves, producing a force proportional to acceleration within 0.5%, for beam deflections up to $\pm 10°$. (c) Model plane engines can be tuned for optimum output while continuously monitoring thrust. (d) This weigh scale is useful for conventional weighing, but can also measure the lift of a helium or hot air balloon. (e) Students have scanned a coin's surface with a strain gage profilometer, displaying real time terrain results in 3D perspective on a computer monitor. (f) Only one full scale reading using a spring balance is required to calibrate this torque wrench in units of ft-lbs or Nm. (g) Small strain gages have fast responses ($< 100\mu$s), and can be used to measure impulse and other transient forces. (h) A strain gage pressure sensor works best with a diaphragm gage. Positive or negative pressures are easily measured. (i) Gages at $\pm 45°$ respond to shaft twist, so this device can measure motor torque.

"Why are beams used for so many strain gaged sensors?"

Mechanical properties of rectangular beams are well understood, so it is straightforward to calculate strain gage signals when a known force is applied to a beam. Rectangular metal, plastic or wooden beams are also easily fabricated from sheet material. Small bars of metal, plastic and wood, suitable for making strain gage sensors are sold at most hobby stores.

Strain on a cantilever beam

A loaded beam has its top surface stretched and the lower surface is compressed, when bent as shown in Figure 5-11. Only the central plane shown by a dotted line (the neutral axis), retains its original length. Strain is maximum on the beam's surface near the clamp, and gages are usually installed at this location to enhance signal output.

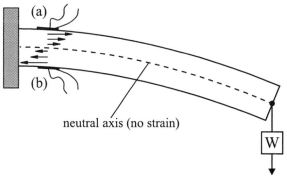

Figure 5-11. The resistance of strain gage (a) will be increased, and that of gage (b) reduced, when the beam is bent as shown. Therefore, a single gage can be used to measure not only the strain magnitude, but the bend direction too. The length of arrows near the beam's root, show strain is positive on the top surface and negative on the lower. This means a single gage attached to either surface can monitor signed magnitudes of an applied force.

Making and testing a strain-gaged tilt sensor

Students at the University of Toronto have used the following device for dynamically levelling moving platforms and for sensing wrist rotation on robotic arms. Sometimes a pair of orthogonal tilt sensors is employed when arm movements are in more than one plane.

     Dimensions given in Figure 5-12 will be used later to compute strain gage signals, but neither beam material nor dimensions are critical.

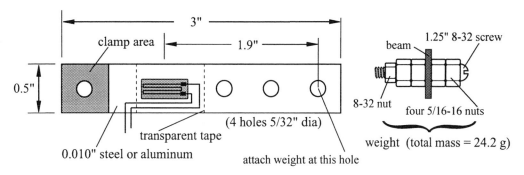

Figure 5-12. Sensitivity of this beam can be changed by placing the screw/nut weight at a suitable location. For maximum signal levels a gage should be mounted close to the clamp, where strain is greatest. An insert in Figure 5-14 shows a side view of the finished tilt sensor. Sensor mounting depends on a user's application, but a sensing beam should always be held at the clamp area indicated in this figure.

Strain gage beam fabrication procedures

[1]   Prepare a beam as shown in Figure 5-12. Springy steel from a 2 lb coffee can (0.010" thick), can be cut with heavy scissors and makes a good beam - wear gloves for this operation. Aluminum gives about three times the sensitivity of a steel beam because of its lower modulus of elasticity, (it's not as stiff).

[2]   Roughen the gage area with 300/400 wet or dry abrasive paper using a random motion.

[3]   Clean the whole beam thoroughly with acetone, (non-oily nail varnish remover is ok.)

[4]   Stick a strain gage to a 4" x ¾" length of transparent tape. Place shiny side of gage to sticky side of tape, because dull side of the gage will be stuck to beam. Make sure the gage is oriented so it lays lengthwise along the tape.

[5]   Stick the tape on the beam with the gage located and oriented exactly where it should go. Solder pads should be remote from the clamp area.

[6]   Peel back one end of the tape until only one end of the gage remains touching the beam.

[7]   Apply cyanoprep™ (chapter 8), or a suitable accelerator/catalyst to the gage backing and allow this to dry.

[8]   Put a small drop of crazy glue where the gage end is contacting the beam. Then lower the tape and gage onto the beam. Simultaneously press down from the top of the tape with your thumb over the gage area. Be careful not to stick your thumb to the beam!

[9]   Rub the gage area continuously with heavy thumb pressure for a minute.

[10]   Peel off the tape very slowly and carefully at a shallow angle. Wipe the gage area with an acetone moistened swab, but don't soak this area.

[11]   Twist two 12" lengths of wire wrap wire (Radio Shack 278-501) in a hand drill, then solder them to the gage pads.

[12]   Wrap one layer of ¾" transparent tape over the gage and solder pads, then bend the wires toward the clamp area as in Figure 5-12. Now wrap a second layer over to secure the wires. Gage wires should always be fastened in this manner to relieve stress on the solder pads. Stress relief also reduces the effects of wire movements on gage readings.

Tilt sensor circuit

Strain gage signals are very small, but are easily amplified using a Wheatstone bridge and an operational amplifier. The complete circuit in Figure 5-13 will fit on a 2" x 2.5" section of solderless breadboard.

Bridge resistors and strain gage leads can be inserted directly into solderless breadboard sockets, and this works fine for many projects. Breadboard contact resistance is about 0.001Ω and wiggling a wire in a socket produces resistance changes of 0.002Ω. So for very small strain gage signals soldering each bridge element gives superior performance.

Metal film 1% precision resistors are essential for the bridge, because regular carbon resistors have poor temperature characteristics and produce unacceptable drift.

120Ω strain gages are recommended because ¼W 120Ω precision resistors are readily available, (eg. *Hosfelt* QW112 four pack 90¢). *Measurements Group Inc.* also sells 350Ω gages but these require two precision resistors for each bridge element, (ie $340\Omega + 10\Omega$).

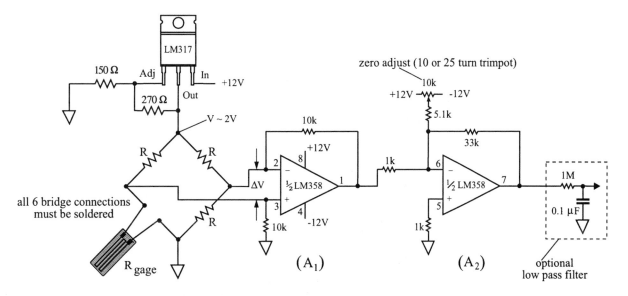

$R_{gage} = R = 120\,\Omega$ (bridge resistors must be 1% metal film)

$$\Delta V = \frac{VG\varepsilon}{4} \qquad \varepsilon = strain = \left(\frac{\Delta L}{L}\right) \qquad V = \text{ bridge voltage (approx 2V)}$$

$$\frac{\Delta R}{R} = G\varepsilon \qquad G = \text{gage factor} = 2.1 \qquad A_1 \text{ gain} = \frac{10k}{0.5R} = 167 \qquad A_2 \text{ gain} = 33$$

Gage: CEA-13-125UW-120 (120 ohms) $28 for 5 gages
Measurements group Inc. Raleigh, North Carolina (919) 365-3800
Special prices for teaching institutions

Figure 5-13. This circuit is suitable for measuring strain gage signals from any transducer from Figure 5-10.

Tilt sensor circuit operation

[1]   Bridge resistors and strain gage leads can be inserted directly into solderless breadboard sockets for most projects [1].

[2]   Build the circuit of Figure 5-13 on a solderless breadboard, then insert the four bridge leads, they go to ground, pins 2, 3 and the LM317 center pin.

[3]   Place the beam in a stable position.

[4]   Turn on circuit power and monitor pin 7 with a DVM. The amplified output will probably be close to either power rail potential (about -12V or +12V).

[5]   Adjust the10k trimpot (zero adjust) until the DVM shows zero volts.

[6]   Flex the beam gently, this will be enough to drive the signal to either rail, so don't press too hard

[7]   Troubleshooting tips are given at the end of Chapter 6, if ciruit problems arise.

[1] When measuring very small strains, first solder the strain gage wires and three precision resistors together as a bridge. This can be done neatly if resistor leads are snipped to ⅛" length. Then attach 4 short solid wire leads (one at each bridge junction). Install this wired/soldered bridge as a single unit onto the solderless breadboard circuit of Figure 5-13.
         A soldered bridge improves drift, and gives fine performance at low signal levels, even on a solderless breadboard.

Tilt sensor calibration

A tilt sensor built to the specifications of Figure 5-12 has a sine response to tilt angle because $F = W \sin \theta$, as shown in Figure 5-10 (a). This sine relationship is confirmed by calibration results in Figure 5-14.

Gage output is nearly linear with tilt angle from -60° to +60°, and this region is very useful for quick angular measurements to about ±2 degrees. This accuracy is adequate for most projects but optimum accuracy over the whole tilt range is possible using the relation $V(\theta) = V(90°) \sin \theta$ or by calibrating the sensor as described in the caption of Figure 5-14.

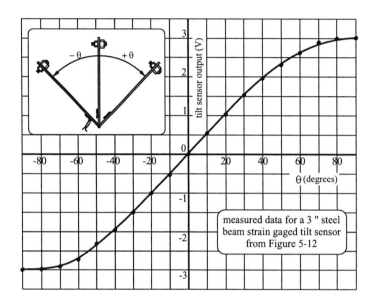

Figure 5-14. These measurements were obtained with a 3"x ½" strain gaged beam made from a 2 lb coffee can, described in Figure 5-12. During calibration the sensor was attached at the clamp area to a builder's plastic circular level (~$15), using double sided sticky tape, (a protractor works just as well but measurements take a little longer). Because the sensor has near linear response from -60° to +60°, it can be used over this range with a precision of a few degrees, (this is adequate for many robotic levelling applications and more than sufficient for wrist rotation feedback control).

Strain gage sensing notes

[1]   Raising the bridge voltage in Figure 5-13 from 2 volts to 5 volts increases tilt sensor gage sensitivity by a factor of 2.5.

[2]   All strain gages are heated slightly by bridge current but metal beams conduct this heat away efficiently. Gages mounted on plastic, fiberglass, wood and composite beams drift more slowly until thermal equilibrium is established. Because heating power increases with the square of the bridge voltage, heating effects are 6.25 times larger when bridge voltage is increased from 2 to 5 volts.

[3]   Aluminum is one of the best materials for strain gaged sensing beams, it is lightweight, easy to machine, does not corrode, and has about three times the sensitivity of steel.

[4]   Full scale output on the tilt sensor can be adjusted to suit a particular project if the fixed 33k resistor at pins 6 and 7 of Figure 5-13 is replaced by a 100k trimpot.

[5]   A tilt sensor or accelerometer made with a dual beam flexure shown in Figure 5-15, is stiffer and less sensitive to spurious vibrations. For small amplitude displacements the load undergoes pure translation, unlike a single beam where it moves in an arc.

[6]   Just for fun, twang your sensing beam and observe the beautiful damped simple harmonic vibratory gage output waveform on an oscilloscope.

[7]   Gage information and costs are given in Figure 5-13, special prices are available for qualified courses at teaching institutions. A strain gage information package is available on request from Applications Engineering, Measurements Group Inc.

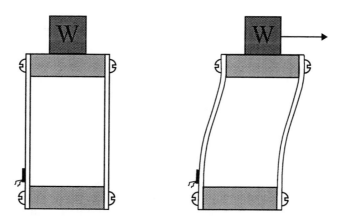

Figure 5-15. The tip of a loaded cantilever moves in an arc (as shown in Figure 5-4a). However, a dual beam flexure shown here undergoes only lateral displacement. This device is suitable for measuring acceleration on robotic vehicles.

Calculating strain gage signals

With a little experience you will be able to make a strain gaged beam for a particular application without doing any calculations. Beam flexibilities can be compared by measuring their tip deflections for various loads, and with a small stock of gaged beams you can measure a variety of parameters.

Calculations for our tilt sensing beam from Figure 5-12, are given below. These are valid for other beam sizes and different materials, using information from Table 8-4. Derivations and explanations of formulas given below may be found in good physics texts or books on stress analysis.

Elastic or Young's modulus of a mild steel beam	$E = 20 \times 10^{10}$ Pa (Nm^{-2})
Active beam length (load point to gage centerline)	$L \sim 1.9"$ (0.0483 m)
Beam width	$b = 0.5"$ (0.0127 m)
Beam thickness	$t = 0.010"$ (2.54 $\times 10^{-4}$ m)
Load = 0.0242 kg, (one screw and 5 nuts)	$P = 0.0242 \times 9.81 = 0.237$ N

Beam surface strain (ϵ) at the gage location is:

$$(5.3) \quad \textit{Strain measured by gage} = \epsilon = \frac{6\,P\,L}{E\,b\,t^2} = \frac{6 \times 0.237 \times 0.0483}{20 \times 10^{10} \times 0.0127 \times (2.54 \times 10^{-4})^2} = 419 \times 10^{-6}$$

Final output voltage at pin 7 is calculated in Equation 5.4, using information in Figure 5-13 and strain from Equation 5.3

$$(5.4) \quad \text{Output signal for } \theta = 90° = \frac{VG\epsilon}{4} \times 167 \times 33 = \frac{2 \times 2.1 \times 419 \times 10^{-6} \times 167 \times 33}{4} = 2.42 \text{ volts}$$

This calculated value of 2.42V is within 20% of actual measured output (3 volts), and is acceptable because modulus information for steel coffee cans is not available. In addition, Eqn. 3 is only valid for a point load, and 5 nuts on a long screw is definitely not a point load. Strain gage beam calculations normally agree with measured values to about ±10% if beam modulus is known accurately.

Conducting plastics for making force and strain sensing resistors

Synthetic rubbers impregnated with carbon and conducting polyethylene have fairly low resistivity, and are widely used in industry as anti-static materials for protecting sensitive electronic components. Conducting wrist straps, floor mats, shoes, shipping bags, computer keyboard pads, and a variety of other commercial products are made from conducting plastic material for anti-static workstations. Some telephones also use conducting fluoroelastomer keypad switches.

Hobbyists can use such materials to make sensors for weighing, touch sensing, monitoring strain or pressure, angle measurements, and for some sensing techniques outlined in Figure 5-10. These sensors will not have the precision or accuracy of a commercial strain gage, because all rubbers have considerable hysteresis due to their molecular composition. However, they can be used to fabricate simple, inexpensive, and very effective first order sensors. Typical accuracy after an appropriate calibration is about ±3% to ±15% depending on the sensor type. When one considers these devices can measure parameters where other alternate devices either fail or are prohibitively expensive, they are valuable additions to an experimenter's armamentarium.

Anti-static black rubber foam is advertised as electrically conducting, but really falls into the semiconductor category when compared with other materials in Table 5-4.

Table 5-4. Material resistivities

	resistivity ρ (Ω.m)		resistivity ρ (Ω.m)
silver	1.6×10^{-8}	silicon	600
copper	1.7×10^{-8}	anti-static foam	~10 to ~250,000
Nichrome ™	100×10^{-8}	hard rubber	~10^6 to ~10^{16}
carbon	3×10^{-5}	Teflon™	~ 10^{17}
germanium	0.5	fused clear quartz	> 10^{18}

Many useful sensors can be made from the anti-static foam highlighted in this table. Black conducting rubber foam is widely used for shipping and storing static sensitive components, and can often be obtained at no cost. Volume resistivities for materials in this table cover a range of 10^{26}, an impressive span for any physical property.

Selecting conductive rubber foam for sensors (the squeeze test)

Although any compliant conductive material shows a decrease in resistance when it is compressed, not all conductive foams make good sensors because of hysteresis (Figure 5-3). Conducting rubber foam suitable for sensors must recover its shape immediately after a hard squeeze between thumb and forefinger. Therefore, squeezing is the first test to apply when selecting foam for sensor fabrication.

Conducting rubber foam is available from large electronic outlets in three thicknesses, ⅛", ¼", ⅜", but is mainly sold in large sheets eg. Velostat™ foam ρ ~ 1200 Ω.m , *Newark* 24" x 36" x ⅛" #46F8208 , $30.48. You can probably obtain free samples of suitable foams from an electronic store by first buying a few items, then explaining your interesting device application to a friendly salesperson. Take along your DMM and a pair of ½" x ½" wired contact plates. Ask to test some of their scrap material, (they receive many shipments in foam and this is frequently discarded). Explain the tests are nondestructive, because it is not necessary to cut or abuse the rubber foam.

First use the squeeze-recovery test to reject any samples that are not perfectly elastic. Next check selected materials with your contact plates using a hard squeeze between thumb and forefinger. This corresponds to a force of about 3200 gram weight or 31 Newtons. The best materials will show < 100Ω and fall into category (3) of Figure 5-16.

Conductive polyethylene bubble foam (usually pink), has both high resistivity and poor elasticity and is unsuitable for making sensors.

Fabricating and testing a Force Sensing Resistor (FSR)

An FSR can be made from two tinned steel plates and a few drops of contact cement. Thin flexible wires are soldered to each plate, then covered with transparent tape for stress relief, Figure 5-16. Contact cement forms a tenacious bond and sensors made in this manner will withstand rough handling.

Contact cement is very volatile and deteriorates quickly when exposed to air, so buy only small quantities in a tube, not a can or bottle. Apply a tiny spot of cement (~1mm dia.) at each corner of a plate with a pin, and allow this to dry for 30 seconds before attaching to the conductive foam.

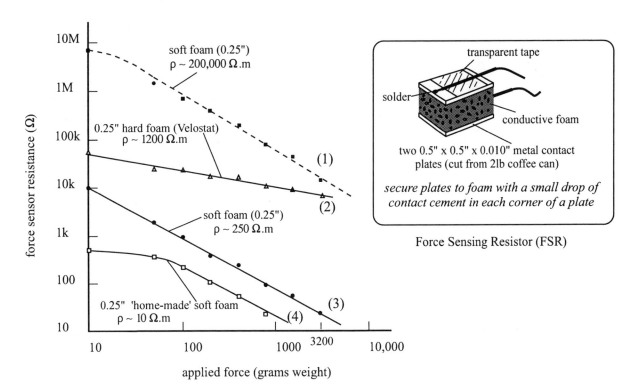

Figure 5-16. Foam resistivity is determined by pressing a pair of metal plates to foam and measuring resistance with a DMM. A hard squeeze between thumb and forefinger produces a force of about 3200 gram weight (31N). Items (1) and (3) are specimens used for protecting ICs during shipment. (2) is a commercial foam sold by *Newark* and *Electrosonic*. Experimenters can make their own low resistance foam as in (4). Items (1) through (4) all recover instantly after deformation and respond well to transient forces.

Thin tinned steel from coffee or food cans is recommended for contact plates because it solders extremely well and unlike copper, does not oxidize. It is available at no cost in a variety of thicknesses, smaller cans are usually made from thinner plate. Always solder wires to contact plates and cover these with transparent tape to prevent wires breaking. Use wire-wrap wire for very small sensors.

Usually an FSR is calibrated for a specific application, but it is easy to check your materials against those in Figure 5-16 using an inexpensive food or postal scale shown in Figure 8-33. These spring balance weigh scales are often equipped with a removable weighing pan. Remove the pan for your tests, then press a sensor to the scale platform and record resistance for various loads. This is a very effective method, easily outperforming a digital scale or beam balance, in both ease of use and cost.

Measuring FSR output signals

Simple divider circuits are adequate for monitoring FSR outputs, these also provide linearization where required as shown in Figure 5-17, all curves are for type (3) foam. Linearity can be further improved by using an additional resistor in parallel with the FSR, (an optimum value can be determined empirically for a particular application). Sometimes it is advantageous to retain a foam's logarithmic response, allowing a sensor to function more like a switch. For these cases a high resistor ($R \sim 10k\Omega$) is appropriate.

Results in Figure 5-17 are repeatable to $\pm 3\%$, making these devices suitable for many robotic measurement applications.

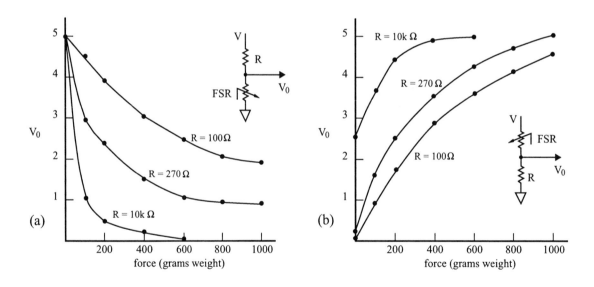

Figure 5-17. Force sensing resistors made from black conducting type (3) material of Figure 5-16, are simple reliable devices with measurement repeatability of $\pm 3\%$ at room temperature. For weighing, pressure or force sensing and other tasks requiring linearly proportional measurements, use a low value resistor ($R \sim 100 \ \Omega$). Linearity can be further improved by adding a resistor in parallel with the FSR. If a high value for R is chosen ($R \geq 10k\Omega$) an FSR will respond quickly to small forces. This is useful when instantaneous threshold detection is required.

FSR response time

Force sensing resistors can respond to rapid changes in applied force. Sensor size, foam resiliency and its resistivity are the main factors governing signal rise and fall times. A ½" x ½" sensor from type (3) foam responds in less than 5 ms to an impulse tap from a small hammer, as shown in Figure 5-18.

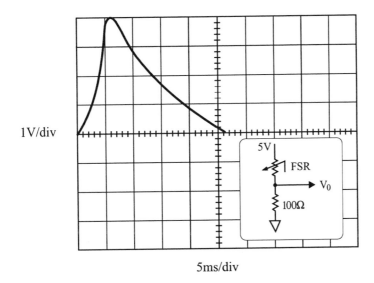

1V/div

5ms/div

Figure 5-18. Conductive foam sensors have fast response times, as indicated by this output from an FSR which was given a sharp tap with a small hammer. Type (3) foam was used for this test. Weights dropped from known heights provide predictable impulses, and can be used for calibrating force sensing resistors.

Squeezing eggs without breaking them

Students have used plate type force sensing resistors, similar to those shown in Figure 5-16, to pick up an egg with robotic fingers, remotely controlled by a human hand. Rubber sleeved FSRs at robot fingertips are used in a feedback loop to ensure the eggshell cannot be broken, no matter how much force an operator tries to apply. In this project the FSR signal was also used to develop a feedback tactile force, so an operator's fingers sense a pressure equivalent to that being applied to the egg. A gripping force equal to 100 gram weight (~ 1N) permits an egg to be held and lifted along its minor axis without cracking the shell.

Measuring finger movements for sign language recognition

Monitoring hand, wrist and finger motions is difficult even for a professional roboticist. Pros have used fiber optic sensors, break point bend monitors, wires, pulleys, accelerometers, linear and rotating potentiometers, rate gyros, ultrasonic ranging, tilt sensors, mercury switches, FSRs, microswitches, Nitinol™, magnetostrictive devices, capacitive sensors, magnetic fields, inductive loops, Hall effect sensors, strain gages and probably other techniques in their quests.

How can an amateur roboticist compete with professionals and their esoterica? The performance gap is not as wide as one might imagine. With a little ingenuity a hobbyist can produce results that compare very favorably with those obtained using sophisticated commercial quality equipment, but at a fraction of their cost.

From a few hand and finger orientations for sign language/finger spelling shown in Figure 5-19 it is apparent that sensors are needed for monitoring finger bending, finger separation, hand and wrist positions. Sensors are usually attached to a closely fitting glove, made from a material that can be stretched to fit various hand sizes. Hot melt glue is recommended for attaching sensors to a glove, because it is one of the best fabric glues, and tiny dabs can be applied with a small, low heat glue gun. Hot melt glue joints are also easily undone with a small soldering iron tip, run at low temperature from a solder iron controller. This is a major benefit of hot melt glues, because a glove may have many sensors, and should not be discarded just because one sensor needs replacing.

Figure 5-19. A few of many sign language hand profiles as seen by a receiver are depicted here. Each letter of the alphabet can also be communicated by finger spelling. A transducered glove is the most convenient method for monitoring hand and finger motions, and sensors should be attached with hot melt glue since this permits their easy removal or replacement.

Monitoring finger separation with a conductive surface foam angle sensor

Although plate type FSRs can be used to measure finger separation as in Figure 5-20 (a), a foam sensor shown in (b) is more comfortable, and also measures the degree of finger separation.

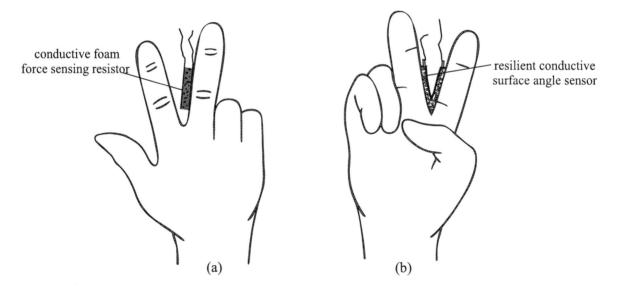

conductive foam
force sensing resistor

resilient conductive
surface angle sensor

(a)   (b)

Figure. 5-20. Force sensing resistors are usually attached to a tightly fitting glove with holt melt glue. Finger separation can be monitored with a plate type FSR shown in (a) but a flexible conductive foam strip is more comfortable for the operator and provides proportional angle outputs. Approximate closure force generated between 1st and 2nd fingers is 120g weight (1.2N) or 4.2 oz.

Making and testing a conductive foam angle sensor

An angle sensor shown in Figure 5-20 (b) and Figure 5-21 can only be fabricated from very soft foam that deforms progressively for angles $\theta > 90$ degrees. *HARD FOAMS WILL NOT WORK*. Ideal compliancy is just a little stiffer than sponge material used for foam stick paint brushes. A satisfactory sensor can be made from type (3) conductive rubber foam, or a fine grained bathroom sponge with a pliable conductive coating on one surface.

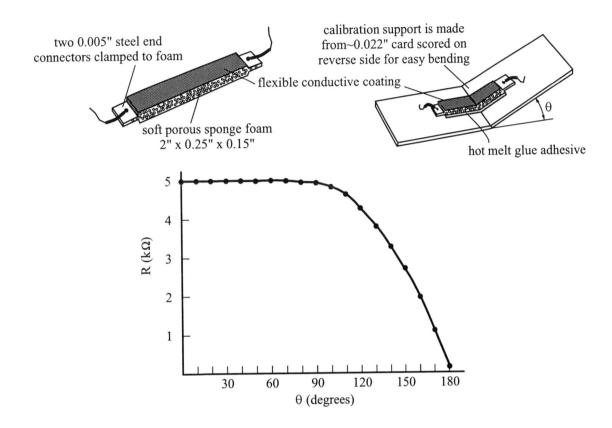

Figure 5-21. This conductive foam sensor *cannot be constructed from hard foam material*. A soft compliant foam is essential because it deforms readily, squeezing into itself as θ increases beyond 90 degrees, (R ~ 5 sinθ for the sensor shown). The sensor's working principle is more easily understood by making a dummy sensor from bathroom sponge foam as illustrated above, and observing foam action as θ is changed.

The operating principle behind the foam angle sensor is more readily understood by attaching a 2" x 0.25" x 0.15" piece of soft bathroom sponge to a card, so it can be bent from $0° \le \theta \le 180°$. During bending over the range $90° < \theta \le 180°$, upper surfaces of both sponge sections squeeze into each other, progressively reducing the sensor's surface contact resistance. Hard foams cannot conform in this manner, remaining parallel to the backing card until θ reaches nearly 180°, at which point sensor resistance drops precipitously.

Because adjacent fingers cannot be separated by an angle greater than about 60°, only that portion of the calibration curve from $\theta \sim 120°$ to $\theta \sim 180°$ is required for sign language monitoring. Note that angular sensitivity of the sensor in Figure 5-21 is not logarithmic, but closely approximates the relationship R = 5 sinθ.

The sensor used in Figure 5-20 was cut from a soft fine-grained packaging foam using a thin blade utility knife. When cutting foam use a ruler guide and make successive shallow cuts, working progressively through the material, rather than trying to cut through in a single slice. Light cuts also permit better control of blade angle, resulting in more perpendicular cuts. A flexible conductive coating made from rubber cement, carbon black and lacquer thinner, is applied to the top foam surface with a toothpick.

Two tinned steel end caps 0.5" x 0.25" x 0.005", cut from a 2lb coffee can with scissors, are given an initial sharp bend at about 90° with pliers, then formed over foam ends with finger pressure. A final crimping with pliers makes a good contact, and wires are soldered quickly to these electrodes to avoid overheating the foam. Coffee cans are painted on their outside and this is a plus for experimenters. Contacts can be soldered where required by scraping to bare metal, leaving an insulating layer elsewhere.

Calibrating a foam angle sensor

The best calibrations are those made 'in situ', ie. by operating a sensor exactly the way it will eventually be used in its final mode. Ideally, foam angle sensors should be calibrated on a glove worn by the final operator. This direct method avoids possible discrepancies that may arise due to gage orientation, and surface adhesion effects during remounting procedures.

Calibration data in Figure 5-21 were obtained by attaching the sensor to a paper card, underscored with a light cut to ease bending. A protractor laying on a table was used to measure angles by sighting along card edges while recording sensor outputs. Use a sheet of white paper beneath the protractor. The sensor can be removed for reuse if it is attached with hot melt glue.

Home made conductive foam, coatings and cements

Only a few inexpensive materials are required to make sensors that compare favorably with commercial devices. Home made force, angle and strain gage sensors can be constructed to suit particular applications allowing greater design flexibility, they are low cost, can be made to any size and are very effective. All desirable sensors have not yet been invented, so this is an area where amateurs can make contributions.

For those who like to dabble, a list of some procedures tested by the author and discarded in favor of recipes described later, are given below. Carbon black was used as the conductive ingredient except where otherwise stated.

RTV (silicone seal)	acrylic caulk
hobby black paint	Zap flex (flex crazy glue)
regular Zap	glycerine
gelatin (similar to Jello®)	white super glue
nail varnish	nail varnish + glycerin
nail varnish + rubber cement	household cement
tuner lube	urethane varnish
Krylon® crystal clear	Vinylite® cement
Letraset® spray	Uhu® bond-all
Tamiya paint	stencil adhesive
Aquadag®	Ailene's stretchable glue
Zap-A-Goo®	Ivory® detergent + water
Wall guard® + detergent	methyl alcohol
water glass (sodium silicate)	battery electrode graphite
pencil graphite	speedometer cable lube
Latex + various additives	rubber cement + copper powder
rubber cement + iron filings	charcoal
activated carbon	laser printer toner

Most of these mixtures when tested were either non-conducting, had high resistance, were too stiff, too brittle, erratic, unstable or did not dry completely.

Carbon black

Conductive carbon black is a key element for home made FSRs, foam angle sensors and stretch resistors. It is a fine jet black powder mainly used for paint pigmentation and in tire manufacture, (tires are non-conducting). Carbon black is formed by burning hydrocarbon products such as acetylene or mineral oils under partial combustion conditions. Diamond, lamp black, graphite, activated carbon used for water purification, and BBQ charcoal are also carbon. However, all these are unsatisfactory for making conductive sensors.

A carbon black powder suitable for home made conductive sensors and obtainable from local distributors, is manufactured under the trade name Conductex® by: *Columbian Chemical Co.* 1600 Parkwood Circle, Atlanta, Georgia 30339. (800) 235 4004 / (404) 951 5700

Rubber cement and lacquer thinner are also required, and the latter can be purchased at hardware stores. Rubber cement used as a paper adhesive works well. Small quantities, in 4oz bottles with brush, are sold by office supply houses and stationers. This cement works better for making sensors than rubber cement sold in tubes for bicycle tire repairs.

Home made conductive foam (for sensors in Figure 5-16)

Small celled foam suitable for sensors is used as a shock absorber or packaging material for shipping purposes, and can be obtained at no cost from many sources. Hardware stores sell a variety of foamed plastic products such as bathroom sponges, foam paint brush refills, foam sanding blocks, refills for sponge head mops, and auto windshield squeegees.

Type (4) conducting foam of Figure 5-16 is made by kneading a thick slurry of carbon black and lacquer thinner into foam as detailed below. Exact proportions of the mix are unimportant because a sensor's output can be subsequently trimmed by gain adjustment during signal amplification.

[1]   Use a piece of glass plate for this work (8" x 8" is fine). Do not use acrylic or other plastics because most of these are dissolved by lacquer thinner. Glass plate is cleaned by scraping off dried unused mix with a steel blade, then wiping with paper towel.

[2]   Cut foam to the desired width and thickness, but make it longer than required so it can be held at one end, keeping fingers clean and dry.

[3]   To about ¼ teaspoon of carbon black add lacquer thinner using a medicine dropper, then stir with a round toothpick to a consistency of soft ice cream.

[4]   Knead mixture uniformly into foam using one side of a round toothpick.

[5]   Squeeze out excess fluid by running the toothpick over the foam in a longitudinal sweeping motion.

[6]   A hair dryer can be used to drive off volatiles. Drying is complete when the blackened foam has recovered to the same dimensions as an untreated portion of foam.

[7]   Treated foam can be attached to metal plate electrodes as in Figure 5-16 and excess trimmed with scissors.

[8]   Gently dab the sensor sides with a slightly moistened paper towel to reduce dusting. Foam sensor sides can be sealed with a thin mixture (~ 1: 20) (rubber cement : lacquer thinner) but optimum flexibility is maintained with no sealant.

Common sense safety precautions: Both rubber cement and lacquer give off toxic, flammable vapors. But these products can be used in small quantities with discretion in a well-ventilated area, without resorting to draconian safety measures. Keep all containers sealed when not in use and dispose of waste promptly.

Conducting coating procedures (for home made sensors in Figures 5-21 and 5-23)

Flexible conductive coatings are used in fabricating foam angle sensors of Figure 5-21, and stretch resistors in Figure 5-23

[1]   Use a glass plate and round toothpicks for all mixing

[2]   Using the rubber cement bottle brush, drip a pea-size volume of cement onto an equal volume of carbon black on a glass plate and mix quickly with a toothpick.

[3]   Add lacquer thinner with a medicine dropper until consistency is suitable for the specific coating application. Stirring should be continuous during this operation.

[4]   Apply mixture to only one side of foam with a toothpick. When making a foam angle sensor, drying can be accelerated with a hair dryer.

[5]   Stretch resistors are made by holding rubber cord ends with thumb and forefinger of both hands. The cord is then rolled through the slurry, twirling both ends to give a uniform coat along the cord. This is much easier than it sounds in print.

Work quickly during these operations or mixtures will dry to an unusable state. Only make small quantities because repeated addition of lacquer thinner to prolong workable time, produces an inferior product.

Commercial conductive coatings

Some commercial conductive paint products listed below are handy for roboticists. It is a lot simpler to paint a nickel print electrical touch plate onto acrylic sheet, than mount a piece of metal. Wires can also be attached or bonded using conductive paint in many situations where soldering is impractical.

Television Tube Koat	# 49-2	*GC Electronics*
Nickel Print	# 22-207	*GC Electronics*
Silver Print	# 21-1	*GC Electronics*
Circuit Works	2220-STP	*Planned Products*
Aerodag® G		*Acheson Colloids*

The first three items are supplied with a brush applicator, Circuit Works is a pen applied paste. Aerodag G spray is a colloidal graphite suspension in isopropyl alcohol, it has good adhesion and can be used for electronic shielding or as a dry lubricant. These are all excellent products but cannot be used to make FSRs, foam angle sensors or stretch resistors because they lack the necessary flexibility. All crack under even small strains.

A home made flexible conducting adhesive

Hobbyists can make a very flexible conductive glue with excellent bonding properties from contact cement and carbon black. This conducting cement is very useful for attaching small wires to electrodes, it is inexpensive, waterproof, and easily made as required.

"How good is home made conducting rubber?"
Sheet resistance of a thin conducting film is specified in Ω/square usually for a thickness of 0.001" (25.4μm). Resistance is independent of area providing it is a square and thickness is specified. A commercial flexible carbon fluoroelastomer spray Electrodag 501™, manufactured by Acheson Colloids has a sheet resistance of 400Ω/sq. Their silver coat Electrodag 504 is 0.1Ω/sq. By comparison, home made conducting rubber for stretch resistors has a sheet resistance of about 3Ω/sq.

Measuring finger bending and large strains

Roboticists must often measure very large strains to duplicate human movements. For example maximum surface strain at finger knuckle skin is about 50% or 500,000 microstrain. This is well beyond the limit of commercial strain gages which cannot be used above a few thousand microstrain for repetitive measurements.

It is easy to check finger skin strain or stretch, by marking two dots on the skin centered over a knuckle and separated by 2cm, as shown in Figure 5-22. Keep fingers flat on a table during the marking process. Now bend the finger to its maximum and remeasure dot separation. It will have increased to about 3cm, an extension of 1cm or a strain $\dfrac{\Delta l}{l} = \dfrac{1}{2}$ *or* 50%

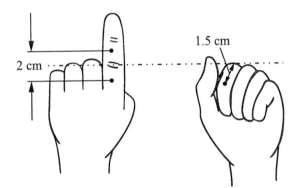

Figure 5-22. Finger knuckle skin strain can be measured by comparing markings in both stretched and relaxed states. Large strains of 50% or more can be dynamically monitored using stretch resistors.

Very large strains have traditionally been measured in ad hoc fashion using flexible tubes filled with conductive paste or mercury. Moving plungers have also been employed to suck or squeeze resistive fluid into a cylinder. Strings and pulleys using potentiometer arrangements outlined in Figure 5-5 are obvious contenders, but these methods are difficult to invoke for a hobbyist. A major impediment to all these techniques, apart from cost, large sensor size and mounting problems, is the difficulty of making dynamic measurements

Stretch resistors

A stretch resistor is simply a length of elastic cord covered with a flexible conductive coating as illustrated in Figure 5-23. If the coating is intimately bonded to the cord its resistance will increase as the cord is stretched, mimicking strain gage behavior.

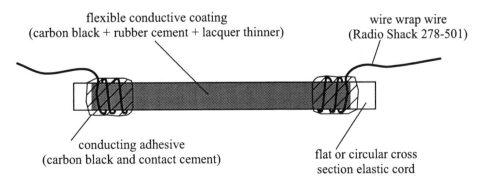

Figure 5-23. A stretch resistor can be stretched to about three times its relaxed length and used to measure strains up to 300%. It is fabricated using procedures given in the text, is easy to use, and has many practical applications

A flexible conductive coating from carbon black, rubber cement and lacquer thinner, previously described for foam angle sensors, is used to make a stretch resistor. A suitable rubber cord is twirled through a carbon black mix to give a uniform coat along its length. Contact cement cannot be substituted for rubber cement in this application because although it has superior adhesion qualities, it cracks at large strains. Thin contact wires are wound around both ends of a stretch resistor, and given a light squeeze with pliers. These connections are then overcoated with flexible conducting adhesive made from contact cement, carbon black and lacquer thinner.

Stretch resistors can be mounted on a glove or other surface using three small dabs of hot melt glue. One at each end, and another at the gage midpoint, (give the gage a slight stretch during this mounting procedure).

Where to buy elastic cord for stretch resistors

Reels of covered elastic thread (~$1/60ft), are sold by arts and crafts stores or sewing centers. This consists of a 0.5mm diameter elastic core with a cross woven double layer polyester sleeve. This material can be stretched about 40%, but if the covering is removed stretchability increases to 300%. Glove sensors made from this material are a bit trickier to manufacture, but have a dainty professional look when installed.

Elastic shock cord available at hardware and auto supply stores, is sold by the foot as 'bungee cords' or hooked 'tie down straps'. These items are a good source of high quality elastic. Only fabric-covered shock cords are useful, because they are made from many small circular or flat elastic strips. If a tie down strap is inspected closely at its hook, the type of elastic content can be established before buying. Very strong stretch resistors can be made from slingshot elastic replacement bands. These powerful stretch resistors are suitable for monitoring heavy body motions.

Stretch resistor performance

Stretch resistors can easily measure strains of 50% and will operate to 300% without damage. A calibration for a 3cm x 1mm diameter sensor is shown in Figure 5-24. These sensors have hysteresis but their measurement repeatability of ± 15% is adequate for operating robotic fingers to pick up objects, and other similar tasks. Note that for strains less than 50% response is almost linear, and no correction is required when a stretch resistor is used for finger bend measurements.

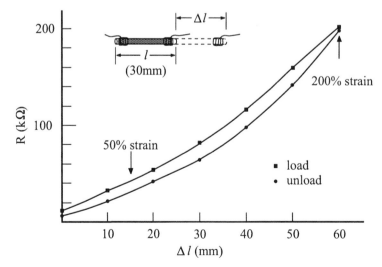

Figure 5-24. This stretch resistor calibration was made using mini-clip ohmmeter leads attached to the conducting layer. Stretch was measured with a ruler. Finer measurements for small extensions can be made using a modified microscope mechanical stage sold by *Edmund Scientific* # C30,058 $69 (Figure 8-31). Despite uncertainties due to hysteresis, stretch resistors work well in many applications. They have been used successfully for monitoring an operator's movements to remotely control robotic joints.

Stretch resistors are readily attached to an operator's glove or other clothing with hot melt glue. They can be used to monitor large dynamic motions and require only a one chip interface circuit, Figure 5-25.

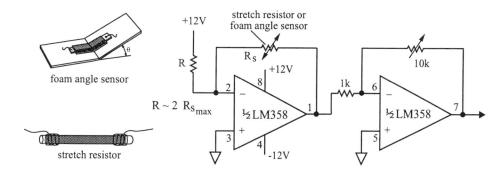

Figure 5-25. Stretch resistors, foam angle sensors and other devices with near linear resistive outputs for applied force, should be run at constant current to preserve their linearity. This circuit is simpler than a conventional constant current source and works well for such devices. To prevent signal saturation make R about twice the maximum sensor resistance encountered during normal operation

Hazards of signal saturation

Vigilance is always required to ensure saturated signals mentioned in the caption of Figure 5-25 are not misinterpreted as genuine measurements. This phenomenon is common in many electronics circuits but can occur in other physical processes too. Even experienced researchers have been fooled at times by failing to recognize the sometimes subtle but insidious features of gradual signal saturation. Always run a full range check on a sensor to check its performance. If you anticipate a linear output yet it starts to flatten out unexpectedly, this may be a sign of saturation. It may be due to a sensor used beyond its prescribed limits, an amplifier that cannot deliver an adequate current, pick-up noise overloading an amplifier, or an op amp output approaching power supply potential. Whatever its root cause, saturation should be corrected, or at least recognized to prevent drawing untenable conclusions from biased data.

Flow measurement

Techniques used for monitoring gas and liquid flow speeds are useful when designing planes, rockets and pneumatically launched projectiles. Some of these methods have other uses too, and can be employed for measuring displacement, or for proximity and security sensing.

Commercial flow measuring instruments run the gamut from simple basic techniques such as paddle wheels, rotating turbines, Pitot tubes and orifice flow meters, to sophisticated electromagnetic mass flow meters, ultrasonic Doppler sensing, vortex shedding meters, Coriolis sensors, vibrating bands and hot wires. Optical methods have permeated many areas of metrology, providing not only alternative techniques but often superior diagnostic performance. Laser Doppler Anemometry or LDA can measure flows non-invasively with high resolution and excellent frequency response, from millimeters a second to supersonic speeds. Commercial LDA equipment has features that boggle the mind but hobbyists can build a laser Doppler velocimeter, and this technique is discussed in a future book.

Sometimes visualization of a flow field is helpful, because point by point mapping is too slow and there is nothing like 'seeing' wind behavior. Small particles of suitable size and density can be used to make flow effects visible, and most people have seen dust devils, those mini swirling tornados that occur around large downtown office towers or in the desert. Helium bubbles, sparks and other methods are used for visualizing flows in wind tunnels, but smoke is favored because of its persistence, visibility, and ease of use. Fog fluid sold by theatrical agencies is a non toxic liquid that forms a dense fog when appropriately heated and is suitable for flow visualization. This medium is used for stage productions, and chilled fog often supplants traditional water vapor condensate from air blown over dry ice (solid carbon dioxide).

Stage smoke juice is handy for amateurs, although many professional tunnels still use kerosene or mineral oils. These are dangerous for hobby use because of their flammability.

Smoke tunnels are used in industry for checking flow conditions over models of aircraft, building structures and automobile shapes. The human mind tunes in very quickly to smoke flow streamlines, and hobbyists will find a home built smoke tunnel is a fascinating and rewarding project, providing snapshots of flow field conditions.

Commercial flow measuring equipment is expensive but *Edmund Scientific* carries some useful instruments at prices from $15.95 (C60,349) to $165 (C34,793). These items measure velocities from zero to 66mph and 0 to 100mph respectively.

A table top wind generator for testing sensors

High accuracy calibrations on wind velocity sensors (anemometers), are usually done in a wind tunnel where low turbulence, laminar flow and variable speed conditions are essential. Flow speed in these tunnels is usually determined with a pitot-static tube or an LDA system. These techniques give accuracies of about ±1% improving to ~ ±0.1% with extraordinary care.

"Wind tunnels suck!"

This truism favored by some aerospace engineering students, has a deeper meaning. In most wind tunnels air is sucked not blown through a test section where models are located, this is done to avoid fan induced turbulence. The Wright brother's wind tunnel employed this principle, and their apparatus can be seen at the US Air Force Museum in Dayton Ohio, one of the finest displays of aeronautical exhibits.

A home made wind generator for testing flow sensors

A small inexpensive wind source is very useful for hobby development work in air flow sensing. A suitable unit can be constructed for less than $10 using a hair blow dryer motor and fan. Retain heater wire for making low resistance power resistors. The thermal cut-out switch can be saved for other hobby purposes.

Hand held blow dryers operate from an ac line but employ dc brush motors because they are cheap, and their speed easily controlled by varying motor voltage. These small motors are designed for nominal operation at 12V. They are run from a full wave rectifying bridge after an appropriate voltage drop across the heater wire. Hand held dryers usually blow at ~30 mph (13.4 m/s) with 12V dc, but can be run up to 30V for short durations. At 30V flow speed increases to 25m/s or 55.9mph.

All fans generate a turbulent air flow due to vortices formed as blades redirect air flow from intake to outlet. Eddies can be a problem but will not seriously affect calibration procedures for our home built instruments, and procedures given later permit sensor calibrations to ± 5% accuracy.

Two-speed electrically powered garden blowers (~ $50) have speeds >150mph. A 1.5 HP Shop Vac™ blows up to 50mph, and its speed can be varied from a dimmer switch. Noise may be a problem if either of these appliances is used for protracted periods.

Blow dryer wind speeds

Performance of a squirrel cage blow dryer fan is plotted in Figure 5-26, and shows wind speed is proportional to motor voltage for exit velocities up to 30mph. Above this speed drag retarding forces load the motor quadratically, (drag increases as velocity squared). Getting speeds beyond ~25m/s is impractical.

A suitable dc power source for running a blower at higher speeds is shown in Figure 4-4. This requires a mandatory isolation transformer for safety. A blower can also be run at variable speeds up to 30mph from a PC switching supply using a 317-type variable voltage regulator shown in Figure 4-2.

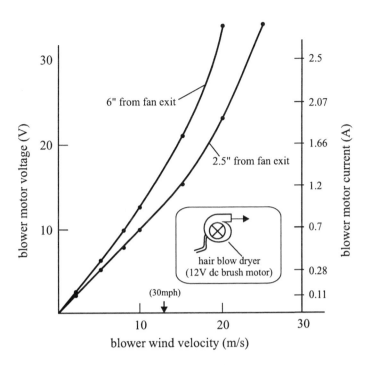

Figure 5-26. Wind speed is linearly proportional to blower motor voltage for velocities up to ~30mph where drag forces start loading fan blades appreciably. This region is useful because wind sensing schemes can be quickly checked with little data reduction. After a squirrel cage or turbine type blower fan has been calibrated, it can be used to characterize the performance of cup anemometers and other sensors described in this chapter.

Absolute wind velocity[2] measurements

Pitot tubes are used for measuring flow speeds in air and liquids, they can be seen at the foremost part of high speed aircraft and as 5" right angled tubular projections below a wing on many small planes. A knowledge of Pitot probe pressure and air density [3] are sufficient to calculate flow velocity. A simple Pitot tube measurement described later is recommended, and is the most accurate method for measuring air flow speed without investing in a commercial anemometer

Calibrating sensors accurately from a moving automobile is tricky because air accelerates as it moves around a vehicle. Errors as large as +30% can occur if a cup anemometer is extended by hand from a car window. Flow patterns around a moving auto are complex because of various projections, side mirrors, wheel wells, bumpers, windshield wipers etc. and these produce a turbulent layer near the vehicle.

---

[2] Velocity is a vector quantity and is different from speed. Because of turbulence we will frequently be unsure of flow direction in our tests, so both terms are used here synonymously.

[3] Up to 58m/s or 129.7mph, air is not significantly compressed when it is rammed into a body, and a Pitot tube can be used without any corrections below this speed. At higher speeds impacted air density increases, and eventually a shock wave is produced when flow speed becomes supersonic. Fortunately corrections can be applied even at high velocities, because a Pitot tube is one of the simplest yet most accurate methods for measuring fluid flow speeds.

A Pitot tube for measuring wind speeds

You can make a simple Pitot tube from flexible plastic tubing sold in hardware stores. Because this device is based on fundamental hydrostatic principles, it will always give a true indication of air flow speed, (below 130mph of course). Mount the tubing as suggested in Figure 5-27, leaving extra portions to reach the flow, a 6 foot length of ⅜" x ¼" clear food grade vinyl tube is fine, add water to fill about 8 inches of tube length. Tube dimensions are noncritical, unless the inside diameter is so small that surface tension forces become significant,

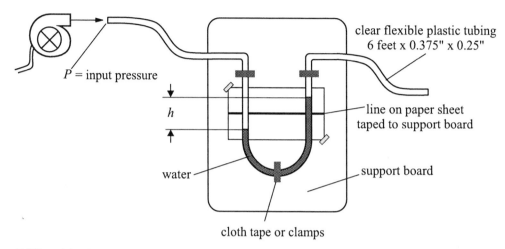

Figure 5-27.   This Pitot tube apparatus gives an accurate measure of wind speed by following instructions given in the text. Smaller diameter tubing will not give larger displacements and may cause difficulties due to capillary action.
       Professional Pitot tubes are made with a tapered or elliptically rounded entrance and prices start at around $200. Pitot-static tubes consist of a small tube with a single forward facing orifice used for measuring dynamic pressure. This tube is placed inside a larger tube that has a ring of round holes an inch or so back from the dynamic entrance. This circumferential set of holes is used to sample static pressure. Static and dynamic pressures can be measured independently or differentially with a Pitot-static tube.

Calculating flow velocity from Pitot tube measurements

[1]   Align a lined sheet of paper so tops of both water columns coincide with the line with no air flow, this makes it easier to measure h.

[2]   Air flow entering the tube in Figure 5-27 generates a pressure P that pushes water so it rises in the right-hand tube arm. This height h should be measured in millimeters with a ruler.

[3]   Compute flow velocity from V(m/s) = $4\sqrt{h}$ (mm water).
       For example if h = 10mm then $V = 4\sqrt{10}$ = 12.6m/s or 28.3mph.

Derivation of the relation in [3] is shown in Eqn. 5.5, for those who wish to try a low density oil instead of water.

$$P = \rho_w g h = \tfrac{1}{2}\rho_a V^2 \qquad V = \sqrt{\frac{2\rho_w g h}{\rho_a}}$$

$$\rho_w = \text{water density} = 1000 \, kg\,m^{-3} \qquad \rho_a = \text{air density} = 1.225 \, kg\,m^{-3} \qquad g = 9.81 \, m\,s^{-2}$$

$$V\,(m/s) = 126.6\,\sqrt{h} \;\; (h \text{ in meters of water}) \quad or \quad 4.00\,\sqrt{h} \;\; (h \text{ in mm of water})$$

(5.5)

[4]   Only a single measurement is required using a high flow speed with the tubing placed about 2.5" from the blower exit. This velocity measurement is used to convert readings from a cup anemometer or a strain gage wind meter to an absolute scale. These latter instruments are described later.

Using pressure gages for flow speed measurements

Experimenters may be tempted to try an electronic pressure gage for monitoring dynamic pressure from a Pitot tube, but this is not as easy as it seems. A piezo-resistive pressure transducer suitable for blood pressure monitoring and barometric pressure recording is sold by *Electrosonic*,$5.50 MPX100GP (sensitivity ~5psi/3.5V).

An air flow of 20m/s generates a pressure $P = \frac{1}{2}\rho V^2 = 0.5 \times 1.225 \times 400 = 245$ Pascals (Pa) or 0.0355 psi, producing an output of 25mV from the transducer. This signal level corresponds to less than 1% of full scale gage output (which is 3.5V), resulting in a pressure conversion accuracy of only 10%.

Measuring wind speed accurately with electronic pressure transducers is expensive. Current pressure gage technology is hard pressed to outperform a Betz water manometer developed in the early part of this century. A superb German-made mechanical instrument, the Betz can resolve 0.01 mm or 40 microinches in a 300 mm water column.

A strain gage wind meter

In chapter 4 drag was shown to be a significant factor in auto gas consumption. At 100mph air drag accounts for 50% of automobile fuel usage and drops to 6% at 40mph because drag scales as velocity squared. Drag force F_D is greater for a truck than a car because a truck has a larger frontal area A and a car is more streamlined, (resulting in a lower drag coefficient,C_D). Drag force in Newtons can be calculated from eqn. 5.6.

$$F_D = \frac{1}{2}\rho\, V^2\, C_D A \ (N)$$

$$\rho_{air} = 1.225\ kg\,m^{-3} \qquad C_D \sim 1.2 \ for\ a\ square\ flat\ plate^4 \qquad A = frontal\ area\ (m^2)$$

(5.6)

[4] Drag coefficient varies with Reynold's number but is about 1.2 for $V<$58m/s and small size plates.

The strain gaged beam from Figure 5-12 fitted with a drag plate makes a convenient, sensitive and precise wind meter. It has the added advantage that gage output can be predicted using eqn 5.6 and previous measurements from Figure 5-14. Remove nuts and screw from the beam and attach a 5cm x 5cm piece of stiff cardboard using double sided carpet tape, center this card on the first hole as before.

Drag force on this cardboard plate at a wind velocity of 10m/s is calculated in eqn 5.7.

$$F = \frac{1}{2}\rho\, V^2\, C_D A \ = \ 0.5 \times 1.225 \times 10^2 \times 1.2 \times 0.05^2 \ = 0.184 \ (N)$$

(5.7)

From previous results shown in Figure 5-14 and commented on in eqn 5.4, our beam gave a measured signal of 3V for a load of 0.237N. With a drag card the signal should be $3 \times 0.184 \div 0.237 = 1.31$V.

A wind calibration of the strain gaged beam fitted with a drag card is shown in Figure 5-28, and in a 10m/s flow it generates a signal of 1.55V. This value differs by 18% from the predicted signal of 1.31V but is acceptable because force is produced not only from wind striking the card, but the lower unclad portion of the beam as well. Gage output in Figure 5-28 is parabolic as implied by eqn. 5.7 and can be translated directly into velocities by fitting measured data to an equation $y = \alpha x^2$ where y = strain gage output, and x = velocity in m/s.

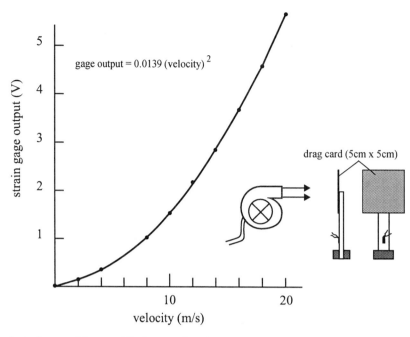

gage output = 0.0139 (velocity)^2

drag card (5cm x 5cm)

Figure 5-28. A strain gaged beam with drag card makes a very good wind sensor. Unlike most cup anemometers it can measure down to zero velocity. The transfer function is parabolic as predicted, and can be used to calculate wind velocity from gage readings. The constant (0.0139) is obtained from a least squares fit of data in this figure.

A rotating cup wind meter

Cup anemometers are still used for some field meterological wind speed measurements. They are simple reliable devices manufactured with both mechanical or electrical readouts. Miniature fan turbines are made that rotate on just a breath of air. These are carefully balanced jewelled bearing instruments resembling muffin fans used to cool electronic equipment. Because of their low friction, these devices spin at rates proportional to wind velocity. An infra red reflective sensor and some electronics can turn a jewelled bearing fan into an excellent anemometer.

Such turbines are hard to find but experimenters can build a very good cup meter for less than $10 using ping pong balls, balsa wood, and an audio cassette tape player motor as shown in Figure 5-29. The device can be built in under an hour and has an accuracy of ±5% after a calibration.

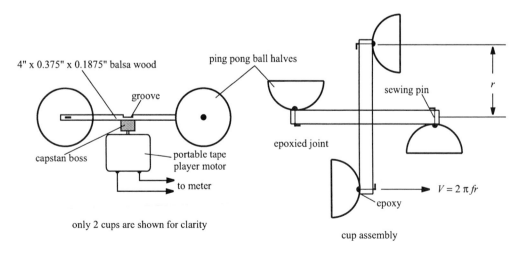

Figure 5-29. This four-cup anemometer can be made in less than an hour. It is a simple reliable device with a linear output above 10m/s (22.4mph). With a freely turning motor this unit operates down to wind speeds of 4m/s.

Ping pong balls cut with a thin utility blade heated in a flame, are attached to balsa shafts by a sewing pin pushed through the wood, bent, then snipped off. Each cup should be cemented with five minute epoxy where it meets wood. Epoxy is also used to join the two rabbeted balsa sections together.

Motors used in portable tape players have low turning torque and are ideal for a cup anemometer. They are equipped with a capstan boss, and five minute epoxy can be used to glue balsa shafts to this boss. Balsa wood is recommended because it is strong, cuts easily with a utility knife and needs no dynamic balancing due to its low density. Metal beams can inflict serious injury if they separate while rotating at high speed, and this hazard is significantly reduced with balsa.

Calibrating a cup anemometer

Calibrate your cup anemometer by measuring motor output for various blower speeds with the cups positioned 2.5" from the blower exit. A plot of motor volts versus blower voltage will be linear at higher speeds, as in Figure 5-30.

One absolute value of flow velocity from the blower is required to complete the calibration. This is obtained from a Pitot tube measurement outlined in Figure 5-27. Set up the apparatus using a clean cut tube end supported on a stand, facing into the flow and 2.5" from the blower exit. Adjust blower speed for a height difference h of about 15 to 20 mm and measure this accurately with a ruler. True wind speed can then be calculated using eqn. 5.5.

Tests show ping pong cups reach about 90% of the wind speed, and this relationship can be used as an additional calibration check. A stroboscope or optical tachometer is used to determine cup rotation rates. Rotational velocity as shown in Figure 5-29 is calculated from $V = 2\pi fr$.

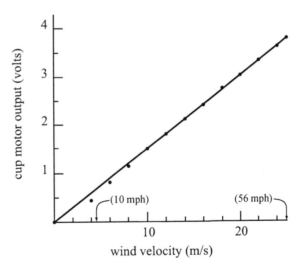

Figure 5-30. Voltage generated by a 4-cup ping pong ball anemometer varies linearly for wind velocities above 10 m/s. Average wind speeds can be measured with a cup anemometer even in fairly turbulent conditions. Faraday's law predicts generated EMF should be proportional to motor shaft speed and is confirmed in this figure.

Some comments on using hot wires for wind measurements

When a tungsten filament is heated its resistance changes as shown in Table 4-7. If the filament is subsequently cooled by a flow, more current is needed to reestablish its original working temperature. This principle is exploited in modern hot wire anemometry and large dynamic operating ranges, with frequency response to 50kHz are possible.

Commercial hot wire probes are easy to use, cause minimal flow perturbation and can be tailored for use in many gases and liquids. These probes use thin tungsten wires as small as 0.0002" (5.08μm) diameter and can measure flows from a few centimeters a second, and have even been employed at supersonic velocities.

Duplicating this performance is difficult for an amateur, because even miniature tungsten filament lamps use heavier gage wire. Miniature lamp filaments are relatively long, resulting in slower response time because of greater thermal inertia. It is also not easy to linearize a hot wire probe's output, but a hand drawn scale obtained from a wind calibration using procedures described earlier, can be attached to an analog meter.

Mini Maglite™ lamps or an equivalent sold by *Radio Shack* #272-1149, 2.5V @ 300mA, $2.99 for 2 lamps, are best for an amateur hot wire apparatus. These bulbs are tiny (0.125" dia x 0.4") and their 0.001" diameter coiled tungsten filaments are only about 0.030" x 0.005", cold resistance ~0.7Ω. Lamp pins fit into solderless breadboard sockets or a 0.1" female header socket making experimentation easier. A lamp's glass envelope can be broken by squeezing it in a screw clamp or vise.

A practical wind speed meter using a 5¢ signal diode

Self heating of a 1N4148 or 1N914 silicon signal diode (~ 5¢), can be used to make a simple, practical rugged, precise and reliable anemometer. The circuit in Figure 5-31 drives 450mA through a diode causing it to overheat, because its nominal current rating is 10mA. Signal diodes tolerate this over current operation remarkably well. They behave in a very stable manner, and can be operated continuously for many days without drift or a change in their calibrated wind sensitivity. They are hot to touch when running, but will not burn fingers. An added benefit accrues from this high temperature operation, the diodes are insensitive to ambient temperature fluctuations and respond only to cooling from wind or a touch.

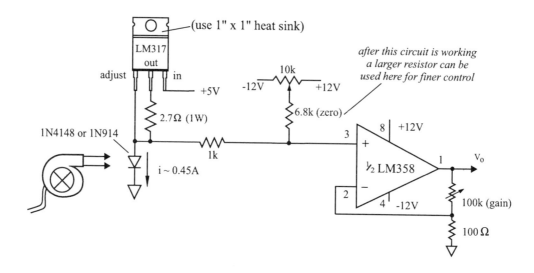

Figure 5-31. This sensor gives an output of 0.5V from a hand wave at 6", it is rugged, very repeatable and has low drift. Diodes can be interchanged, requiring only a zero adjustment and a full scale setting from a reference wind generated by a hair blow dryer run at a fixed voltage.

Figure 5-32 shows a diode wind sensor's response up to 25m/s or 55.9mph. The region below 2m/s has been omitted from this plot but is stable and very useful for low speed measurements down to zero velocity.

After initial warm up, sensor outputs are repeatable to better than 2% over several days continuous operation and response time is ~5 seconds from 25 m/s to 2 m/s and 7 seconds from 2m/s to 25m/s. For smaller velocity excursions response is about 1 second.

Diodes of the same type conform to the calibration curve of Figure 5-32 to better than ±5% from 5 to 25m/s, when zero and gain are adjusted for the same outputs. Diodes from the same batch usually agree to better than ±2%.

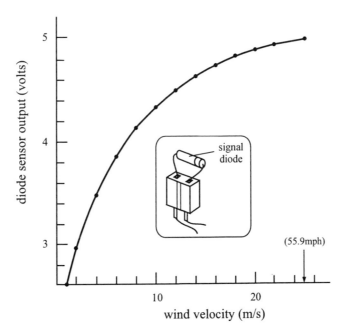

Figure 5-32. Electrically heated 1N4148/1N914 signal diodes give a repeatable, stable, logarithmic output from zero to over 60mph. Very low velocities are easily measured but have been omitted from this plot because of their near-linear nature. A signal diode sensor can be conveniently mounted on a solderless breadboard, or supported in a small section of female header strip as shown in the figure inset.

A multi-component strain gage balance

Lift, drag, and roll forces on a model placed in a wind tunnel or a flowing stream of liquid can be measured simultaneously using a strain gage balance illustrated in Figure 5-33. A metal rod or 'sting' is used to hold a model for testing, and a streamlined shroud is placed around the balance itself to shield it from wind or other forces. Balances are always placed downstream in a flow, or can be installed outside a wind tunnel by taking a sting extension through tunnel walls, floor or ceiling.

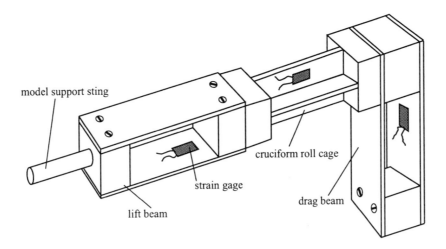

Figure 5-33. This three component balance can be calibrated using weights to apply known forces while measuring gage outputs independently. Three separate channels are required for simultaneous data acquisition, or outputs can be multiplexed.

Temperature sensors

Many techniques are available for measuring temperatures but no single method spans the range from absolute zero (-273.15K) to thermonuclear temperatures (5×10^7 K). Expansion properties of gases, liquids and solids are used for measurements from low to medium temperature regimes. Good mercury or organic thermometers based on thermal expansion are useful from -200°C to 500°C. These are recommended for home calibrations because of their simplicity and reliability.

Thermistors, thermocouples, RTD's (Resistance Temperature Devices), thermal gages and solid state sensors are among the most convenient devices because they give electrical outputs. They are also small, rugged, accurate, relatively inexpensive and cover temperatures of most interest to hobbyists and experimenters. Measuring instruments using these sensors are manufactured for temperatures from absolute zero to the melting point of many metals ~3000K. Optical techniques supplement these measurement capabilities from thermal infrared (~700C), to maximum temperatures.

Thermocouples and thermopiles

Any two dissimilar metals placed in intimate contact generate a small voltage when heated and this is called the Seebeck effect. Iron and Constantan (used in wire wound resistors), give the largest signal at $60\mu V \,°C^{-1}$. Even a combination of stainless steel and wire wrap wire (which is silver plated), produces about $1.4\mu V \,°C^{-1}$. All thermocouples have low output impedances depending on their wire gage, but are usually less than 30Ω except in long lengths of fine wire.

A thermopile is an assembly of thermocouples connected in series to produce higher voltage. This composite unit is subsequently blackened to absorb radiation, and can be employed for many types of radiometry. Thin film thermopiles are rugged, reliable and have an almost flat response to radiation from ultraviolet wavelengths to more than 50μm in the far infrared. They may contain more than 100 thermocouples, and are invaluable for remote temperature sensing when used with telescopic optics.

Pyroelectric materials also have a flat spectral response but give a signal output only when the radiation level is changing. They are used in passive infrared security sensors, and can detect movements of a 98 °F human body against a steady thermal background. A PIR security sensing element may be used by hobbyists for low temperature thermography as shown in a future book in this series.

Hobbyists can make their own thermocouples by twisting two dissimilar metal wires together, then drawing an arc at low voltage (~20V), and high current (~20A). Use a graphite rod connected to one side of an ac supply and touch the twisted wires attached to the other ac line, separate the two and an arc will form fusing the two metals. A graphite rod can be salvaged from a heavy duty 1.5V cell and may be sharpened to a point at one end with a file. Common sense and elementary safety considerations dictate these operations need an isolation transformer to reduce the risk of electrocution from a raw 115V supply line.

Commercial thermocouples are either welded or fused using a capacitive discharge that produces a short duration high current discharge of controllable power. Direct current arcs are difficult to control and are inferior to alternating current for amateur junction fusing operations.

Optical pyrometry

Hot tungsten filament radiation has been studied extensively, because it can be used to approximate black body radiation up to the melting point of tungsten, 3400°C. This feature is used in optical pyrometers from about 700°C to characterize a hot body of unknown temperature. Either a visual or photoelectric comparison is made against a glowing tungsten lamp filament of known temperature. Other optical techniques compare energy from two or more infra red spectral bands, and can be used for monitoring temperatures up to 5000°C and beyond. At extreme temperatures (>20,000K), all matter becomes a plasma of free charged particles in equal quantities, and optical methods must be used for determining temperature because no physical probe can tolerate such conditions.

Measuring temperature at high altitudes

Measuring air or gas temperatures at low densities is difficult, because of poor heat transfer between gas molecules and a physical measuring device. This is why hobbyists must use thermometers or thermistors with discretion when determining temperature at high altitudes. For example, atmospheric temperature at an altitude of 125 miles is about 960°C/1760°F. A thermometer cannot be use to measure this temperature, because only a few gas molecules are available to heat a thermometer bulb. Consequently, the thermometer will stabilize at an equilibrium state established by radiation effects. This will be very cold at night and a little warmer when illuminated by the sun.

Similar difficulties abound in thermometry and an experimenter must always be alert to ensure a valid measurement of temperature is obtained. 'Temperature' at very low gas densities depends on how fast gas molecules are moving. This 'kinetic temperature' is impossible to measure accurately with a conventional thermometer or even the smallest thermocouple probe. Temperature in these cases is obtained indirectly either from pressure measurements, or by ionizing the local atmosphere with an electron beam. An electron beam produces a glow just like the aurora borealis, and this glow is then analysed spectrally.

Temperature measuring methods

Typical capabilities of some important temperature measurement techniques are summarized in Table 5-5. These represent current state of the art performance in high quality commercial instrumentation.

Table 5-5. Thermometer characteristics

	range	response time	accuracy	repeatability	resolution	linearity
glass thermometer	-200 °C +500 °C	≥5s	≤0.1 °C	≤0.1 °C	<0.01 °C	0.1% FS
thermistor	-100 °C +400 °C	>0.5s	±1% FS	0.1 °C	0.1 °C	<0.5%FS
thermocouple	-270 °C +2320 °C	>1s	±0.5 °C	0.5 °C	0.1% FS	0.1% FS
RTD	-250 °C +1000 °C	≥5s	0.1 °C	0.1 °C	0.1 °C	0.1% FS
LM135/235/335	-50 °C +150 °C	>5s	≤1 °C	0.5 °C	0.5 °C	~1%
bimetallic dial gage	-75 °C +530 °C	>30s	~1% FS	~1%	~0.3% FS	1%
thermal gage (strain gage)	-195 °C +260 °C	<0.01s	1C FS	0.1 °C	0.1 °C	<1 °C FS
optical pyrometer	+700 °C >+5000 °C	5s	~1% FS	0.5%	0.2%	~0.5%

High quality sensing elements and break point linearization are used in top quality commercial temperature monitoring instruments. Few amateurs can match this performance because only economies of large scale production make such procedures affordable. Hobbyists can make inexpensive sensing equipment to equal these tabled standards over a range of at least -80°C to +250°C, by using procedures described in this chapter.

The basic principles of any technique listed in Table 5-5 can be exploited by experimenters, and much can be learned by building equipment based on these methods.

An exotic thermal sensor used in a heat sinking missile may cost $10,000 or more and require cooling to below -150K. Even temperature controlled ovens are too expensive for most hobbyists. However, with a little effort and small expense, it is possible for experimenters to produce and measure temperatures from -80°C to over +250°C (-148°F to 482°F). Procedures are given below that enable hobbyists to test their own sensor inventions over this regime to an accuracy of ±1°C

A 500°F pop-can oven

If you have ever observed how long a domestic kitchen stove oven takes to heat up and cool down, you probably realize this thermal inertia contributes to temperature uniformity inside the oven. This long time constant also makes a kitchen oven inconvenient for hobby work, where temperatures must be cycled rapidly during calibrations.

A small low power oven that heats and cools rapidly is shown in Figure 5-34, it is safe for an operator, because even at 300°C the can's exterior may be touched quickly without burning, due to its low thermal capacity. About 3kJ of thermal energy is stored in the oven's mass at 300°C, equivalent to operating a 100W lamp for 30 seconds. But because of its low thermal mass and large surface area, the oven cools surprisingly fast and this energy is stored and dissipated safely. Support legs may be held continuously while the oven is running, and thirty seconds after removal from the lamp by holding the legs, the oven may be clasped in a hand.

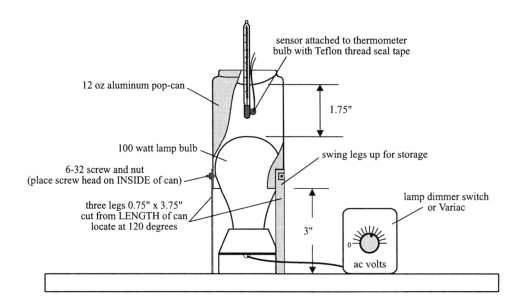

Figure 5-34. A pop-can oven heats and cools rapidly. It is safe, convenient for testing, and suitable for calibrating small sensors accurately. The oven is designed to operate up to 250°C but will reach temperatures over 300°C if the lamp is run from a Variac™ at 125V. A nearly linear temperature gradient (~2.5°C/mm at 200°C) exists between lamp and can roof. Effects of this gradient are substantially reduced if a sensor under calibration is attached to a thermometer bulb or a small thermal mass.

Two cans[5] are required to make the oven, and these may be cut with heavy steel scissors. Excellent edges are possible with a little practice. Holes can be pierced with an awl or a sharp nail and subsequently enlarged by rotating needle-nosed plier tips in the hole. Make a top hole no larger than necessary to avoid convective heat loss. Aluminum foil can be placed around an oversized hole, but an opening just large enough to fit the thermometer and sensor wires is best. If made to dimensions given in the figure, an oven's temperature and warm up time can be predicted from Figure 5-35.

[5] The ubiquitous aluminum pop can is an excellent example of good engineering design. Its apparent simplicity understates the care and attention to detail required to make this quintessential utilitarian product. Less that 40% of a can's 15g mass is used for walls that are sometimes only 2 mils or 50μm thick, and its strong base accounts for 25% of total weight. Aluminum is the most abundant metal on earth with a specific gravity of 2.69 a melting point of 660°C and a specific heat capacity of 0.900Jg^{-1} °C^{-1}. It is also the most pure commonly available metal.

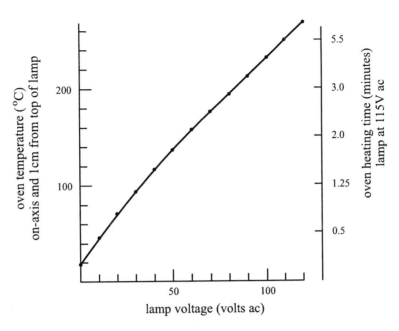

Figure 5-35. A pop-can oven heats to 250°C in about five minutes with a 100W lamp at 115V ac, and cools to room temperature in one minute after being removed from the lamp. Temperatures over 300°C are possible if lamp voltage is increased to 125V with a Variac™.

An almost linear temperature gradient of ~2.5°C/mm occurs from lamp to can roof at 200°C, as indicated in Figure 5-36 (a), and this scales approximately with temperature as suggested by a plot at 50°C shown in Figure 5-36 (b). This gradient is of small consequence when a sensor is attached to a thermometer bulb or a piece of metal. Gradient effects are also be reduced if a sensor is oriented so its long axis is parallel to the can roof.

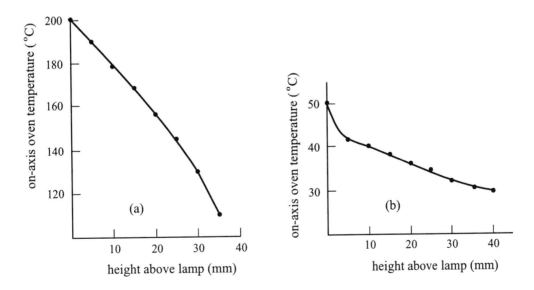

Figure 5-36. Temperature variation from lamp to can roof depends on operating temperature, but its effects are reduced when a sensor is attached to a thermometer bulb, or mounted with its major axis parallel to the floor.

Calibrating thermal sensors

If a sensor responds linearly to temperature it requires calibrating at only two known temperatures, preferably close to the sensor's operational end points. For example National Semiconductor's LM335 has an uncalibrated error of 5°C over its -40°C to 100°C operating range, but improves to ±1°C with a calibration. A 1N4148 signal diode sensor after calibration at water's ice and steam points is accurate to ±1°C and linear from -80°C to +250°C. This diode is recommended for hobby use because of its low cost, and predictable, precise, repeatable performance.

Fixed point reference temperatures

"Water freezes at 32 °F and boils at 212 °F."
If this were only true under all circumstances, temperature calibrations would be a snap. Unfortunately a few caveats exist, and a little care is necessary to get accuracies of ±0.2°C at these two reference temperatures.

Ice water fixed point

Ice taken from a freezer is not necessarily at 0°C or 32°F. It may be -10°C/14°F or colder or warmer, depending on the thermostat setting, room temperature and even the type of water.

A reliable ice-water bath at 0°C is made by first crushing ice to ~½" sized nuggets, placing them in an insulated Thermos™ jug, adding tepid water, then agitating periodically for five minutes. If distilled water is used for both ice cubes and topping up, this ice-water mixture will be within 0.2°C of zero Celsius. Ice can be broken in a heavy plastic bag on a carpet using a hammer.

Boiling water or steam point

Distilled water only boils at 100.00°C if its surface is in equilibrium with water vapor (steam) at normal atmospheric pressure, (101325 Pa or 760 mm of mercury or 1013.25mb). A reasonably accurate estimate of pressure can be obtained from a local weather station or a reliable barometer. A first order boiling point correction for atmospheric pressure p is given by BP = 100 + 0.0367 (p-760).

World maximum and minimum atmospheric pressures are given in the 1989 Guinness Book of Records as 1083.8mb and 870mb (812.9 and 652.6 mmHg) corresponding to water BP's of 101.9°C and 96.1°C, a spread of almost 6°C or 10.8°F. [Example: BP = 100 + 0.0367 x (652.6-760) = 96.06°C]

Boiling point depression at high altitudes is more common. At 5000ft normal atmospheric pressure is 635mm Hg and water boils at 95.4°C. Experimentalists in Denver or mountain top dwellers are usually aware of these problems if they like hard-boiled eggs.

Cold temperature calibrations

Freezing mixtures down to -21.3°C/-6.34°F can be made by mixing table salt (NaCl) with crushed ice or snow. Ice can be pulverized with a hammer, carpet, and plastic bag as described previously.

Dry ice or solid CO_2 has a boiling point of -78.2°C/-108.8°F and is listed under 'dry ice' in some Yellow page directories. It is also retailed by theatrical stage equipment suppliers. A price of 70¢/lb for scrap dry ice is reasonable, but some manufacturers will often donate a 1lb bag for a worthy cause. Take along an insulated container, this stuff is very cold and disappears overnight in a home freezer. It also freezes flesh instantly. One pound blocks of dry ice can be made from compressed CO_2 in 60 seconds using an attachment available from *Cole Parmer*, or *Edmund Scientific* C71,533, $74. Hi-dri siphon compressed CO_2 gas, used for cooling environmental chambers, is available from gas suppliers.

Home refrigerator freezing compartments operate down to about -15°C. A sensor enveloped in plastic food wrap can be calibrated in a home freezer by immersion in a bath of rubbing alcohol, using a thermometer for temperature comparison measurements.

Using a glass thermometer

What could be simpler or inspire more confidence than to attach a sensor to a glass thermometer's bulb, then record both outputs over a range of temperatures? Well this is surely simple, but not necessarily accurate over a large temperature range.

The most accurate glass thermometers are 'total immersion' types. This means the whole thermometer must be immersed in a fluid to measure a liquid's temperature. Many other thermometers use a circumferential band marker signifying the immersion point. Cheap thermometers just leave one guessing.

It is evident that if a 'total immersion' thermometer's bulb is being roasted at 400°C and the remaining column section sits at 20°C a serious reading error exists, because the unheated column should also be heated to 400°C to comply with initial certification procedures. Immersing a thermometer completely for every measurement is tedious and sometimes impractical, so corrections must be applied to compensate for any unheated column length.

Full correction of exposed stem cooling becomes tricky if great accuracy is needed, and parameters such as glass type, indicating fluid formulation and the stem's thermal gradient must be known. For most industrial and hobby work an accuracy of about 0.2°C is adequate, and this can be achieved using eqn. 5.8.

(5.8)

$$T = T_m + \alpha l (T_m - T_s)$$

T = true bulb temperature, T_m = measured temperature, T_s = midpoint temperature of liquid column

α = 0.00016 for mercury
α = 0.001 for organic fills (alcohol etc)
l = exposed column length in degrees

If you think it is too much trouble to make corrections, let us look at a total immersion mercury thermometer with only its bulb stuck into our pop-can oven. Assume a column of mercury equal to 258°C protrudes from the can. A measured temperature of 258°C and a mid stem temperature of 20°C really correspond to a true bulb temperature T = 258 + 0.00016 x 258 (258 - 20) = 267.6°C. A discrepancy of almost 10°C or 17 °F[6].

Correcting an organic fill thermometer is even more important, since a reading of 137°C with an exposed column length of 130°C corresponds to an uncorrected error of 15°C.

A very good mercury thermometer (-10°C to +400 °C) costs about $20 at a laboratory supply house or from *Cole Parmer*. Organic fill candy thermometers operating from 75°F to 400°F (23.9°C to 204.4°C) retail for less than $10 at hardware stores.

Linear sensors are easily calibrated at two fixed reference points as described previously, but nonlinear devices and new inventions are best calibrated in contact with a thermometer bulb, as shown in Figure 5-34. This procedure is recommended for all important measurements.

[6] To convert Celsius to Fahrenheit or vice versa:

     [1]   add 40
     [2]   multiply by 5/9 or 9/5
     [3]   subtract 40

This works for all temperatures because -40 °C = -40 °F, the only temperature where both scale values agree.

An important note on using a glass thermometer at high temperatures

Because water expands when freezing it can easily break apart large boulders or even lift the side of a house. Expansion forces are in fact virtually irresistible. It should therefore be no surprise to learn that a glass thermometer will break if heated so the liquid column expands beyond its upper limit.

Attaching sensors to a glass thermometer bulb for testing

Sensors must be properly attached to a thermometer's bulb because it is made of very thin glass and easily shattered. Mercury is often used as an indicating fluid, and this is a dangerous element outgassing toxic vapors for decades, even at room temperature.

Consequently, a sensor should only be fastened to a glass thermometer with Teflon thread seal tape. Available from plumbing or hardware stores in 40ft x ½" rolls for ~$2, this thin (2-5 mil), pliable, stretchable, high and low temperature tape, can be used to wrap sensors gently but firmly onto a thermometer bulb. It flexes to give a thermometer bulb protection, and is easy to apply and remove.

High temperature wiring

Above100°C the insulation qualities of many common wire coverings deteriorate, and if used at 115V ac can be hazardous. Fortunately most temperature sensors run at low voltages, and heating introduces only minor insulation difficulties.

Insulation on general purpose PVC hook-up wire is rated at 100°C and Teflon at 200°C, but these ratings are conservative. For low voltage work a hobbyist may use Teflon at 300°C for extended periods, providing the wire core is not stressed so it tends to cut through the insulation. Magnet wire can be used at 250°C or higher, and even wire wrap wire survives this temperature if lines are not twisted together. Give each wire a separate wrap with Teflon tape, where it exits the oven.

Sensor connections should be made by wire wrapping onto sensor terminals because soldered joints lose strength above 200°C. Wire wrapping tools are shown in Figure 7-5.

Solid state temperature sensors

National Semiconductor manufactures a series of precision solid state temperature sensors fulfilling most hobby needs, and their specifications are listed in Table 5-6. They are small, low current, easy to use zener devices, with an internal amplifier.

Table 5-6. Solid state temperature sensors for hobbyists

	operating temp (continuous)	operating temp (intermittent)	accuracy (uncalibrated)	accuracy (after calibration)	price
LM135	-55 °C +150 °C	200 °C	0.5 to 1 °C	0.3 to 1 °C	$4.30 *Electrosonic*
LM235	-40 °C +125 °C	150 °C	1 to 3 °C	0.5 to 1.5 °C	$4.70 *Electrosonic*
LM335	-40 °C +100 °C	125 °C	2 to 6 °C	1 to 2 °C	$1.99 *Hosfelt*

These National Semiconductor temperature sensors have good basic accuracy, which can be improved with a single fixed point temperature calibration at room temperature. An LM335 is the most popular device and is more readily available. It is inexpensive, and after a simple calibration is more accurate than most cheap thermometers

All sensors in Table 5-6 have a slow response to temperature changes. When immersed in a flowing liquid an LM335 has a time constant of about three seconds, increasing to ~8s in an air flow, it should be given good thermal contact where possible.

In its untrimmed operational mode any sensor from Table 5-6 has a linear output of 10mV $°C^{-1}$, and this may be amplified if needed as shown in Figure 5-37 (a). These three terminal devices do not have the same pinouts as regular transistors but are configured as in Figures 5-37 (b) and 5-37 (c).

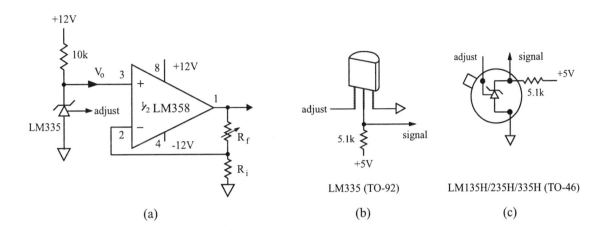

(a)   (b)   (c)

Figure 5-37. Untrimmed operation of these sensors gives reasonable accuracies. A simple room temperature calibration, and one extra trimpot as shown in Figure 5-38, improves things by a factor of about two.

Outputs from all three NSC thermal sensors can be trimmed for greater accuracy. This is usually done at room temperature, after allowing the sensor to reach thermal equilibrium while sitting close to a thermometer. Using the circuit of Figure 5-38 a sensor will provide linear, reliable and accurate performance over its specified operating range, after trimming as suggested.

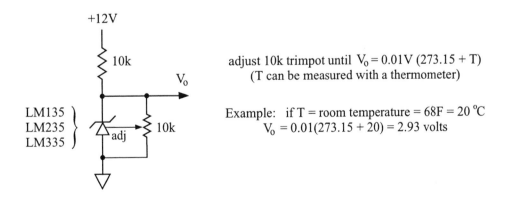

adjust 10k trimpot until $V_0 = 0.01V (273.15 + T)$
(T can be measured with a thermometer)

Example: if T = room temperature = 68F = 20 $°C$
$V_0 = 0.01(273.15 + 20) = 2.93$ volts

Figure 5-38. Trimming a National Semiconductor thermal sensor is straightforward, giving enhanced accuracy for only a small effort. Additional amplification is usually only necessary for precise differential temperature measurements, or when the full scale output must be set at a specific level.

A transistor temperature sensor

Inexpensive TO-92 temperature sensing transistors with good accuracy are sold by *Newark* for 63¢. These devices are manufactured by Motorola as MTS102 and MTS105, and have a linear temperature response from -40 °C to +150 °C. A small silicon NPN transistor may be used instead of these special parts, and performance of a 2N3904 is illustrated by a calibration curve shown in Figure 5-39.

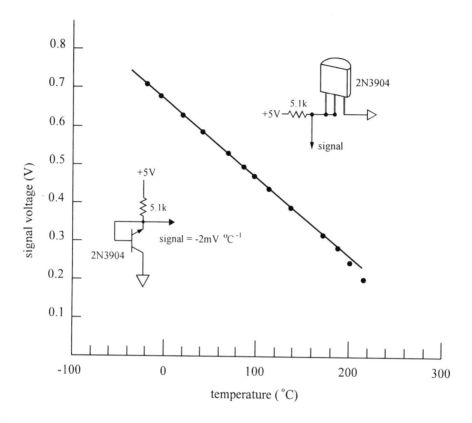

Figure 5-39. Small NPN transistors or MTS102/105 thermal sensors are inexpensive, with good linearity and repeatability. Signal amplification as applied in Figure 5-37 elevates the performance of these simple sensors to equal that of an LM335.

A 5¢ temperature sensor with excellent precision and linearity from -100°C to 250°C

It was shown earlier that a 1N4148 signal diode has excellent stability and repeatability, when used as a wind sensor. These desirable characteristics can also be employed to make a simple but very effective temperature sensor, when a signal diode is run at a forward current of 1mA.

A 1N4148 or 1N914 can be used over a larger temperature range than either LM135, LM235, LM335 or MTS102/105 and has comparable or better linearity, precision and repeatability. In addition these signal diodes have a faster time response, are more readily available and less expensive.

If a 1N4148 is initially trimmed as outlined in Figure 5-40, then its forward voltage drop varies with temperature, as shown by the plot in this figure. Results are repeatable to ±0.1% and V_f can be amplified using the circuit in Figure 5-37.

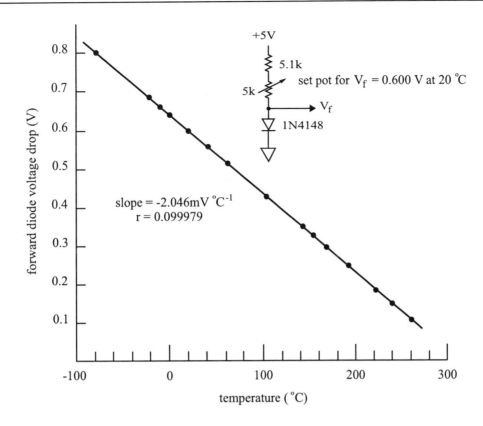

Figure 5-40. This simple, inexpensive temperature sensor is very stable and has an output repeatability of 0.1%. It also has faster response to temperature changes than either an LM335 or MTS102/105. After calibration at 0°C and 100°C, signal diodes can be used for measuring temperature from -80°C to 250°C with an accuracy of ±1°C.

Not all 1N4148 diodes have a slope of exactly -2.046mV/°C as shown in Figure 5-40, but tests show slopes usually differ by no more than ±2% from this value. If a signal diode is biased for 0.600V at 20°C, it can be used for temperature sensing to ±3°C from -80°C to 100°C without further calibration.

If 1N4148 sensors are calibrated at ice and boiling water points as described earlier, a sensor accuracy of ±1°C can be attained over a -80°C to 250°C range.

Capacitive Sensors

A device that can monitor force, pressure, displacement or strain is always useful to a roboticist. A home made sensor with such capabilities, and no thicker than this sheet of paper is described below.

Presently no commercial capacitive sensors suitable for hobby use are available on the surplus market. This is because many other types of sensors can monitor all parameters of interest, consequently commercial users have been slow in embracing capacitive technologies. This situation is now changing as users become aware of the outstanding performance of capacitive sensor products, which are often unmatched by other methods.

A capacitive sensor used for high precision weighing may have less than 4% of the creep and hysteresis present in even a high quality strain gage. For example up to 999,999 identical parts can be weighed and counted using a capacitor scale with an error of only 1 in 750,000. Some of the best pressure transducers also employ capacitive techniques, operating in a highly linear manner over wide pressure ranges with good accuracy, excellent resolution and repeatability. One of the foremost competitors to the venerable Betz manometer mentioned earlier, is based on a pressure sensing capacitor. Such devices can resolve differential pressure changes of 1 part in 100 million.

Capacitive sensors are now sold for measuring acceleration, weight, humidity, tilt angle, displacement, pressure, and rotation angles to 1 arc second. Simple proximity sensors exploit capacitive coupling between one plate of a capacitor and another body, and touch sensors also employ this principle for alarms and appliance switching. Most microphones in use nowadays also use capacitive technology, where one plate of a capacitor is permanently energised by an electret. These sensors are now so common that prices have fallen below $1.

A parallel plate capacitor in Figure 5-41 is one of the simplest electronic devices and its capacitance depends only on plate area, dielectric medium and plate separation. Capacity varies linearly with common plate area, and this feature is used to make variable vane rotary capacitors, or liquid level and tilt sensors. Separating the plates of a capacitor changes capacity in a nonlinear fashion, but in some modern instruments plate separation is kept constant by null-servoing. Servo current measurements are then used to determine plate displacement with excellent linearity, repeatability and high instrumental precision.

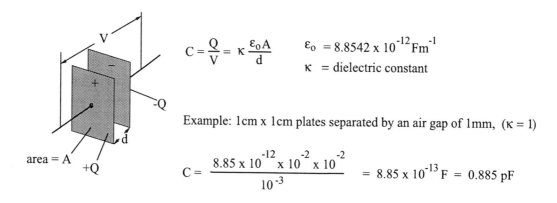

$$C = \frac{Q}{V} = \kappa \frac{\varepsilon_0 A}{d} \qquad \varepsilon_0 = 8.8542 \times 10^{-12} \, \text{Fm}^{-1}$$

$$\kappa = \text{dielectric constant}$$

Example: 1cm x 1cm plates separated by an air gap of 1mm, ($\kappa = 1$)

$$C = \frac{8.85 \times 10^{-12} \times 10^{-2} \times 10^{-2}}{10^{-3}} = 8.85 \times 10^{-13} \, \text{F} = 0.885 \, \text{pF}$$

area = A

Figure 5-41. Capacitive sensors based on variable plate area, plate separation or dielectric content, are basic simple mechanical devices for hobby applications. Capacitance between cylindrical electrodes varies as $[\ln(R/r)]^{-1}$ and as $(rR)/(R-r)$ for a spherical capacitor. These geometries can also be used for sensing.

It is not essential that both plates of a capacitor be planar. In fact a spherical electrode is more forgiving if unavoidably tilted during finicky measurements. This feature is handy in Atomic Force Microscopy or AFM, a burgeoning field with an impact in science comparable to the electron microscope. Hemispherical AFM micro probes can be used for non contact sensing of even soft surfaces, giving displacement resolutions of 10^{-11}m (3.94×10^{-10} inch) or better.

Techniques for measuring small capacitances when mapping molecular features, have kept pace with improvements in capacitive sensor miniaturization. Resolutions of 10^{-7} pF are possible with appropriate guarding techniques. Although circuitry required to measure such minute capacitances accurately is not particularly complex, it must be carefully shielded. Stray capacity from a mere hand wave can easily produce a full-scale meter deflection.

Miniature capacitive gages are fabricated using thin film photo lithographic and etching processes, and tiny teeter totter plates made with such techniques can measure forces as small as 3.58×10^{-8} oz (10nN).

Capacitive transducers have some appealing properties for sensor designers. They dissipate virtually no energy (their resistance is often > 10^{10} Ω), so self-heating common to most other gages is absent in capacitive gages. Air capacitors are also thermally insensitive over a wide temperature range, this eliminates the requirement for temperature compensation, that becomes increasingly difficult to apply for high precision measurements. Output from a capacitive sensor can also be predicted from basic units of voltage and length, making these sensors absolute devices from a technical perspective.

The capacity of an air gapped capacitor increases about 80 times when water is used as a dielectric. Some other materials also have dielectric constants greater than 1 as shown in Table 5-7.

Table 5-7. Dielectric constants

	dielectric constant κ		dielectric constant κ
vacuum	1.00000	glass/quartz	3.75 - 6.75
air	1.00059	mica	7.0
Teflon™	2.1	acetone	21
polyethylene	2.3	methyl alcohol	33.1
epoxy	3.6	water	78.3
waxes	2.1 - 4.5	titanates (Ba,Sr,Mg,Ca,Pb)	15 - 12,000
oils	2.2 - 5.3		

Even slight contamination makes water electrically conducting, so capacitor plates must be insulated if water is used as a dielectric. Dielectric constants change little with frequency, typically <10% from 60Hz to 1MHz. Temperature has a bigger effect but is <1% °C^{-1} for most materials and 0.3% °C^{-1} for air. All insulators break down when stressed at very high voltages and dielectric strength also varies with test procedures. For dc potentials, Teflon's dielectric strength is ~60 x 10^6 V/m (1523V/mil) and air is ~ 3x 10^6 V/m. Mica is an excellent natural high temperature dielectric withstanding 5500V/mil.

Capacitive techniques for hobby use

State of the art instrumentation specifications always whet an engineer's appetite. Hobbyists must be content with lowered expectations, and rely on ingenuity to implement such methods at the 'homemade' level. A roboticist can build a rain gage sensor, tilt monitor or a pneumatically operated pressure sensor as suggested in Figure 5-42. These capacitive gages are straightforward in their operation, and their varying output capacitances can be measured with a circuit given in Figure 5-45.

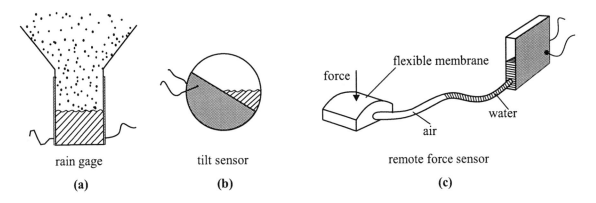

rain gage   tilt sensor   remote force sensor
(a)   **(b)**   **(c)**

Figure 5-42. (a) A capacitive rain gage has its sensitivity increased if a funnel is used to collect water from an area larger than the measuring cell. (b) If capacitor plates are insulated with varnish or epoxy, this tilt sensor can be filled with water, which has a high dielectric constant (κ ~ 80). (c) When the flexible membrane is squeezed, air travels along a flexible tube forcing water between capacitor plates of a sensing cell.

A slimline capacitive force sensor

A low profile (~ 0.005" thickness) capacitive force sensor can be assembled from readily available materials as detailed in Figure 5-43. It can be used for finger tip force sensing, or wherever cramped space inhibits using a large sensor. This sensor responds to even light finger loads (<0.5g / 0.02 oz), yet still gives useful signals at pressures > 20 psi, (5lb loading for the sensor in Figure 5-43).

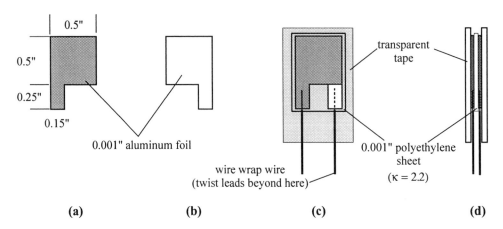

Figure 5-43.
Figure 5-43. A very thin capacitive force sensor is made by sandwiching 0.001" polyethylene sheet between two identically shaped pieces of 0.001" aluminum foil. The assembly is held together with ¾" wide transparent tape that also holds wires in electrical contact with the foil legs. Legs should be offset as shown in (c), confining capacity changes to the ½" x ½" sensing pad area.

Output signal repeatability for this gage is about ±10% for uniform loading, and a gage may be taped to a glove for finger-pressure sensing applications.

Making a capacitive force sensor

Dimensions for a capacitive force sensor are not critical, but a gage made according to Figure 5-43 will respond as shown in Figure 5-44. A polyethylene sheet dielectric is recommended even though plastic food wrap (eg Saran Wrap™) is thinner (0.0005"), and probably has a similar dielectric constant to polyethylene. Plastic food wrap is more difficult to handle because of static charge, and develops creases more easily.

[1]   Aluminum foil pieces can be cut with manicure scissors but a sharp razor blade, rule and glass plate[7] do a better job.

[2]   After cutting, a foil section can be lightly taped to an oversized piece of 0.001" polyethylene (removable tape is helpful for this operation).

[3]   Tack a wire to a foil leg with a small piece of transparent tape then overlay this combination with a length of ¾" transparent tape.

[4]   Flip the foil and plastic assembly over, adding the second piece of foil as in [3]. Orient this foil section as shown in Figure 5-43 (c). Separating legs in this manner restricts capacity changes to the ½" x ½" sensing region.

[5]   Trim off surplus tape and polythene but do not twist wires together at this time if you want to measure the capacitance of your gage on a DMM. Doing so adds appreciable static capacity (~ 20pF per foot of twisted wire wrap). Capacity introduced by twisting wires together in the finished sensor is of no consequence, because this is bucked out by zeroing gage output, using the circuit in Figure 5-45.

Polyethylene sheet is available as painters' plastic drop sheets from hardware stores. The cheapest variety is also the thinnest and usually 0.001" or 1 mil. Polyethylene bags used by customers to pack their own produce at grocery stores are satisfactory, also of 1 mil thickness, and just the right price.

[7] A glass plate is useful for many cutting, and mixing operations, it is easily cleaned and not scratched significantly by a utility knife or razor blade. Surprisingly, a sharp edged blade is not dulled appreciably when cutting on glass, so paper and balsa can be cut with precision. Blades are easily honed to perfection on a piece of 400 grit paper pressed onto a glass plate and can be repeatedly sharpened, extending blade life considerably. All types of tape can be cut to custom widths as narrow as 0.020" (just peel off after cutting). A sheet of quad-ruled paper attached beneath the glass makes a useful cutting guide. 'Float' glass (made by pouring molten glass onto liquid tin), is flat to ~ 0.00015" (~7λ) per inch and makes a good surface plate for marking parts before machining or cutting.

Calibrating and using a capacitive force sensor

A small spring-type food or postal weigh scale can be used to calibrate a capacitive gage, using the same procedures outlined in text following Figure 5-16. Because this gage has a high sensitivity to small forces, coins can be used to provide forces less than 100 grams, (25¢ weighs ~5.7g and a one cent coin ~ 2.5g)

Cut a ½" x ½" pressure plate from a 0.010" coffee can, and flatten it with pliers. A metal shim can also be flattened between two heavy flat steel plates, either in a vise or with a sharp hammer blow. Hammering metal directly creates a dimpled surface. This pressure plate is taped to the gage head and centered over the sensing region.

Tape the gage head periphery and pressure plate to the scale platform for calibration. Measurements are made by coin loading, or pressing the sensor area while reading the scale needle value. Uniform loading is obtained by pressing the point of a Philip's plastic handled screwdriver to the center of the pressure plate.

Carefully constructed gages have similar characteristics and perform as shown in Figure 5-44, after appropriate circuit adjustments for zero and gain.

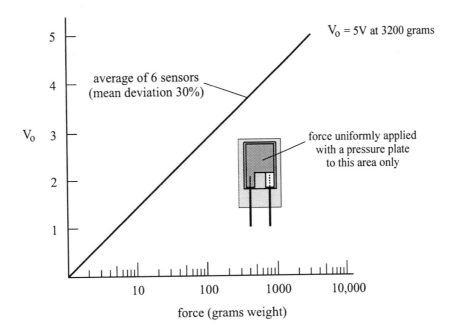

Figure 5-44. A spring food or postal scale can be used to calibrate a capacitive force sensor. Coins provide incremental loading below 100g. (25¢ ~ 5.7g, 10¢ ~ 2.3g, 1¢ ~ 2.5g). A ½" x ½" steel pressure plate taped to the gage head confines load effects to the sensing area during calibration.

Measuring capacitance

Sensor capacitance (~130pF), is measured with the circuit of Figure 5-45. Fixed capacitors (50pF, 100pF, 200pF) can be used for preliminary circuit testing. Using components specified in Figure 5-45 the circuit will measure capacitors to 1000pF. Larger capacitors can be handled by proportionally reducing the 4049 oscillator frequency as indicated in Figure 5-45

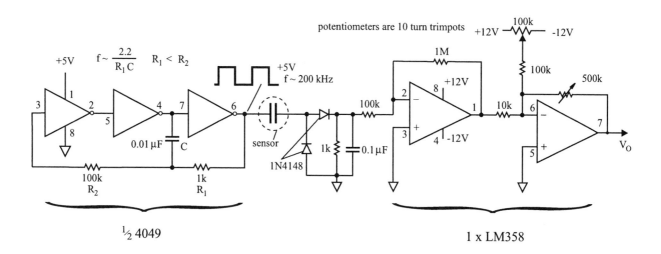

Figure 5-45. This circuit can be modified to measure very large capacitors, but as shown handles values to 1000pF. It has a fast settling time, low drift and is accurate to ±2% using the components specified. A capacitive foil sensor's output may require several seconds to return to zero after maximum pressure has been applied. This is due to mechanical relaxations between the capacitor plates. Circuit response time can be checked using fixed capacitors recommended in the text.

Although the sensor can be used without any special precautions, you will find it is sensitive to pick up when touched directly with bare fingers. Gage signal wires and the gage head itself can be enclosed in a grounded shield if needed, although this is unnecessary for most measurements.

Gage wires can be twisted in a drill, then neatly shielded by inserting them in a creased ½" x 12" length of aluminum foil, this is much tidier than braid, which is difficult to get in very small sizes. The gage head can also be covered in foil and the complete assembly connected to power supply ground, by twisting a bare wire onto the foil shield at the circuit board.

If you wish to measure your sensor's capacity with a DMM this should be done before twisting gage leads together. This is necessary because every 12" of twisted wire wrap wire has a capacitance of 20pF. A sensor made to the dimensions of Figure 5-43 has a calculated capacity of 129pF, if the aluminum foil sections are in intimate contact with the polyethylene dielectric. Gage capacity can be measured using pressure plates on either side of the sensing area. Do this test on an insulating surface, applying a load with an insulated rod, because a bare thumb adds stray capacity.

Capacitive rotation sensors

Variable rotary capacitors make excellent infinite resolution rotation sensors, and their capacity can be measured using the circuit in Figure 5-45. Variable capacitors are manufactured in a range of sizes from fingertip size trimmers, to massive high voltage capacitors employed for tank circuits in powerful transmitters. Capacity does not vary linearly with rotation angle for all variable capacitors, because some designs have eccentrically moving or specially shaped plates.

Small trimmer capacitors are made with multiple conducting plates separated by plastic film dielectric sheets. Maximum linear capacity variation for these devices is about 7pF to 100pF for 180° of rotation. Trimmers are usually less than ½" diameter and are hard to turn, but can be mechanically coupled for rotation sensing with a little care. Because trimmer capacitance is easily affected by conducting materials, shaft couplings should have an insulating break section.

Capacitors salvaged from old analog radios are ideal for sensing applications. Air gapped multi vane types have the best stability, and some are equipped with shaft bearings, gear trains and anti-backlash pulley drives. These units make superb sensors because of their mechanical integrity and repeatability. Variable capacity from 40 to 1000pF is possible in a 2" x 2" x 3" volume for an air capacitor, and this includes a geared drum drive covering the full capacity range in three revolutions.

Vendors of surplus electronic equipment or large old radios are the best sources for suitable rotary capacitors. Even small portable analog radios have good quality rotary capacitors that can be used for rotation sensing. Currently *C & H Sales* stocks an anti-backlash 5 independent section capacitor (each section is 8.5 to 30pF), #VACAP9400 $5.00. These sections can be connected in parallel to make a unit with larger capacity. Some useful capacitors for angular sensing are shown in Figure 5-46.

Figure 5-46. Rotary variable capacitors may be salvaged from old radios and used for angle sensing.
[A] This 20-100pF 180° multi plate capacitor from a small table radio has very smooth operation and is usually equipped with a pulley wheel. [B] Two sets of plates (20-200pF) and (20-200pF), with the same plate areas but different plate separations. [C] This air gapped capacitor 40-1000pF (2x20-500pF) has a 3 turn geared drive with anti-backlash mechanism and is ideal for accurate rotation sensing. Similar capacitors are found in old style radios. [D]180° 15-50pF ceramic trimmer.

Robot obstacle avoidance using capacitive proximity sensors

Walls and other obstacles in a moving robot's path can be detected by a variety of techniques. Ultrasonics, optical probes, touch sensors, inductive, pressure, piezoresistive, microwave, stretch/foam resistors and capacitive methods are all useful.

Capacitive sensors are easily made and will detect the presence of electrically conducting and non-conducting material. A pair of 3"x3" aluminum foil plates placed side by side on a flat surface and separated by a gap of ½" detects a hand at 10", wood at 6" and a sheet of paper at 3". Flexible capacitor plates are readily fabricated by attaching aluminum foil to one side of a 0.001" polyethylene sheet with glue stick adhesive. These plates can be cut to size and shape with scissors, then attached in any configuration at convenient locations on a vehicle.

Capacity is measured using the circuit of Figure 5-45 but with a 10MΩ feedback resistor replacing the 500k trimpot at pins 6 and 7. If your sensing application requires 5V logic signals a comparator can be added to condition V_o as suggested in Figure 3-22.

Unlike tactile sensors, most other proximity sensors have operational behaviour quirks and will false trigger under certain conditions. Capacitive proximity sensors also require a little fiddling to make them operate reliably, but for some applications this may be worth the effort. Optical methods are probably the most dependable obstacle sensing techniques, and are covered in a future book.

Magnetic sensing methods

Magnetism and magnets

A magnetic field is established about any current carrying conductor (Figures 2-27, 2-45) and a similar field is formed around each electron orbiting the nucleus of an atom. Ferromagnetic materials such as iron, cobalt, nickel, dysprosium and gadolinium are readily magnetized by aligning atomic dipoles associated with these orbiting electrons. Once adjoining moments are coupled they remain so, even after the magnetizing force is removed. Groups of 10^{17} to 10^{21} atoms naturally agglomerate into tiny magnetic colonies called domains, and act like small magnets that can be aligned by an external field.

Permanent magnets are made by aligning atoms in domains within ferromagnetic alloys. This requires a net energy input and some of this energy is stored in the resulting magnetic field. If one magnet is used to make another magnet, it loses some of its energy in realigning atoms in the new magnet. Similarly, if like repelling poles of two magnets are forced together, the magnetic field strength of both magnets will be decreased *(don't do this unless you want to ruin your magnets)*.

Over very long periods, thermal effects gradually misalign atomic spins and a magnet becomes demagnetized, although this may take centuries in modern materials. This demagnetization process can be accelerated by heating a magnet, causing magnetic dipoles to assume random orientations, eventually resulting in a zero net magnetic field.

Spontaneous magnetization ceases to exist above the Curie temperature T_c which is 770°C for iron, 358°C for nickel and 1120°C for cobalt. Most permanent magnets have high proportions of some or all these materials, but demagnetization may occur at 500°C even for high temperature cobalt-samarium magnets. Most powerful permanent magnets currently sold are neodymium based. These start to lose their magnetic properties well below T_c sometimes as low as 100°C, so it is wise not to overheat individual magnets or those used in motors etc.

Mapping magnetic fields with 5 minute epoxy and iron filings

Small particles of iron sprinkled on a sheet of paper held close to a magnet are useful for mapping field lines surrounding magnetic poles. A permanent record of a magnetic field pattern can also be made by using a slurry of iron filings and clear 5 minute epoxy

Suitable filings can be made by rubbing a large nail on a file. These filings can be stirred into clear epoxy on a piece of white card using a toothpick, then smeared into a uniform thin layer. If the slurry coated card is placed over magnet poles or the region of interest, filings will align themselves with the field lines. Remove the card when a suitable image has formed, then allow the epoxy to cure and form a permanent field map for later scrutiny.

Weaker magnetic fields may be mapped using a slurry of iron filings and regular epoxy heated with a hair dryer to a thin consistency. This lower viscosity coating is then used as described previously.

Examining magnetic fields with rotary search coils and electron beams

The basic principles of analog meters, electric motors and generators are all based on Faraday's induction law. Television tubes, CRTs and particle accelerators use a reciprocal effect to bend the paths of moving charged particles.

Small coils rotated in a magnetic field can be employed to measure magnetic field strength and are called rotating coil gaussmeters. These are fundamental devices, so if rotation rate $\omega=2\pi f$, coil area A and number of coil turns N are known, then field strength can be determined by measuring the induced current $i=E/R$. In this relation R is the coil resistance and $E=NAB\omega \sin\omega t$ (N=# of turns of area A). In SI units magnetic flux density B should be in Wb/m^2 or teslas (T) and A in m^2.

(Note: If an analog meter is used to measure the coil voltage $E_{max}=NAB\omega$ and $E_{avg}=2\,E_{max}/\pi$).

If you plan to build a rotating coil gaussmeter, don't forget to include slip rings for signal and power leads, or do your measurements quickly for a prescribed amount of wire twisting.

Charged particles moving in a magnetic field experience a force whenever they move at an angle to magnetic flux lines. This force cannot change a particle's speed but does alter its velocity vector direction.
Electrons with a velocity V can be made to move in a circular path with an appropriate magnetic field (the resultant trajectory radius is $r=m_e V/(qB)$. In SI units $m_e=9.11\times10^{-31}$kg , B is in Wb/m^2 and $q=1.6\times10^{-19}$C.

Even a small ceramic button magnet causes color distortion when placed near a TV screen or computer monitor because electrons are redirected and excite the wrong color phosphors. If a very powerful magnet is used for this test, it may magnetize some interior portions of the CRT, so be careful. If this does happen you can degauss or demagnetize those parts by waving the same magnet in a random fashion, first close to the offending part then gradually moving it further away. This procedure continuously changes the direction of unwanted magnetization, eventually reducing it to zero.

Trajectories of moving electrons are easily altered by magnetic fields and over long paths even the earth's weak field will deflect high energy electrons by a significant amount. The aurora borealis and aurora australis are natural manifestations of such particle steering. In most TV sets and color monitors electrons are accelerated to about 25keV, reaching about 30% of light speed. Such high speeds reduce perturbation by stray magnetic fields, but equally importantly excite screen phosphors to give a really bright picture.

Magnetic sensors

Although in principle electron deflection and rotary, flip or snatch search coils can be used by hobbyists for magnetic sensing, other techniques are available which are easier to apply. For example, Hall effect and magnetoresistive sensors are inexpensive devices widely used in industry and readily available for amateur use. They are easily incorporated into projects, and complement other techniques described in this book.

Magnetostrictive force sensing

When a body is placed in a magnetic field its dimensions change very slightly as the field reacts with each atom, this phenomenon is called magnetostriction. A reciprocal effect takes place if a body is subjected to stress or strain that produces magnetization changes, and is termed magnetoelasticity.

Magnetostrictive effects have been used to make knock sensors for auto engines, and for position sensing. However magnetostrictive and magnetoelastic devices are less common than resistive sensors, capacitive sensors, strain gages or piezoelectric devices. Pointing out that just because an effect exists and is a promising measurement technique, it may not be widely exploited if existing methods have a well-established record or are less expensive.

The Hall effect

Many sensing techniques lie in limbo for decades, until technological improvements vault them into the practical measurement arena. For example, Edwin Hall of Harvard University observed in 1879 that moving electrons follow curved trajectories in solids when placed in a transverse magnetic field, similar to electrons in a cyclotron. Like so many great discoveries viewed in hindsight, the Hall effect appears so obvious we can look back retrospectively and wonder why it was not discovered much earlier.

Hall's discovery was interesting and of fundamental significance, but of minor practical importance until Hall semiconductors became available around 1950. Hall effect sensors are now used extensively in current measurements, as gaussmeters, wattmeters, modulators, high frequency multipliers, proximity detectors, position limit switches, and for commutation in brushless dc motors. Millions of Hall devices are currently used annually in auto ignition systems, tachometers and for speed controls.

Most first year college physics texts give good descriptions of Hall's discovery, that shows a voltage develops across a current carrying conductor immersed in a magnetic field B as in Figure 5-47.

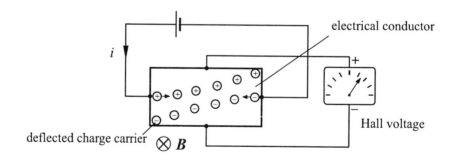

Figure 5-47. Moving charge carriers are deflected by a transverse magnetic field B directed into the paper, producing a voltage across the conductor. For ordinary metals this Hall voltage is minute and difficult to measure using modest field strengths and currents. Integrated semiconductor sensors have moved the Hall effect from laboratory to the industrial environment, by producing signals 10,000 times larger than those obtained in copper.

If a 0.5" x 0.04" copper plate is used as the conductor in Figure 5-47, then a 50 amp current with a powerful magnetic field (B = 10,000 gauss), will generate only a few microvolts of Hall potential. Silicon devices can increase this Hall voltage to millivolts, making fabrication and use of practical sensing devices feasible.

An inexpensive Hall sensor for hobbyists

A simple reliable non-contacting Hall effect sensor can be made for less than $1, as shown in Figure 5-48. Specifications are provided with the chip, when purchased from *Hosfelt Electronics*. This small sensor is easy to use, and the LED will light whenever a ceramic magnet's south pole is brought close to the chip's branded side.

If we can light an LED this same signal could be used to move a Mack truck, so Hall switches are very useful devices. Of course mechanical micro switches and reed relay elements can also be used for proximity sensing, but these often need some additional applied force. In addition, all mechanical switch signals must be debounced before they can be used for computer inputs. A UGN-3013 [8] is a no-contact Hall effect switch and its output is internally debounced. It is also smaller than currently available micro switches. This chip can switch 20mA at more than100kHz, a prime consideration for its use in automobile distributor systems. Hot rodders really appreciate the elimination of mechanical bounce at high revs, because this causes pre-ignition and misfiring, sapping performance. Removing mechanical spring-based points from a distributor also permits auto makers to extend tuneup intervals and reduce pollution.

[8] *Sprague* designed and previously manufactured the UGN-3013 and other Hall sensors described in this section. These devices and newer versions are now made by *Allegro Microsystems Inc.,* 115 Northeast Cutoff, Box 15036, Worcester, MA 01615. (508) 853-5000, FAX (508) 853-5049.

Allegro offers a UGN-3113 replacement for the UGN-3013, but UGN-3013U chips are still available from *Hosfelt* #25-204, 45¢.

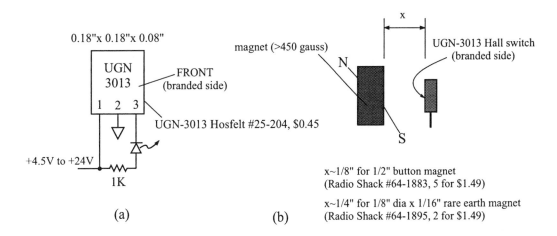

Figure 5-48. A solid state magnetic Hall switch is very useful for non-contact proximity switching. This inexpensive Hall switch gives a bounceless output with rise and fall times of 150 and 400ns. It operates from 4.5V to 24V drawing less than 5mA. In the circuit shown above, a UGN-3013 can sink 25mA at a 100kHz rate.

A magnetic flux of about 450G(gauss) from a ceramic magnet is required to turn on a UGN-3013 and the chip can tolerate any magnetic field strength. Operating distance (x in Figure 5-48), is extended with more powerful magnets. Even adhesive backed flexible magnetic tape (Radio Shack #64-1890, 99¢ for 30"x ½") is adequate for triggering, when a chip is backed by a soft iron plate. Operating with magnet tape is finicky and this material must contact the chip for reliable switching.

Linear Hall effect sensors

Linear Hall sensors give an output proportional to magnetic field strength and can be used to sense cam lobes or notch gaps as suggested in Figure 5-49. When properly adjusted, movement of individual ferrous gear teeth can be monitored with a linear Hall sensor.

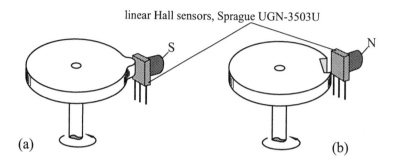

Figure 5-49. Linear Hall sensors such as the UGN-3501 or UGN-3503 have a sensitivity of about 1mV/G and a UGN-3604 is rated at 0.05mV/G. Such chips can be used to detect a cam lobe, notches or even individual gear tooth movement in ferrous metals. These low power non-contacting sensors do not load the target and millions of such devices are used in automobiles and elsewhere, a testament to their performance, reliability and ease of use.

Note that Hall switches such as a UGN-3013, are inadequate for sensing such minute flux changes shown in Figure 5-49. However, a 1" length of paper clip wire is reliably detected with a UGN-3013 using the arrangement of Figure 5-50.

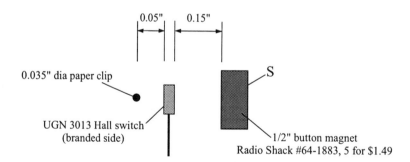

Figure 5-50. A 1" length of a paper clip can be detected by a UGN-3013 Hall switch if a ceramic magnet is positioned behind the sensor as shown. Non-contacting security detection schemes can employ embedded Hall switches and mating steel plates for sensing door closure etc. Because such devices have no moving parts, exert zero effective force, and do not use mechanical contacts, they are ideal for such tasks.

Measuring rotational motion with a Hall switch

Small ceramic or rare earth magnets attached to rotating elements, can be used for triggering a Hall switch as in Figure 5-51. Magnets should be securely attached, and positioned symmetrically to preserve dynamic balance at high speeds. A knife edged cone attached to a magnet may be used to channel field lines as shown in the figure. This reduces interaction times between magnet and sensor and improves rise and fall times. Switching times are always shortest when a narrow rectangular source is moved parallel to a similarly shaped sensor.

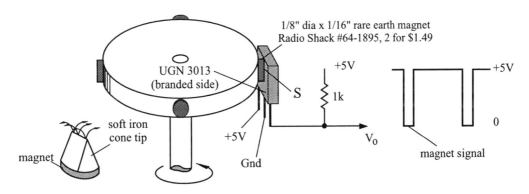

Figure 5-51. Tiny but powerful rare earth magnets attached to a rotating shaft or disk permit non-contact, high speed rotation measurements with bounceless outputs using a UGN-3013 Hall switch. Sensor rise, fall and dwell times can be reduced by attaching a soft iron knife edged cone to each magnet.

Turning on underwater apparatus with Hall and magnetic reed switches

Submersible vehicles and dive computers are favorite robotic projects, and since these packages are completely sealed they must be started externally. A Hall switch can be employed for this purpose using the method outlined in Figure 5-48. For 'turn-on', a magnet is attached to the exterior of the package by small steel strips or a large holed steel washer that permits most of the magnetic flux to reach a Hall switch inside the box. Strips from a 0.01" steel coffee can, or stainless steel washers are suitable for this purpose.

Glass encapsulated magnetic reed switches can also be remotely activated by a magnet as shown in Figure 5-52. Reed switches have an ON resistance <0.1Ω and are suitable for switching low power lines directly to run a submersible package. The switch shown in Figure 5-52 is conservatively rated at 0.5amps.

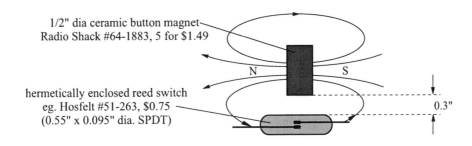

Figure 5-52. Conducting leads on a reed switch are made from ferrous material allowing magnetic flux to flow freely to both contacts. These are attracted together because they become north and south poles. Whenever a permanent magnet is used to activate a magnetic reed switch, it should be oriented so its north and south poles are disposed along the switch length, as in this figure.

Magnetoresistivity

Elements such as nickel and iron are magnetic because electron spins in these materials create small magnetic fields. In the presence of an external magnetic field some realignment of spins occurs changing the resistivity. This resistance change is a few percent in permalloy, a nickel-iron alloy with an initial relative permeability of $\sim 10^4$.

Easy-to-use inexpensive magnetoresistive sensors are available, enabling hobbyists to measure magnetic fields or construct their own sensing devices. Details on monitoring rotary motion, displacement, angular position or current are given below.

Philips Semiconductors manufactures a series of KMZ10 magnetic field sensors with open circuit sensitivities from 1.5 to 22 (mV/V)/(kA/m) or 0.6 to 4.4mV/gauss, when operated at 5 volts. Philips has also published a useful application handbook describing their magnetoresistive sensors:

Philips Data Handbook SC17, 1993, Semiconductor Sensors. *Philips Semiconductors,* Integrated Circuit Div. 811, East Arques Avenue, Sunnyvale, CA 94088-3409 (800) 227-1817 ext 900   Fax (408) 991-3581

Noninvasive current measurements with a KMZ10B magnetic field sensor

KMZ10B sensors are the same size and have comparable sensitivities to linear Hall effect sensors shown in Figure 5-49. They can be obtained from *Hosfelt* #25-207, 49¢ and specification sheets are included with a purchase.

If a KMZ10B is positioned 2mm from a current carrying wire as in Figure 5-53, it responds with a linear output of 1.5mV [9] for each amp flowing in the wire. If the sensor is moved so it touches the wire (r=0) its output increases to 3.5mV/amp [10]. Circuit continuity is not compromised for these measurements.

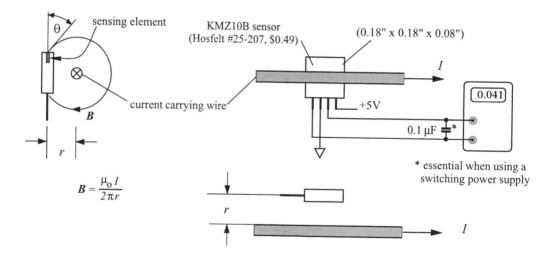

Figure 5-53. This versatile sensor can be used to measure current flowing in a wire from magnetic flux generated by the current. A KMZ10B can be used for checking auto headlight operation or for monitoring current flow in robotic experiments. Because a wire need not be cut, this sensor is very useful for many types of non-invasive current monitoring.

---

[9] KMZ10B sensors have an internal modulator and are designed for operation from batteries or linear power supplies. If a switching power supply is used, always connect a 0.1μF capacitor across sensor signal outputs as shown in Figure 5-53.

[10] A wire carrying 10amps has a magnetic field of ~40 gauss when r = 2mm, this is about 70 times the earth's field strength.

Measuring the earth's magnetic field with a KMZ10B magnetoresistive sensor

If a KMZ10B is connected as in Figure 5-53 and then turned in all possible directions, it will respond with a maximum output V_{max} when it is aligned with the earth's field, or a minimum reading V_{min} at 180° to the earth's field. These values correspond to 3 dimensional sensor alignments not only those obtained in a plane. Because the terrestrial magnetic field is known at any location on earth [11], a KMZ10B sensor's sensitivity can be derived from these readings.

A test shows V_{max} = 3.4mV, V_{min} = 1.5mV with a sensor sensitivity ½(V_{max} - V_{min})/0.58 or 1.64mV/gauss. Philips quotes KMZ10B sensitivity at (4mV/V)/(kA/m). Therefore, when powered at 5V the device sensitivity is 4x5x0.0796 = 1.59mV/gauss [1gauss = 0.0796 kA/m] [12]. This is close to the measured value of 1.64mV/gauss.

This simple observation shows raw output from a KMZ10B can be used as a rough indicator of the earth's magnetic field strength. In addition a KMZ10B output could be used as a crude compass without any additional amplification. A far superior compass can be made using a KMZ10B by following mechanical and electronic instructions given in *Philips SC17 Data Handbook.*

[11] Local university geology departments can provide this information. In Toronto B = 0.58 gauss, dip = 73 degrees, declination is 8 degrees west.

[12] This conversion applies only in air. In general the relationship between flux density B (wb/m^2) and magnetic field strength H (amp·turns/m) is complex, and B cannot ordinarily be expressed as a simple analytic function of H.

Some useful magnetic calibration techniques

Precision magnetometers are usually checked with Helmholtz coils. Helmholtz found that two identical solenoids with the same current and common axis produce an axial interval of uniform magnetic field intensity, when they are separated by a distance equal to their radius R. Using this coaxial coil arrangement hobbyists can duplicate the precision magnetic calibrations of professional laboratories.

A single coil wound on any nonmagnetic tube can also be used to test a KMZ10B sensor, because the field will be fairly uniform over the sensor's short length. Using a 7.5cm diameter plastic food container with 10 turns of #26 AWG wire, an axial field of 0.6 gauss is achieved for a 346mA coil current, as shown by equation 5.9.

$$I = \frac{2BR}{\mu_o N} = \frac{2 \times 0.6 \times 10^{-4} \times 0.0375}{4\pi \times 10^{-7} \times 10} = 0.346 \; amps \qquad (5.9)$$

Tests with this 3.75cm radius coil give an on axis reading with a KMZ10B sensor of ~1mV. This agrees with our previous measurements of the earth's field.

Many readers probably will have realized a rough measure of KMZ10B sensitivity can also be gleaned from the single wire current measurement procedure depicted in Figure 5-53. The left side diagram in this figure outlines the magnetic field geometry and it is straightforward to show that for r = 0.08" B is approximately 10G. This field strength corresponds to a sensor output of 15mV or 1.5mV/G. Don't forget to apply a sine correction for θ, the angle at which B is inclined to the KMZ10B sensing element, (θ~30° in this case).

Electronic compasses and magnetometers

It is tempting to consider using an electronic magnetic compass for determining a robot's heading. Presumably coordinate position will be more certain if this information is combined with data from accelerometers and rate gyros.

This is seldom a good idea inside buildings because of difficulties mentioned in Question 36 of Chapter 2. Outdoor measurements can unfortunately be influenced by local deposits of iron bearing ore in seemingly pristine surroundings. Buried water, electrical conduit, and old iron sewer lines are often pervasive under parking lots, a favorite venue for roving robot trials. However, the technology associated with compasses and magnetometers is fascinating, and when used with circumspection these instruments play an important role in navigation.

Traditionally, fluxgate magnetometers have been chosen for delicate magnetic sensing. Even sophisticated rocket payloads often use fluxgate magnetometers as simple reliable attitude sensing devices, providing supplementary back up data to GPS, gyros, accelerometers etc. Savvy suborbital rocket engineers never forget the earth's magnetic field is comfortingly always there.

An interesting alternative to a fluxgate magnetometer is described in the Semiconductor Sensor Handbook SC17 from *Philips*. A practical circuit and mechanical fabrication details are given for constructing a compass using their KMZ10A1 sensor. This sensor has more than five times the sensitivity of a KMZ10B.

Information on building fluxgate magnetometers may be found in back issues of electronic magazines such as *Radio and Electronics* (currently titled *Electronics Now*) Dec 1988, Sept 1994, Sept 1996, Oct 1996, *ETI (Electronics Today International)* September 1994. Fluxgate magnetometers are still favored for electronic compasses by connoisseurs.

Clamp-on ammeters

Hinged clamp-on laminated iron toroids are used for low frequency ($\leq$500Hz) ac current measurements. Rugged hand held clamp-on probes operating to 1000V and 600A are made for industrial use. These devices will slip over a 1.3" diameter cable and give an instant current indication to about 2% accuracy. AC clamp-on probes operate on the standard transformer principle shown in Fig.5-54 (a). A multi-turn secondary winding is used to sense primary current derived from either a single wire as shown, or multiple turns for greater sensitivity. *Amprobe, Fluke* and *Wavetek* are active in this measurement area and prices range from $100 for a 250A/600V analog ac probe to more than $400 for an ac/dc 600A/1000V digital instrument.

DC measurement with a clamp-on probe is obviously a trickier proposition but is accomplished using Hall effect or magnetoresistive sensors as in Figure 5-54 (b) and (c). Current from a single wire shown, induces a magnetic field in a suitable toroid, and this field is sensed either by a Hall or magnetoresistive sensor. Note the large clearance required for a KMZ10B chip, which must be oriented perpendicular to the field lines, as shown in (c).

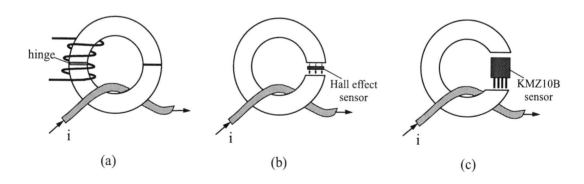

Figure 5-54. (a) Commercial ac clamp-on probes operate to 660A/1000V using a multi-turn secondary to sense a primary current. Hall effect and magnetoresistive sensors for monitoring dc current are shown in (b) and (c). *Electrosonic* stocks a linear analog Hall sensor (SS49 Honeywell/Micro Switch), with a sensitivity of 1.25mV/gauss over ±400G, $2. This device is suitable for use in (b). A single primary wire can be used for (a) (b) and (c), but multiple turns proportionally increase the sensitivity of these devices. Sensors in both (b) and (c) can also be used as null detectors by applying a bucking current through a multi turn secondary winding placed on the same core. This buck current can be made very small if many turns are used.

Magnetic cores, windings and electromagnets

Laminated iron cores are commonly employed for industrial ac clamp-on current meters, and powdered iron or ferrite toroids are used for both ac and dc measurements. Cores should have low retentivity or remanence, so the core field strength tracks primary current variations. A core must also not saturate over the primary current range. Incipient saturation only affects linearity, but when a core is truly saturated gross measurement errors will occur.

Both powdered iron and ferrite cores are suitable for most hobby applications, although iron is usually preferred because it can be cut with a hacksaw. However, there are zillions of ferrite toroids on the surplus market and they are suitable for making (b) and (c) type probes in Figure 5-54. Ferrite is extremely hard and brittle but can be cut with an abrasive tile cutter disk. These cutting disks will go through a small ferrite core in seconds, so watch your fingers. For dc and low frequency applications a steel link from large size steel chain sold at hardware stores, also makes a satisfactory core.

Not all flux from primary or secondary turns is trapped by a core, so windings should be spaced evenly around a toroid for best performance. An analogous effect exists in a 'C' type electromagnet shown in Figure 5-55, where flux loss is indicated by dotted lines. Placing equivalent windings at the poles is not only more convenient, it usually produces a higher pole flux.

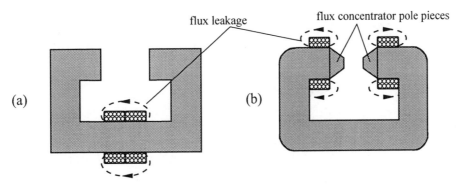

Figure 5-55. Windings on powerful electromagnets are easier to install, and more effective when placed at the poles. Flux leakage, indicated by dotted lines is inevitable with any winding. Sharp corners on a soft iron former contribute to flux loss and a core is usually smoothed as shown in (b). Flux concentrators may be used to increase magnetic intensity in the gap, because the same number of field lines is forced to flow through a smaller area, thereby increasing flux density (Wb/m^2).

Record breaking magnetic fields (high and low)

Using holmium pole pieces, the Francis Bitter Magnet Laboratory at MIT has produced a continuous magnetic field of 35T (1 tesla = $1 Wb/m^2$ = 10,000 gauss). Fields exceeding 100T are predicted using high temperature superconductors, but many technical problems must first be solved. Enormous forces try to rip apart coil windings on powerful electromagnets, and contemporary high temperature superconducting wires are very brittle, making things tough for magnet designers.

With pulsed operation, fields exceeding 70T have been achieved, and winding pressures of more than 190,000psi are generated during operation. Copper has a tensile strength of only 40,000psi and several cases of exploding wires have been reported. Even more powerful fields have been produced on a one-shot basis, by using explosively driven collapsing coils. Although over 200T has been produced with this technique, rebuilding a magnet and replacing test specimens becomes tedious after a few shots.

Magnetization of permanent magnets requires a field strength of 3 to 5 times a material's intrinsic coercivity. Consequently, it is likely that electromagnets may always be the record holders. A capacitor bank of ~2.5 million joules (2.5MJ) is required to magnetize current formulations of rare earth magnets, and practical reasonably priced permanent magnets are still limited to ≤2T.

Massachusetts Institute of Technology also holds the record for the weakest measured magnetic field at 8 x 10^{-15} T in their shielded room, a formidable but less dramatic achievement.

A $5 magnetoresistive rotation sensor

An angular position sensor with linear response over ±60° can be constructed from readily available materials as shown in Figure 5-56. The completed sensor has infinite resolution and gives reliable, repeatable measurements over a ±90° span. This device uses a Philip's KMZ10B sensor and its output is only perturbed if other magnets are placed nearby.

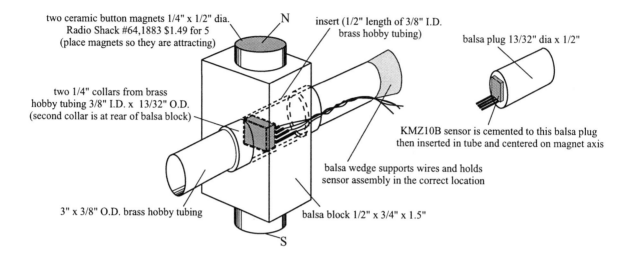

Figure 5-56. This rotation sensor can be built and tested in a few hours following instructions given in the text. It gives reliable and repeatable readings. Because incrementally sized brass hobby tubing is made to exact specifications the unit has low rotational friction. The sensor may also be used in a gravity sensing mode as a tilt sensor.

Brass hobby tubing, available from hobby stores is designed so incremental sizes fit beautifully one inside another. When assembled with such tubing the outer balsa block will spin for at least 5 seconds when given a sharp flick with a finger. Check tubing purchases to ensure they are a good sliding fit before leaving a hobby store. Polishing an inner tube with fine steel wool gives a better fit if required, but adding lubricating oil increases rotational friction.

A non-magnetic 10 gram weight can be added to one end of the block assembly, turning the completed module into a gravity seeking sensor with stiction <2°, (comparable to a builder's angular level).

Directions:

[1]   Cut the balsa block from a ¾" x ½" soft balsa beam with a thin utility knife. Use successive light perpendicular cuts gradually working through the wood.

[2]   Cut brass tubing in the sketch, using a hobby razor saw or a hobby tubing cutter. Finish all ends against a file held flat on a table. Rotate tubing frequently to preserve perpendicularity.

[3]   Hold the 3" tube from [2] in a drill press or a hand drill chuck, then chamfer the free end to a sharp edge at 45° using a fine file or sanding board. Clean the interior edge carefully with a utility blade. The final edge should be razor sharp, so handle it carefully.

[4]   A 'perfect' hole is drilled centrally through the balsa block using the sharp edged tube from [3]. Use a drill press if possible and back the block with scrap balsa to protect the tube's edge on break through. This operation also produces a 'perfect' balsa plug. You will be amazed at the quality of the hole and mating plug. They are genuinely close to 'perfection'.

[5]   Press a ½" length of 13/32" O.D. brass tubing into the block, be sure this insert is also finished with clean ends. Check the tubing from [3] slides and rotates freely in this insert.

[6]   Bend leads on a KMZ10B at right angles to the chip, so when it is attached to a balsa plug the assembly passes snugly through ⅜" O.D. tubing. Trim leads on the KMZ10B to ¼" length.

[7]   Attach sensor to balsa plug with superglue (CA).

        Note: crazy glue wicks into balsa down the wood's grain, so if you have not made the plug as recommended in [4] you will have to use epoxy for attaching the sensor.

[8]   Support plug and sensor assembly in a small vise or with modelling clay etc. Then solder wire wrap wires onto the top of each sensor lead. Do this quickly and neatly, using a heat sink if you are a novice solderer. Attach a small piece of masking tape to protect leads from shorting on the brass tubing.

[9]   Label each lead at its free end, then use a slow speed drill to twist wires, starting 1" from the sensor.

[10]   Slide the plug with sensor into its tube. The plug may require light sanding but don't overdo it, you want a nice sliding fit.

[11]   Slip the completed tube from [10] inside the balsa block insert, then tape on the two collars.

[12]   Manoeuvre the sensor so it is centered on the magnet axis. After it has been correctly positioned, use a balsa wedge to cinch wires to the tube preventing sensor movement,

[13]   Attach both magnets with either double sided tape or epoxy, Then you're ready to power-up and test your sensor.

Point form instructions make even a simple task sound complicated, just follow Figure 5-56 if in doubt.

Calibrating a magnetoresistive rotation sensor

File a 0.4" semi circular clearance on the centre of a 6" diameter plastic protractor, (these are sold for 50¢ at dollar stores). Attach this modified protractor to the balsa block with double sided tape, you may have to rotate it later so don't use glue. Support the sensor on a few books to clear the protractor, holding the tube down with masking tape. Stick a pin into modelling clay on the table as a fiducial angle marker, and remember to put a 0.1μF capacitor across signal leads if you are using a switching power supply.

　　　Your sensor when wired as in Figure 5-53 should give an output similar to that in Figure 5-57, and results will repeat to 0.5% or better over a series of runs.

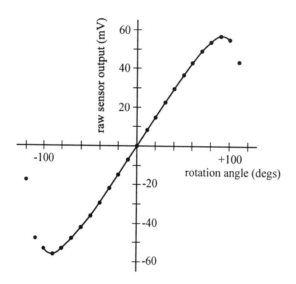

Figure 5-57. Response for the rotation sensor of Figure 5-56 has a useful output over ±90°, shown by the solid line. Two intervals of about 20° indicated by unconnected data points, cannot be used for reliable observations. Results shown in this figure are repeated after 180° so the sensor may be used for continuous rotation measurements if desired.

Inductive displacement sensing

Linear Variable Differential Transformers (LVDT's) are used in weigh scales, load cells and for measuring small displacements. Their outputs are bipolar around a null location that can be set to any convenient position. LVDTs are used extensively in industry because they are accurate, reliable and their outputs are not temperature sensitive. Many units are linear to 0.05% with resolutions of about 0.001", but this can be reduced to 4×10^{-9} inch (0.001μm).

　　　It is unfortunate that reasonably priced LVDTs are not currently available for hobbyists. Even on the surplus market these wonderful transducers are prohibitively expensive, eg. *Servo Systems Co.* PR-266 $129, ±0.050" stroke, 0.000025" repeatability. Making a long stroke LVDT is tricky and labor intensive if uniform sensitivity is needed. Many windings are required and these are distributed unevenly along the device, or windings are complementarily tapered. When LVDTs become available at reasonable prices, a Philips 5520/5521 chip ~$5, can perform all requisite drive and signal pick up conditioning, and these sensors will be widely used by hobbyists.

Operating principles of an LVDT are shown schematically in Figure 5-58. In a practical unit three or more coaxial solenoids are supported on a common axis, and a magnetically compatible core is used to link flux between the coils. Because the core is essentially unrestrained, it can have zero friction and provides infinite displacement resolution. Overall length of LVDTs varies with total stroke range, (30" for ±10" stroke and 5.5" for ±0.5" are typical values).

Many commercial units have self-contained electronics and are highly prized because of their superb performance. Perhaps the only reasonable LVDT source for roboticists is an old digital bathroom scale, circa 1980 because newer versions use strain gages.

An RVDT is the rotary equivalent of an LVDT. It has infinite resolution like our balsa wood/hobby tubing rotation sensor, but operation is normally restricted to a ±40° linear range with an upper limit of ±60°. An RVDT relies on rotation of a magnetically compatible cardioid shaped cam for signal coupling between stator coils, using the same principle as an LVDT in Figure 5-58.

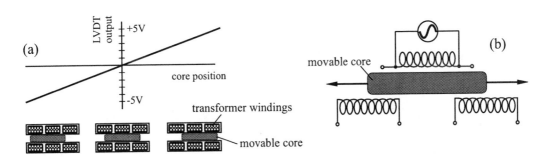

Figure 5-58. Essential mechanical parts of an LVDT are shown in (a), with the core at 3 distinct locations. Corresponding transducer outputs for these core positions can be read from the linear output curve. In (b), inductive coupling to each of the lower coils depends on the movable core's position as suggested in sketch (a). An LDVT has virtually no transverse sensitivity, and an extremely long life, because the core does not contact transformer coils. Commercial units have a frequency response up to 10KHz in small stroke sizes, and can withstand shocks of 1000g for 10ms.

A practical long stroke displacement sensor with infinite resolution

We have seen how RVDT performance can be inexpensively mimicked by a simple rotation sensor shown in Figure 5-56. The reader might wonder "Is there a simple inexpensive equivalent for an LVDT?".

Fortunately the answer is yes, and its features are outlined in Figure 5-59.

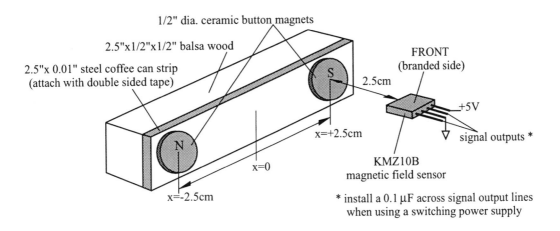

Figure 5-59. This simple infinite resolution displacement sensor uses a KMZ10B magnetoresistive sensor to measure symmetrical and linear field variations between two magnetic poles. A steel strip is used to complete one side of the magnetic circuit and simultaneously hold the magnets.

Hobbyists will find the arrangement in Figure 5-59 is very flexible, and suitable for experimentation. For example, a linear field is still maintained at 1" magnet separation and can be used for increased measurement precision over this shorter interval. Even greater precision may be achievable with smaller spacings. Larger total ranges can probably be achieved with more powerful magnets, but these options have been left for interested readers to explore.

A calibration for the displacement sensor of Figure 5-59 is plotted in Figure 5-60. Output linearity is good over ±1" and the unit can be used for sensing over a 6cm range.

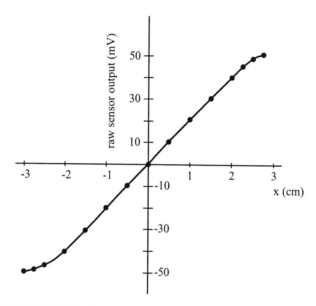

Figure 5-60. Calibration of a KMZ10B displacement sensor shows it has a linear output over ±2cm and a useful range of ±3cm. Data was recorded with a DVM directly from the sensor, with no additional amplification. Remember to place a 0.1µF capacitor across the signal lines if a switching power supply is used.

Future magnetoresistive sensors

Philips can take credit for pioneering developments in practical Anisotropic Magnetoresistive (AMR) sensors, such as the KMZ10B. As we have seen, these are extremely useful devices and can be used in many applications. One drawback with AMR sensors is a tendency to 'flip' their outputs in high magnetic fields. You will experience no problems with the composite sensing units described in this chapter, because they are designed to work within the specified field limitations.

New and improved sensors are now on the market which eliminate flipping, and with 12 times the sensitivity of a KMZ10B when both are operated at 5 volts. These are GMR or Giant Magnetoresistive Ratio sensors manufactured by:

Nonvolatile Electronics Inc. 11409 Valley View Road, Eden Prairie, MN 55344-3617
(612) 829-9217 (800) 467-7141 Fax (612) 996-1600

Prices for *NVE*'s sensors start at $3. Minimum order is $100, and sensors are in 8 leg surface mount format[13]. NVE hints at even higher sensitivity devices in the future, great products with good hacker potential.

---

[13] Any surface mount chip can be wired to a compatible header (usually 0.1" spacing), for hobby use. Work with a small soldering iron tip and a steady hand - practice on an inexpensive chip first, (eg. 74LS00D, *Hosfelt* 59¢). Two methods for attaching surface mount devices to a larger platform are shown in Figure 7-17.

Using eddy currents for braking, propulsion and sensing

Eddy current braking is easily observed experimentally as shown in Figure 5-61, using an aluminum disk attached to a tape player motor boss with glue or double sided tape. When the motor is operated at 6 volts, it rotates at 2380 rpm and draws 30mA. This current increases by a factor of 4 when the magnet is moved to within 0.050" of the disk, and rotation speed drops by a factor of two.

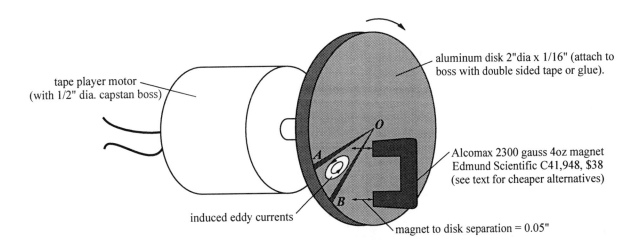

Figure 5-61. A rotating plate is made up of radial segments such as *OA*. Each segment behaves like a length of wire and generates a current as it passes through the magnet's field. Other segments such as *OB*, provide a return path for current from *OA*, and circulating eddy currents are formed as indicated, which oppose disk motion.

If radial slots are cut in the disk eddy current amplitudes will be reduced, because return current paths are then eliminated. Slotting a disk serves the same purpose as laminated cores in regular ac transformers. In high voltage pulse transformers bundles of wires or flat strips are often used as cores to reduce eddy currents, and powdered iron toroids also do an excellent job of reducing eddy current circulation.

A horseshoe magnet is not essential for the demonstration in Figure 5-61, since even small ceramic button magnets have some braking effect. A 2"x1"x¼" rare earth magnet from a brushless motor flat coil actuator provides impressive braking power. This magnet will almost stop the rotating disk although motor drive current rises to over 200mA in its attempt to keep turning. Table 5-8 shows how motor loading and braking forces are affected by two common permanent magnets.

Table 5-8. Eddy current braking using permanent magnets

motor voltage	motor current	(rpm)	magnet type
6V	30mA	2380	no magnet
6V	120mA	1060	1.75" Alcomax ('C')
6V	200mA	100	2"x1"x¼" rare earth

These results show rotation speed and motor current are significantly altered when a conducting spinning disk in Figure 5-61 is subjected to powerful magnetic fields.

Powerful ceramic magnets can be salvaged from speakers or dc brush motors and are also sold at hardware stores, electronic surplus outlets and flea markets. Rare earth permanent magnets from motors, speakers, dc brush motors, linear coil actuators and magnetic suspension systems are flooding surplus markets, and prices are continually falling. A neodymium/iron/boron 0.9"x0.5"x0.156" rare earth magnet, with a 4lb pull is available from *Hosfelt* #75-121 $2.49. *C and H Sales* has a good selection of magnets in many sizes, a small rare earth with 5lb pull #MAG9450 is $2.

Electromagnetic propulsion for the classroom

An excellent tabletop project can be constructed, based on the fact that counter EMFs from eddy currents generate a force on the conducting element. For example, a conducting disk in Figure 5-62 will be repelled along the solenoid axis by induced eddy currents. Use of this principle has been investigated by the military for longitudinal acceleration of metallic 'smart rocks'. At speeds of 18,000 mph such projectiles could be placed in an earth orbit, and escape velocity is just a little higher (25,000 mph), requiring about twice the launch energy under ideal conditions.

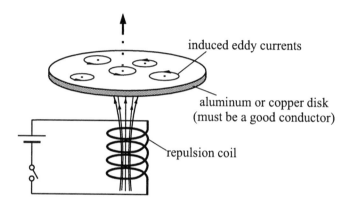

induced eddy currents

aluminum or copper disk
(must be a good conductor)

repulsion coil

Figure 5-62. Repulsion forces due to induced eddy currents on a conducting body can be used to accelerate projectiles to very high velocities. A controlled linear accelerator using multiple discrete solenoids is a good project for a one or two student group.

Students can accelerate a shell made from aluminum hobby tubing, by using coaxial coils strategically positioned along a length of plastic pipe. Optical or inductive sensors track a projectile's location, and this information is used to apply current to appropriate coils as required. Circuitry is either hard wired to coils, using position sensor signals directly for switching, or a software loop can be incorporated. Computer control is more satisfying, because acceleration and velocity parameters can be displayed, adding a touch of class. Fine tuning for optimum drive performance is also quicker and more flexible with a microprocessor controlled loop.

Microwave ovens and eddy currents

Eddy currents generate heat due to resistive dissipation losses in conducting elements, and this is why transformers overheat without laminated cores. This characteristic has been exploited for decades in powerful radio frequency induction heating furnaces, which are used to melt special metals. RF heating of metals under vacuum conditions provides a contaminant free environment, almost impossible to reproduce with other techniques. Microwave ovens use the same eddy current principle, but these are relatively recent additions to household kitchens.

Using eddy currents for strain measurements

Eddy currents usually conjure up a picture of an overheated transformer, or a levitated train zooming along at 300mph. However, eddy currents can also be used for sensing movement, strain, torque etc., and are commonly exploited in metal detectors used by treasure hunters on a sandy beach.

    A passive target in Figure 5-63 is shown with a few eddy current profiles, resulting from pulsating magnetic force lines generated by a coil. If the target is mechanically deformed by some external force, these eddy current paths will react by changing their shapes slightly. Commercial sensors are made that detect these small changes, and such instrumentation can monitor target distortions as small as $0.002\mu m$ ($\sim 10^{-7}$ inch).

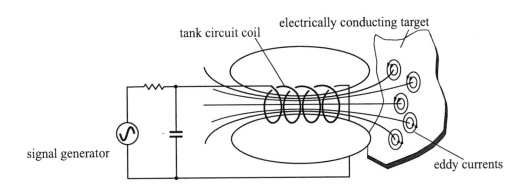

Figure 5-63. Target movements of less than $0.01\mu m$ can be resolved by non-contact monitoring of changes in eddy current fields. These field changes are reflected back into the exciting coil, changing its inductance and the tank circuit's natural frequency.

An ac bridge can be used to detect eddy current fluctuations, and these show up as impedance shifts in the bridge of Figure 5-64. Using two coils makes it easier to isolate inductance changes resulting from eddy current perturbations, and the coils are normally placed close to a target for good magnetic coupling.

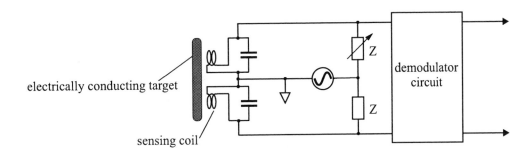

Figure 5-64. An impedance bridge (shown here schematically), is used to detect target movements that cause small inductance changes in both sensor coils. A demodulator provides a quasi dc output corresponding to target strain, torque or position.

Impedance changes may be monitored directly with a frequency counter, if target movement or deformation is large. Interested readers can measure target position by using a single resonating LC circuit coil, and a good frequency meter. A CMOS 4046 phase-locked loop is also suitable for measuring these small frequency shifts.

　　　　Two differential coils shown in Figure 5-64 give better performance and greater sensitivity, because a bridge can be nulled for an undisturbed target condition. A nulled circuit also permits the use of higher subsequent signal amplification without saturation.

Piezoelectric sensors

When a piece of natural quartz is squeezed, a voltage is developed across its squeezed faces, Figure 5-65a. The electric charge generated is called piezoelectricity from the Greek word piezo, meaning to squeeze or press. Conversely a voltage applied across the same faces alters the crystal's dimensions as in (b), causing it to grow or shrink depending on applied voltage polarity. These effects are the most direct conversions of mechanical force to electrical charge and vice versa, presently known.

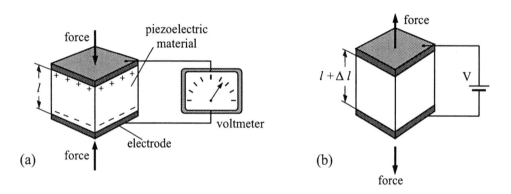

Figure 5-65. (a) Squeezing a piezoelectric material generates a temporary voltage across the pressed faces, and if electrode faces are pulled apart the charge polarity will be reversed. A DVM or other high impedance meter can measure transient voltages generated by a piezoelectric transducer or PZT. (b) Applying voltage across a PZT produces a stretch Δl as indicated, or an almost equivalent shrinkage if polarity is reversed. Hysteresis can be 15% in some cases and piezo materials also creep a little after voltage has been applied or removed. Very high electric fields can depole a PZT, so any applied electric field should be limited to $\leq 15 V/mil$.

Piezoelectric transducers can be used for sensing minute movements or can generate very small displacements, they have infinite resolution and show no wear even after 10^8 cycles. Piezoceramic materials made from lead zirconate titanate or barium titanate are used for exotic tasks such as micro positioning, laser tuning, accelerometers, micro force sensing or as simple audio beepers. Piezo BBQ lighters develop 10kV sparks at the push of a button and ultrasonic cleaners, humidifiers, watch alarms and even some loudspeakers or microphones use piezoceramics too.

　　　　Piezo movements can act in microseconds, and when used as bulk devices PZTs are stiff and can move loads of 10,000 lbs, but only over very small distances, (usually <<1mm). Piezo elements are very high resistance devices and any applied charge dissipates very slowly. Therefore an electrically stressed piezoelectric specimen slowly regains its original dimensions when the electric field is removed, except for hysteresis losses. Piezo elements can only monitor ac forces because voltages generated by mechanically stressing a PZT also dissipate, even if the applied force remains unchanged.

Piezoelectric disks and benders

Hobbyists can obtain piezoceramic bender materials as disks from *Radio Shack* 273-091 $2.49, *Edmund Scientific* C35,200 $5.25 for six, or by dismantling audio buzzers from *Hosfelt* #13-144 and #13-325 $1 each. These items are usually in the form of a 0.010" brass plate 1" to 1.5" dia., with a circular piezoceramic area on one side of about 0.75" diameter, as in Figure 5-66.

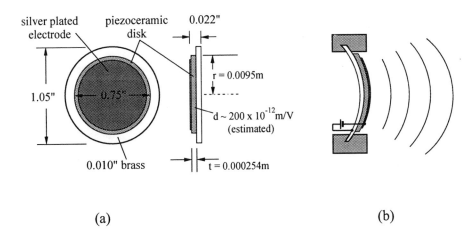

(a)   (b)

Figure 5-66. (a) Most surplus piezo disks have dimensions similar to those shown in this figure, and some disks have extra piezo areas for generating additional audio tones. Lead zirconate titanate and barium titanate are widely used piezoceramic materials in commercial products. (b) When used as audio buzzers, disks are usually clamped around their outer circumference. If a voltage is applied across such a clamped piezo disk it will bow symmetrically, pushing air forward as suggested in this figure. A rapid alternation of the applied voltage produces an audible or ultrasonic sound wave, depending on disk dimensions. Optimum efficiency, and usually the most intense movement, occurs when a disk vibrates at its natural resonant frequency.

A piezoceramic bender

A piezoceramic disk can be used to deflect a laser beam if it is clamped on only one edge as shown in Figure 5-67 (a). A solid state visible laser module is suitable for this test eg: *Hosfelt* #75-174, $54.95, 5mW at 670nm, 3V dc at 85mA, with an adjustable collimating lens. This laser is small (0.5" dia. x 0.75"), and will operate from two 1.5V cells, it is not eye-safe so should not be viewed directly or by reflection.

A small piece of plastic shaving mirror attached to the disk with modelling clay will reflect a beam as shown in (a). Beam movement is D=3.3mm (0.13"), as measured on a graph paper screen placed 1m from the mirror.

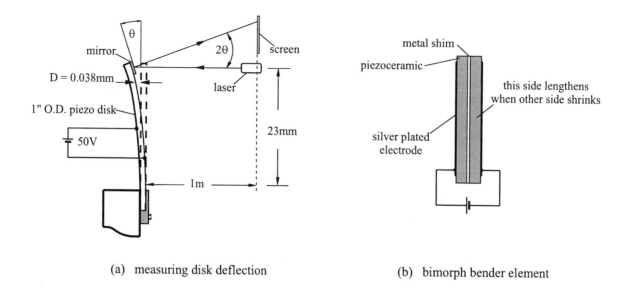

(a) measuring disk deflection   (b) bimorph bender element

Figure 5-67. (a) A disk clamped at one edge acts as a linear bender, and its tip displacement can be determined by measuring the deflection of a laser beam. (b) Commercial bimorphs are formed by bonding two piezoceramic coatings with opposite polarizations to a metal interface. Bimorphs deflect about twice as much for the same voltage as an equivalent disk (or unimorph), shown in (a).

When a mirror moves through an angle θ, a reflected beam is rotated by 2θ. Therefore, disk displacement D at the mirror's location is 3.3/2 x 23/1000 = 0.038mm or 0.0015" [14]. This movement can be increased by applying a higher voltage, but only up to about 150V [15].

Bimorphs are more commonly used by professionals because of their lower drive voltage and increased stiffness. Unfortunately no surplus bimorphs are currently available and new prices are high. Overall, piezo benders are inadequate for deflecting a laser beam through large angles, and galvanometers or speaker mounted mirrors are usually employed for those tasks. Optical galvanometers are discussed in Chapter 6.

[14] Mirror displacement in Figure 5-67 can be calculated approximately if the piezo disk is assumed to bend in a circular arc of radius $R = t/2\epsilon$. Strain (ϵ) due to piezo expansion in the disk plane, is calculated from the applied electric field $E = \epsilon/d$. In our application $E = V/t$, so $\epsilon = Vd/t$. Numerical values for r, t and d are listed in Figure 5-66 and applied voltage V is 50 volts, from Figure 5-67. A value for the polarization charge coefficients parallel to the disk face is taken as $d = (d_{33} + d_{15})/2 \sim 200 \times 10^{-12}$ m/V.
Using this information mirror displacement $D = r^2 Vd/t^2 = 0.014$mm. This calculated displacement is lower than the measured value of 0.038mm, but is not unreasonable because exact properties of the surplus piezo disk used are unknown, and interaction with its brass backing plate is problematic.

[15] If an electric field of E~15V/mil is exceeded the material may lose its piezoelectric properties due to depoling. At high applied voltages a piezo element may also arc to the metal base, forming a low resistance short circuit or arc track. This often renders the element unusable. Piezoelectric properties can also be destroyed by overstressing, since this also generates high voltages. Fast high voltage spikes from piezo elements can easily 'blow up' a DMM or other solid state device, if an element is rapidly placed under high stress.
Unlike natural quartz, piezoceramics must be poled or polarized to activate their piezoelectric properties. Suitable raw crystalline ceramics are heated to a temperature where they fuse without actually melting, (this is called sintering). They are then placed in a high dc field to align their electric dipoles. Heating accelerates the poling process but is done with care because many piezoceramics depolarize at ~200°C. Care must also be exercised when soldering leads onto piezo elements to avoid destructive overheating.

Ultrasonic proximity detection

A portable 40kHz transmitter and receiver that reliably detects small obstacles at 0 to 24" or a wall at 8 feet, can be constructed on solderless breadboards. Circuits for both modules are shown in Figure 5-68 and these can be built on individual 2.5" x 3.5" sections of protoboard, then powered by 9V batteries secured to boards with rubber bands. This portable format is convenient for checking out an operating environment prior to robot trials. A regular 6.5" breadboard can be cut with a hacksaw and used to make both transmitter and receiver. Completed units are shown in Figure 7-15.

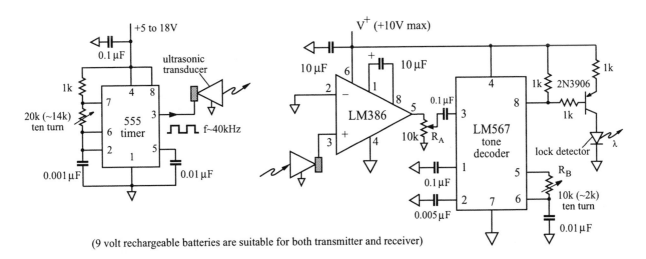

(9 volt rechargeable batteries are suitable for both transmitter and receiver)

ultrasonic transmitter   **ultrasonic receiver**

Figure 5-68. Either circuit with its battery fits on a 2.5" x 3.5" solderless breadboard, and operates with most 40 kHz ultrasonic transducers. A set of suitable transducers is sold as a pair by *Hosfelt* #13-180, $3.50. Transmitter or receiver operating current is ~15mA, but receiver current rises to ~35mA when in lock. Continuous operating time is >6 hours on fully charged 9V nickel cadmium batteries. A visible LED is used as a lock indicator, and lights whenever a viable ultrasonic signal is detected. Both circuits constructed on protoboards, are shown in Figure 7-15.

Ultrasound waves from the transmitter are produced by movement of a lightweight aluminum cone attached to a thin slab of piezoceramic. An ac voltage from the 555 alternately compresses and expands this slab causing the cone to vibrate, producing sound pressure waves. An ultrasonic receiver functions in a reciprocal manner, as sound waves received by a cone excite a piezo slab, which in turn generates an ac voltage.

40kHz aluminum transducer cones have a diameter d= 6.5mm, and an ultrasound wave at this frequency f has a wavelength $\lambda=v/f$ =8.6mm (v=sound velocity ~344m/s at 20°C). We can estimate the ultrasound beam divergence angle θ for a 40kHz transmitter cone using Rayleigh's criterion, ie. $\theta=1.22\lambda/d$=1.22x8.6/6.5=1.6 radians = 92°. This is a very large beam divergence by optical standards, where a HeNe laser beam has a divergence of 1mrad or 0.06°. It also explains why Polaroid's sonar camera focussing systems and ultrasonic tape measuring units employ large diameter transducers. Such devices have a beam divergence $\theta \sim 15°$.

Large beam divergence means ultrasound cannot be used to resolve fine details in a room's surroundings. If human vision ($\theta \sim 0.02$ °) had the same resolving power as a 40kHz ultrasonic system, letters on this page would only be legible if they were 40 feet high!

Optimizing output from an ultrasonic transmitter

Maximum ultrasound output is produced when a transmitter's piezo element is driven at its resonant frequency. Resonant operation also gives the most efficient conversion from electrical energy to mechanical force. Optimizing transmitter output is straightforward if a microphone is used to monitor the ultrasound level as shown in Figure 5-69.

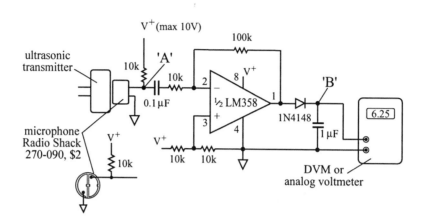

Figure 5-69. A small electret microphone can be used to optimize output from an ultrasonic transducer. The microphone should be positioned for a maximum output and then monitored with an oscilloscope at point 'A'. Alternatively, if additional amplification is provided as shown, an analog or digital voltmeter can be used for monitoring at point 'B'. An analog sound level meter *Radio Shack* 33-2050 $31.99 is also very convenient for measuring an ultrasonic transmitter's output. This Radio Shack unit is equipped with a jack for simultaneous oscilloscope monitoring.

Typical microphone monitor values are given below when both modules in Figure 5-68 are operated from 9V NiCd batteries (ie. 7.2V).

0.6V peak to peak at point 'A'.   *Measured with an oscilloscope.*
6Vdc at point 'B'.   *Measured with a DVM, (DVM reads ~3.5V with no transmitted signal).*
0dB on 100dB range.   *Measured with sound level meter #33-2050.*
1V peak to peak.   *Measured with an oscilloscope on output jack of a sound level meter*

These values were measured with a microphone closely coupled to the transmitter and positioned for maximum signal output.

Receiver adjustments for obstacle detection

An ultrasonic obstacle detection system shown in Figure 5-70 can detect a pencil at one foot, or a wall at 8 feet. A 3.5" x 1.5" paper card placed between transmitter and receiver reduces direct signal coupling between modules, and makes obstacle detection more reliable. If both units are placed on a book, they can be conveniently carried and pointed at targets as required. Detection of an obstacle is indicated visually by an LED lock light, so the system does not require any additional diagnostic equipment.

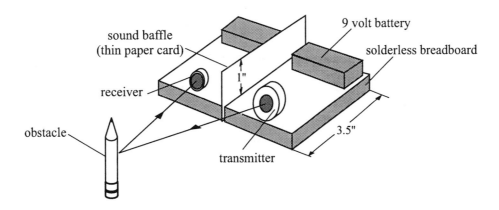

Figure 5-70. Two 9 volt rechargeable batteries can be secured to circuit boards with rubber bands. A ½" length of #22 AWG tinned solid wire soldered to the free end of each battery lead can be poked into the power bus, eliminating the need for a power switch. Completed units are shown in Figure 7-15.

Set R_A of Figure 5-68 to pin 5 potential, then adjust R_B until the lock detector LED is lit. At this setting a wall can be detected by reflection at more than 8 feet. If the units are separated and pointed at each other in a direct line of sight, lock is established over 30 feet. When used in a small room the receiver signal must be attenuated using R_A, to prevent spurious reflections from giving a continuous lock condition.

Adjustment procedure summary for obstacle detection

[1]   Set transmitter for maximum output using a microphone as in Figure 5-69.
[2]   Place transmitter and receiver side by side with a baffle as in Figure 5-70 and ~24" from a wall.
[3]   Adjust R_A and R_B until LED lights continuously (lock condition).
[4]   Point transmitter and receiver into an obstacle free area and adjust R_A until LED goes off.
[5]   Repeat [2] and light will go on.

System will now detect a pencil on a string at 1 foot (30cm) and a wall at 24". Other ranges can be set using these same procedures and longer ranges are obtainable by running the transmitter at 15 volts (two nine-volt rechargeable batteries in series).

Ultrasonic communications

An ultrasonic data link can be established using the transmitter and receiver from Figure 5-68. Transmitter output is modulated by connecting a transistor switch in Figure 5-71 to pin 5 of the 555 timer of Figure 5-68.

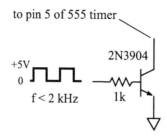

Figure 5-71. TTL level data input to the base of this transistor will modulate the transmitter of Figure 5-68. Direct line of sight conditions give reliable data transfer at 2kHz when transmitter and receiver are separated by 10 feet. Range increases to 20 feet for a 1kHz rate. These distances may be further increased by operating the transmitter at 15 volts. Spurious reflections from furniture, walls, people etc. interfere with data transmission at long ranges, but a bounce link is possible under static conditions.

Polaroid's ultrasonic ranging system

Polaroid Corporation has employed ultrasonic ranging as a focusing aid in their camera systems for many years. It is a sophisticated system using a single 50kHz 1.5" diameter electrostatic transducer as both transmitter and receiver. Operating at 4 discrete frequencies, their unit can measure distances from 6" to 30ft with ¼" resolution.

Hobbyists quickly realized potential applications for this ranging system, and Polaroid responded by providing all essential components in OEM kit form. A $99 kit includes 2 sensors and 2 boards, plus full technical specifications, circuit diagrams and other interesting and useful data. Details can be obtained from:

Polaroid Corporation Technical Assistance (800) 225-1618

When interfaced to a computer, a Polaroid ranger system can be used for mapping objects in a three-dimensional field. Two stepper motors are connected together orthogonally to move the ranger's transducer in a rectangular raster of horizontal and vertical steps. Range sampling at known elevation and azimuths allows a 3D perspective of what is 'seen' by the ranger to be displayed on a video monitor.

Polaroid's transducer emits and receives energy most efficiently within a circular cone angle of about 15 degrees. This is very good by ultrasonic standards, but gives an acuity of only 16" at a range of five feet. The spatial resolution of an optical ranging system is far superior because light wavelengths are about 10,000 smaller that those of ultrasound. However, optical ranging is also more complicated and expensive. Currently, reasonably priced ranging options are sparse for hobbyists, but professionals have difficulties in this area too.

Some comments on using ultrasonics for robotics

A whispered word can be heard anywhere in a quiet small room, so it is not surprising that high level 40 kHz ultrasound is detectable in the same manner. Sound at 10Hz or 40kHz ricochets from any object. Hard surfaces provide good low loss bounce signals, and soft open textured materials soak up sound energy.

Anechoic chambers, have their walls lined with fiberglass wedges because sound waves interact with this material and are reflected and absorbed at each successive bounce, Figure 5-72. Light traps act in a similar way but use polished black glass because of their smaller wavelengths ($\lambda \sim 550nm$).

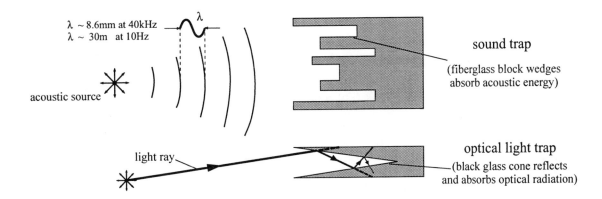

Figure 5-72. Sound waves are reflected from hard surfaces such as wood or ceramic tile with little energy loss, (that's why tenors love to sing in a tiled bathroom). However, fluffy porous surfaces break up sound wave fronts, and their pressure energy is dispersed and dissipated by interference and absorption. A highly polished cone made from black glass traps light very effectively, and is a good approximation to an optical 'black hole'.

Because we cannot 'see' how sound behaves in a room, ultrasonic behavior can at times be puzzling. It is usually a good idea to check out a new operating environment with a portable transmitter and receiver, unless you are confident of your robot's detection capabilities.

Polaroid did not make their ranging system complicated just as a development project. Ultrasound ranging is tricky and a Polaroid system is well worth the expense, if time spent in making a sonar ranging system work is factored into costs.

A piezoelectric touch sensor

Versatile and sensitive touch sensors can be fabricated from piezoelectric disks as shown in Figure 5-73. Wire feelers soldered to the tip of a clamped disk respond to even the slightest touch. Sensor sensitivity can be adjusted electronically with this circuit, and set at any desired level with the 5k potentiometer.

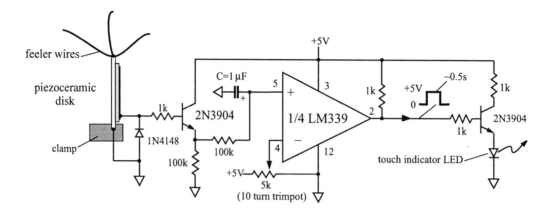

Figure 5-73. Feeler sensitivity is determined by the LM339 comparator threshold level, set with a 5k trimpot. A 1µF electrolytic capacitor and a 100k resistor act as a signal stretcher, producing a rectangular output signal pulse for ~0.5s, with each sensed touch. Feeler wire length should be tailored to avoid excessive oscillation when twanged. A 6" length of #20 AWG wire or a shorter length of smaller gage is satisfactory. Leave insulation on feeler wires except where they are soldered to a disk. An LED provides a visual indication whenever a feeler wire has been touched. Smaller sensors can be fabricated by cutting piezo disks on a metal shear.

This type of sensor is superior to a lever actuated microswitch because it is more sensitive and detects force from any direction. Several feelers can be soldered to a single disk, then trimmed to appropriate lengths providing omnidirectional sensing capability.

Cutting piezoceramic disks for making small piezo touch sensors

Piezoceramics are extremely hard and brittle and will crack if severely flexed. However, several parallel strip unimorph benders can be cut from a single piezoceramic disk using a foot operated metal shear. A 30" metal shear easily slices through both the brass and ceramic of disks shown in Figure 5-66. A rapid cutting motion is necessary, producing clean edges on both sides of a cut. Local metal shops specializing in ducts, galvanized flashings and eavestroughing have suitable metal shears for cutting piezo disks.

A pair of large scissors will also cut piezo disks, but these leave only one very clean edge and a curled waste section with almost no ceramic still attached. Dremel mounted water-cooled abrasive disks do not seem to be a viable option for making small unimorphs, because ceramic is vibrationally and thermally shocked from its brass base.

Small piezo strips are used in the same manner as a complete disk, and are easier to mount. A small bracket made from a 0.01"coffee can soldered to one end of a strip touch sensor, makes a convenient clamp.

A simple ac piezo accelerometer or impulse sensor

If the transistor's output at pin 5 of figure 5-73 is monitored, a damped sinusoidal signal will be observed when a feeler is twanged. This feature can be exploited to make a simple accelerometer by using a tip loaded piezo disk or beam. The device will only respond to changes in acceleration (unlike our strain gaged beam of Figure 5-14), but accelerometers of this type have a high frequency response and are widely used in industry for measuring impulse, dynamic forces, shock and vibration.

Dynamic performance of a piezo accelerometer is often measured on a shake or vibration table, which is essentially a platform supported and driven by a large loudspeaker type movement. Powerful shake tables with water cooled coils and massive magnets, are used to test the mechanical integrity of rockets and their payloads before a launch. These tests are essential because onboard components and associated hardware items sometimes undergo tremendous stress during launch, and anything that can shake apart usually does so.

Hobbyists can use an ordinary loudspeaker to characterize the performance of a home made piezo accelerometer. Select a high compliance speaker because it will have a large free displacement. Speaker displacement can be estimated by gently pushing the cone to determine its maximum travel, (when the salesperson is not looking)!

Speaker cone displacement as a function of drive voltage can be established by first applying known dc voltages and measuring how far the cone moves. These movements are measured with a magnifying glass and a rule, or by the laser technique described in Figure 5-67.

If a speaker cone is not heavily loaded then its motion will be simple harmonic when operated at low frequencies. Cone amplitudes are proportional to applied voltage. Acceleration a is calculated from amplitude A and oscillation frequency f ie: $a_{max} = \omega^2 A$, ($\omega = 2\pi f$, and A = maximum cone displacement). Acceleration forces are maximum when cone movement changes direction. Don't forget to orient a piezo disk's plane so it is perpendicular to the speaker's axis.

For low g accelerations dc accelerometers are often calibrated on a spin table, where g forces are easily controlled by varying rotation rate. If you wish to duplicate such test procedures, remember to install slip rings so power lines and signal wires can be connected to the accelerometer as it rotates. Sometimes it is possible to avoid slip rings by using long test wires and allowing these to twist for a certain number of revolutions.

Astronauts are spun in a large centrifuge to experience the effects of g forces, so they may learn how various physiological indicators precede a blackout (which can occur above ~5g). Small high speed laboratory centrifuges costing less than $2000 can reach accelerations of more than 15,000g [16] in ten seconds, (1g = 9.81ms^{-2}).

[16] Pea-sized metal spheres have been spun at 1,000,000rpm to generate forces equal to more than 5,000,000g. However, according to the Encyclopedia Britannica these achievements are dwarfed by a magnetically levitated rotor spinning in a vacuum of 10^{-6} torr, that produced incredible accelerations of more than 1,000,000,000g (10^9 g).

A piezo microbalance

If extra mass is added to an oscillating piezo disk its resonant frequency will be lowered, and this effect may be used to make a very sensitive weigh balance [17]. For 1cm diameter disks operating at a few MHz, shifts of a few hertz correspond to mass changes of about 10 nanogram (3.5×10^{-10}oz).

Monitoring the resonant frequency of a vibrating element is one way to make piezo devices responsive to dc forces, and hobbyists may find this a fertile area for exploration. Miniature vibrating tuning forks can also be used for this purpose, and an electromagnetic tine driver is described in a future book.

[17] Balances employing this technique have been used to measure erosion rates of materials for the proposed NASA space station. Engineers have found atomic oxygen eats away at carbon composite materials, severely compromising their mechanical integrity in distressingly short times. Because atomic oxygen is a major atmospheric component gas at high altitudes, protective methods must be developed to protect space structures against this very reactive gas.

Miscellaneous sensors

Polymer films respond to changes in relative humidity by changing their electrical characteristics, usually resistance or capacitance. *Electrosonic* sells a Philips RH sensor 691-90001 for $10.

Hobbyists can monitor humidity accurately in other ways too. For example, atmospheric condensate forms on a cooled plate at the dew point, and a polished cooled surface can be monitored to detect a decrease in reflectivity when condensation occurs. Students have used this principle to construct a thermoelectrically cooled relative humidity meter, but other similar ideas are also viable.

Liquid crystal displays are not only used for laptops. Someone astutely saw their potential for monitoring temperature, and they are now widely used for that purpose. A simple extension of this LCD temperature measurement technique resulted in a battery monitor now used by Duracell and Eveready. Battery current is used to heat a resistive strip that has an LCD temperature indicator. Heating is proportional to current, so an LCD's color is a rough guide to a cell's remaining energy capacity.

SQUID (Superconducting Quantum Interference Device) sensors using Josephson junctions push state of the art capability in many fields. Large organizational backing is needed to develop such devices but their operating principles are not difficult to understand, and may suggest new ideas or alternate uses.

Because bodily functions are fundamentally electronic, SQUIDs are used for noninvasive interrogation of brain activity, as with MRI scanning. However, SQUIDs can also be used to measure current, resistance, voltage and other electrical or magnetic parameters at very low levels. Some researchers are employing SQUIDs to detect gravity waves or minute geological perturbations. Temperature changes of 10^{-12} °C can be measured with a SQUID.

SQUIDs have been manufactured and employed commercially for over 25 years. Currently their sensitivity is about 5×10^{-15}T, enabling them to measure fields about 10^{18} times smaller than the earth's magnetic field.

Faraday rotation of a polarized beam has been used for decades, but employing thin films for magneto-optical imaging is a new wrinkle. Very high magnetic field strengths are required to rotate a light beam's plane of polarization, but such fields are commonplace in many laboratories.

Charge transfer screens are popping up again in new applications, and the latest versions operate through a 1" store glass window for touch screen shopping. Charge transfer can be used for proximity detection of even metal items that are not grounded, opening another avenue for hobbyists.

Power generated by piezo elements when they are flexed can be fed back to control bending. This principle is employed by Active Control eXperts of Cambridge, MA to actively damp a smart ski produced by K2 Corp.

Fiber optic sensors have some capabilities not offered by other sensors, and home built fiber sensors for hobby uses are described in a future book. More sophisticated optical fiber sensors employing an internal Bragg grating are now manufactured and sold commercially by EPC, Toronto, Canada.

Acoustic sensing systems similar to those used for submarine sonar systems, are also used for ultrasound body scanning. Non destructive testing (NDT) of composite aircraft wings uses ultrasound, and similar techniques are employed for monitoring the physical integrity of large structures. Gas flow rates, and gas temperatures can be determined by monitoring ultrasound velocities.

Advances in infra red methods over the past decade have vaulted uncooled detectors to a prominent place in thermal imaging techniques. Pyroelectric and bolometric devices are now widely used by the military and law enforcement. Several consumer products are also available at reasonable prices. Lost children can be spotted by their thermal images from the air, even in heavily forested areas. State of the art thermal sensors are still expensive but thermal imaging procedures may be used by amateurs.

Earthquake precursors and small shifts in the earth's surface can be detected with a laser interferometer. Such devices are sophisticated, but their principles are easily understood and may be applied by amateurs. On a smaller scale, commercial instrumentation can measure displacements with a resolution of 0.1μ" (0.0025μm) at a distance of 2.5".

Several more pages could be filled with brief descriptions of useful sensors. For example, miniature accelerometers, rate gyros, position sensors and chemical sniffers are available. Eventually such microsensors will become more prevalent, and prices will drop accordingly, making these devices affordable for hobbyists. Tiny radar units may soon be marketed for ranging and detection. Many biological sensors are currently being investigated, opening up new areas for amateur experimentation.

Optical sensors

It was pointed out at the beginning of this chapter that optical sensors occupy a special place in the sensing arena, and they will be dealt with in a future book. Optical techniques are among the easiest for amateurs to implement, offering greater hacker potential than almost any other technique.

Inventing sensors

Most sensing techniques are disarmingly simple when their operation is explained, and one may muse *"I could have invented that!"*. No person or organization has a monopoly on ideas, so beginners should not be reticent to test or suggest a new notion, no matter how bizarre it may seem. Nifty ideas abound in the sensor arena, and readers are encouraged to seek out new devices and learn their principles. Very few highs match the thrill experienced when a working sensor is made from one's own design ideas.

Force Transducers and Mechanisms

Transducers play an important role in robotic projects, helping to solve many difficult mechanical problems. Lifting, rotating, propelling, levitating, braking, damping and other actions can all be generated from suitable transducers. Motive forces for such actions are mostly derived from transducers based on magnetic, capacitive, electrostatic, piezoelectric, chemical, thermal, pressure, or hydraulic principles.

Not all transducers have been invented yet, and a look at the animal kingdom shows we can still learn a great deal by mimicking nature. Artificial muscles operating with animal efficiency[1] and performance will be a giant step forward for robotics, because they are flexible and possess a good power to weight ratio. Artificial animal muscles are being studied by several investigators, but such transducers have many years of development ahead before they reach surplus stores.

Birds can fly hundreds of miles on just a few ounces of fuel[2], using their onboard navigation systems to arrive with pinpoint accuracy. Bats can detect and catch 1mm insects using ultrasonic wavelengths of ~3mm (100kHz). The military would pay millions for an equivalent capability at any wavelength in the same sized package. A human can easily swing a 40lb suitcase onto a shelf, but designing a 150lb perambulating robot to perform the same action is a daunting task.

To be fair, no animal other than man has reached the moon or can communicate over a million miles of empty space. Given time and resources humans will probably duplicate nature's achievements in all areas. After all, life has been developing for millions of years, but intensive human engineering efforts have been applied for only a relatively short time, and are improving at an exponential rate.

[1] Most human muscles have a power to weight ratio of ~10W/kg, but a bird's flight muscles at 100W/kg are far superior. By comparison a top of the line model aircraft engine has a power to weight ratio of ~2300W/kg, and a top quality electric motor ~1850W/kg. A good example of the latter is an AstroFlight FAI Competition electric motor. This unit has an efficiency of 80% at 16,000rpm. When operated from a 27 cell NiCd pack, it uses 60A at 25V, delivering 1000W of shaft power - impressive performance! *AstroFlight Inc. 13311,Beach Ave. Marina Del Rey, CA 90292 (310) 821-6242.*

    Note that neither fuel, battery weights or other accoutrements are included in these ratios. These latter items and overall efficiencies must be considered when drawing parallels between animal and mechanical power plants.

[2] A one pound seagull can cruise at an airspeed of about 20 mph, consuming about 3% of its body weight per hour, (~ 4oz on a 100 mile trip). Food to energy conversion efficiency for animals, is roughly 24% as discussed in Chapter 4.

Mechanical devices

In many situations a transducer acts only as a prime mover, and must be coupled to a load by some contrivance that applies the transducer's force in an effective way. For example levers, pulleys, capstans, brakes, clutches, yokes, dashpots are vital secondary elements often employed in conjunction with transducers to extend their capability. Some of these devices will be discussed in this book where immediate elaboration is necessary, but the topic is too extensive for complete coverage here, and readers are referred to the following books.

"The How and Why of mechanical movements" Harry Walton, 1968 Popular Science E.P. Dutton and Co. Inc. New York. Library of Congress Catalog Card # 68-31227.

This very useful and practical book with excellent illustrations, explains operating principles of most fundamental mechanisms. Levers, pulleys, balances, engines, differentials, transmissions, brakes, clutches, gyros, accelerometers and hydraulics are all described in a user friendly fashion. Despite many printings it is almost impossible to find a new or used copy of this popular book, but it is still stocked by many good libraries.

Another helpful text for hobbyists, with excellent descriptions and drawings of numerous inventions, gadgets, nifty tricks and mechanical procedures is listed below. This book is one of the best in the practical mechanical genre.

"Mechanisms and mechanical devices sourcebook", Nicholas P. Chironis ed. McGraw Hill 1991 ISBN 0070109184, ~$70.

Rotary to linear motion conversion

Solenoids, galvanometers, loudspeakers, electric clutches, motors, resolvers, linear actuators, relays and many more devices may be classified under this heading. Tiny stepper motors that move wrist watch hands and huge hydroelectric generators are extreme examples of rotary transducers. Rotary motion can be converted to linear movement by cam, yoke, screw, or piston action as in Figure 6-1, but only a few transducers produce longitudinal motion directly. Low friction rack and pinion longitudinal slides are available from *Edmund Scientific*, 12", A61,285, $72 or 24", A61,286, $92. These stages have a vertical load capacity of 5lbs. Surplus priced rack and pinion parts, and dove-tailed slides are sold by *C&H Sales*.

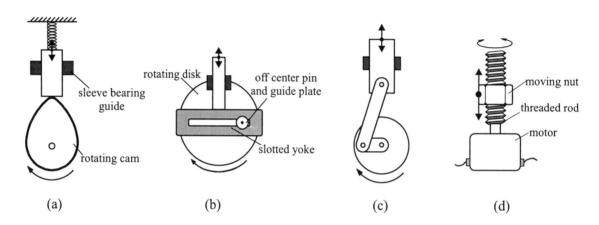

(a)   (b)   (c)   (d)

Figure 6-1. Operating principles illustrated in these diagrams can be used by hobbyists for converting rotary to linear motion.   (a) Cam action is versatile, and stroke format (linear, sine, triangle etc.) is modified by selecting an appropriate cam profile and a suitable follower.   (b) End dwell times may be altered by using different slot shapes on the basic yoke design shown here.   (c) Used in every reciprocating gas engine, this rugged low friction mechanism is probably the most widely used rotary to linear converter.   (d) Millions of floppy and hard disk drives, printers, feeders etc. employ this screw principle for generating linear motion.

Loudspeaker action

Loudspeakers or voice coil actuators move in a linear manner, with a stroke usually less than 0.5". But in spite of travel limitations, loudspeaker action is one of the more widely applied transduction techniques. Figure 6-2 (a) illustrates how a current carrying wire in a magnetic field is subjected to a force indicated by the arrow, and Figure 6-2 (b) shows the essential features of a loudspeaker in cross section. Because of radial symmetry each speaker coil element undergoes a similar force to that in (a), and coil movement is up or down depending on current polarity. A loudspeaker's coil movement is constrained to be truly longitudinal by a flexible corrugated support spider, usually made of an open weave varnished fabric.

Speaker action has no backlash, low hysteresis and coil travel is proportional to drive current, providing continuous and precise movement resolution. A stiff but light conical diaphragm acts as a moving piston to generate sound pressure waves, and with good design this can move very rapidly before mechanical limitations cause it to buckle against air pressure. For hobby purposes, even faster coil action and longer coil strokes are possible if this conical diaphragm is removed.

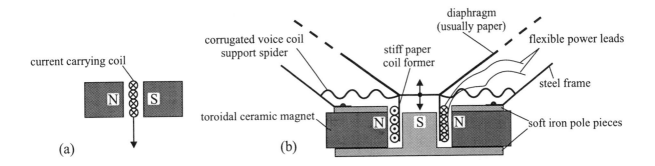

Figure 6-2. (a) A force acts on any current carrying medium in a magnetic field, and is a fundamental operating principle of many linear and rotary electrical actuators. (b) Modern permanent magnet loudspeakers often use a large doughnut shaped ceramic ferrite magnet with soft iron pole pieces to produce a high flux density in the coil gap.

Applied electromagnetics

Magnetic transducers for compact disk players

Electromagnetic action is versatile, and can be as gentle as the touch of a feather or powerful enough to shake-test a giant rocket. It is ideal for delicate positioning tasks, and compact disk technology relies on coil actuators to keep a laser beam aligned and focused for information retrieval. Figure 6-3 (a) shows schematically a single beam CD optical arrangement, and Figure 6-3 (b) a two-axis electromagnetic position controller.

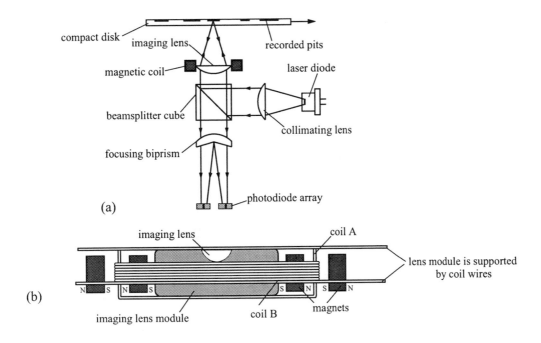

Figure 6-3. (a) A single beam optical arrangement for reading pit information from a compact disk uses two photodiode bi-cell detectors to decide when the imaging lens is in optimum focus. This detection scheme can also tell if the laser beam is correctly laterally positioned on the pit track. (b) Position and focus information from the photodiode array is used to control two sets of magnetic coils, servoing the imaging lens for both focus and tracking. Coil A translates the lens module laterally for tracking, (ie.in and out of the paper plane). Coil B moves the lens up and down for focus adjustment.

179

Compact disks are not flat by optical standards [3], so focusing on this wobbly target moving at ~1m/s, while simultaneously steering a lens to detect a 1μm (40 microinch) wide speck, is a tough acquisition and servo problem. It is doubtful whether CD technology would be possible without using magnetic coil actuation to steer and focus a lens onto such tiny moving bits of information.

[3] Some CDs have a bow of 0.5mm (500μm) across their surface, yet they still play perfectly. A 4" diameter λ/10 (0.05μm) optical flat retails for about $500, so it is fortunate active focusing described in Figure 6-3 works. Otherwise, CDs would be very expensive, heavy, and fragile.

Linear propulsion using magnetic forces

A magnet accelerated through a solenoidal coil is a rugged device and can be used as a smart rock, or as a propulsion element for small vehicles. Using the principle outlined in Figure 6-4, students designed and built a high speed magnetic bullet train that traversed a 1m by 1.5m rectangular track with appropriately curved bends.

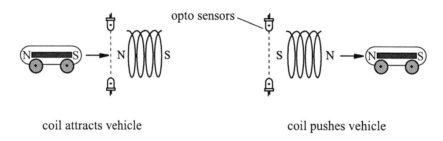

coil attracts vehicle   coil pushes vehicle

Figure 6-4. A magnet can be pushed or pulled through solenoid coils by carefully timed current pulses. Current triggering and switching may be synchronised by interrogating optosensors used to detect the magnet's arrival at each coil.

Contoured half sections of PVC pipe were used for the track and a 3"x1" dia. four-wheeled balsa wood train, carried a 1.75"x0.25" dia. Alnico magnet (*Edmund Scientific* C40,418 $10). Magnetic coils, strategically placed around the track were sequentially switched under computer control, using location information provided by optical sensor pairs. Fast acceleration, high speeds and direction reversals were possible during operation.

The computer also generated real time color images of both train movements along the track, and a 3D tunnel perspective of the train on a split screen. Top projects employing PC's always include high quality graphics routines, in addition to good control algorithms. These final touches turn an excellent project into one with outstanding appeal.

Rotary transduction forces using coils and magnets

Principles used for many years in electric motors are also employed in other areas, solving many difficult motive problems. Stereo systems with their current high fidelity owe much to magnetically driven loudspeakers, and compact disk technology might still be on a back burner without coil actuated transducers. These and other impressive technical developments rely on basic electromagnetic repulsive and attractive forces reviewed in Figures 6-2 and 6-5. Magnetically induced forces can be extremely powerful and are fundamental to all electric motors and many other devices. Laser scanners and magnetic head positioning systems for floppy and hard disk drives also depend on fast, accurate movements provided by magnetic transducers.

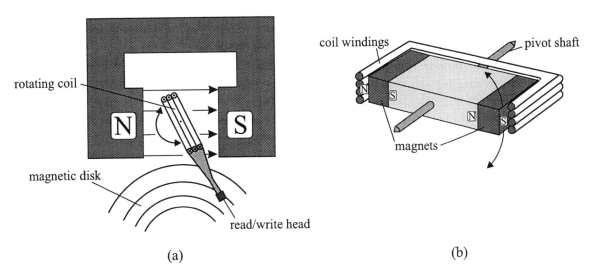

(a)   (b)

Figure 6-5. (a) Rotation forces similar to those employed in electric motors can be used for magnetic head positioning on hard or floppy disk drives. (b) A pivoting member equipped with outrigger magnets is also very useful for generating fast and precise rotary action.

Flat coil actuators

Large linear movements in early floppy and hard disk drives were usually derived from dc or stepper motor action coupled with a screw converter shown in Figure 6-1 (d). However, a rotating arm on a flat coil actuator from Figure 6-6 also provides fast and precise positioning for disk drive heads over small distances. Actuators are often preferred to motors for hard drives, because unlike motor driven screw mechanisms they do not suffer from backlash, and are electrically more efficient (a big plus for laptop computers). Steppers and dc motors always require a rotary to linear converter but a carefully designed coil actuator provides near-linear positioning directly and often with superior speed.

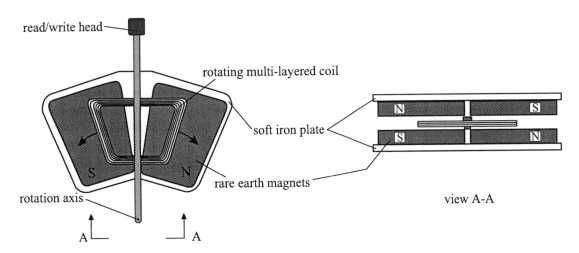

Figure 6-6. Flat coil actuator arms used for data retrieval in hard disk drives only move over a small arc, but do so at high speed. Rare earth magnet assemblies from old hard drives and similar devices are very powerful, and inexpensive at surplus prices; *C&H Sales* 0.6" x 0.62" 5lb pull #MAG 9450, $2 each or 25 for $40. Rare earth magnets made from neodymium-iron-cobalt or samarium-cobalt often have magnetic field strengths of 8kG, or about 100 times the strength of a refrigerator magnet. Alnico (aluminum-nickel-cobalt) magnet strength is usually 3kG or less.

Very fast write and retrieval times are possible with a flat coil actuator, by using rare earth magnets in a sandwich format as in Figure 6-6. Coils often rotate only a few degrees but this motion is amplified by a lever arm, enabling the read/write head to embrace all tracks on a hard or floppy disk. Features used for coil actuators in Figure 6-6 are also employed in some dc brushless motors. Such powerful motors are built into many laptop and regular hard drives. Brushless motors of this type use several segments shown in Figure 6-6 to form a circular armature rotating around the rotation axis indicated. These motors usually employ Hall sensors for sensing rotation, and electronic commutation, thereby eliminating the need for mechanical brush contacts.

One side benefit accruing from hard drive failures in personal computers, is the proliferation of excellent rare earth magnets on surplus markets. These magnets are often cemented to a soft iron backing plate as a complementary pair, and are invaluable in levitation projects for supporting deadweight. Auxiliary electromagnets operating under positional feedback control then compensate for residual weight, providing additional lift as dictated by optical sensors. This static/dynamic lifting duo is very effective for maintaining even large loads at a prescribed height.

Combinations of permanent and electromagnets have been employed in many industrial applications, where a permanent magnet provides a partial holding force supplemented when required by an electromagnet. Such combinations are used in fast acting solenoid valves and for speed enhancement in dot matrix printer head hammers.

Galvanometers for optical writing and scanning

An optical galvanometer is simply a mirrored assembly that can be rotated, and predictably positioned by electronic control signals. A single galvo merely reflects a beam of light so that it moves in a straight line, but galvanometer capabilities are increased significantly when they are used in pairs. For example, a pair of galvos can be arranged as in Figure 6-7 so that a laser beam may be directed to any portion in an X, Y plane by using appropriate mirror tilts.

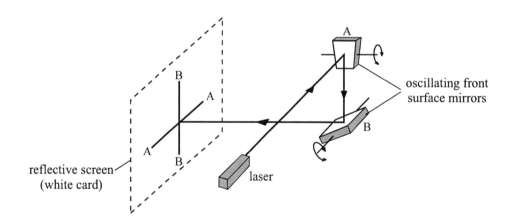

Figure 6-7. Two orthogonally disposed galvanometer mirrors can be used for 2D writing on a screen, and any figure can be drawn with a moving laser spot by using appropriate mirror rotations. Mirror A draws the line AA when rocked about its axis, and line BB is similarly traced with mirror B.

If a laser spot in Figure 6-7 is moved rapidly over a reflective surface its image appears as a continuous line to the human eye, because retinal response is slow. This visual persistence is the same effect that fuses consecutive movie or TV frames into seemingly continuous motion.

It is simple to draw diagonal lines with an X, Y galvanometer spot tracing system, by driving both mirrors from equal amplitude sine wave signals. If a phase shift is then applied to one galvo input, a diagonal line projection will be transformed into an ellipse. Such effects are Lissajous patterns, often described in electronics courses because of their utility and eye appeal. These figures are useful for determining phase differences and frequency ratios by applying suitable signals to X and Y oscilloscope inputs. Modern electronic instrumentation has diminished the usefulness of Lissajous figures, but hobbyists might enjoy reproducing them, using a computer driven laser galvanometer system.

State of the art galvanometers are sophisticated, with servo driven positional feedback, and non-linearities as low as 0.03%. Most employ precision bearings to support a rugged rotatable armature that carries glass mirrors up to 1" x 2" x 0.125, or even larger sizes on special order. Such high performance comes at a price, and even one of the least expensive commercial galvos sold, is beyond the pocket of an average hobbyist (and you will always need *two* for serious work). At the time of writing no second hand galvanometers are available from surplus suppliers, but many are still used in supermarket scanners, so this situation may change. Fortunately, good quality home made galvanometers are not too difficult to build from instructions given in this chapter.

Some commercial galvanometers have resolutions of 1 arc second and response times of 10μs or better. They are employed extensively for laser marking, a procedure where a pulsed focused laser beam writes letters or symbols by localized spot burning on even the toughest materials. Using a set of X, Y galvos a two dimensional design can be projected onto a target with a precision of ~25μm (for laser to target separations of ~10cm). Projected galvo images can be very large, as in laser light shows, or as tiny and precise as an integrated circuit.

Avoiding unnecessary distortion when using an X,Y galvanometer

A pair of galvanometers mirrors should be arranged as shown in Figure 6-7, to reduce pincushion distortion in a scanned figure. Distortion is at a minimum when an incident beam *i* and reflected beams *r*, of Figure 6-8 (a) are in the same plane, sweeping out a straight line image with mirror rotation. However, when an incident beam and pivot axis are coplanar, mirror rotation generates a bowed line image as in Figure 6-8 (b).

Galvo induced optical distortion is usually manifested as either a barrel or pincushion effect; ie. rectangular grid lines are squeezed inwards or outwards. Projecting squares of various sizes is one of the easiest methods to check for image distortion when using X,Y galvos. With a correctly oriented set of mirrors distortion is minimized, and visually undetectable in projected lines or figures.

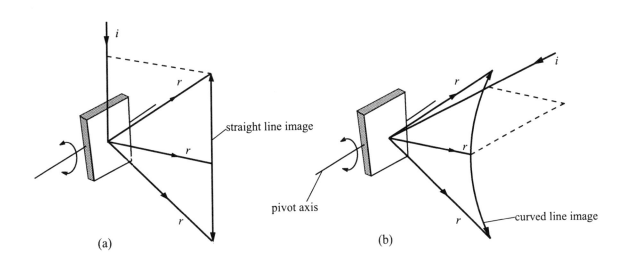

Figure 6-8. (a) Mirror rotation produces a straight line scan when incident (*i*) and reflected beam (*r*), are in the same plane. (b) If an incident beam is not perpendicular to the pivot axis a curved line image will be traced by mirror movement. A pair of X,Y mirrors should be arranged as illustrated in Figure 6-7 to avoid the distortion shown in (b) above.

Laser writing

Using assembly code or a high speed C program, it is possible to write all 26 letters of the alphabet on a screen in a few seconds, (a good test for speed readers if one or two letters are deliberately omitted). Hardware elements for a computer controlled lettering system are shown schematically in Figure 6-9.

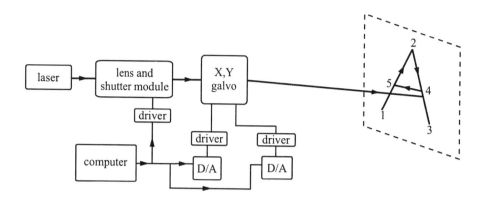

Figure 6-9. Laser letter writing with galvanometers requires a high speed shutter for beam blanking at position *3*, otherwise beam retrace from *3* to *4* adds extra brightness. Beam blanking is also required when a new letter is started, or to increase writing speed for letters such as *E, F, Q* etc. Each letter must be drawn in about 1/20 second or less to appear as a seamless solid image with no flicker.

More information on selecting appropriate lenses, interference filters, optical detectors, and eliminating background radiation for optical scanning are given in a future book. Details on a tabletop modulated laser beam scanning unit using a photomultiplier, modulator and associated circuitry are also given in the same place.

A loudspeaker based high speed optical shutter

A fast shutter is required for blanking a laser beam, to prevent overwriting described in Figure 6-9. This can be done with a loudspeaker driven blade shown in Figure 6-10. Beam occlusion occurs when a sharp knife blade attached to a speaker (from which the cone has been removed), moves rapidly upwards blocking the laser beam. Shuttering speed is enhanced by focusing the laser beam to a small diameter at the knife edge, because a small beam is covered more quickly. A 2" or 3" speaker is suitable for shuttering, and toroidal ceramic magnet types as shown in Figure 6-2 work well. Only a high compliance speaker gives the required translation at high speed, but fortunately many inexpensive speakers have this characteristic.

Piezo disk motion can also be used to blank a laser beam, but requires a switching voltage of ~150V, (see Figure 5-67 and accompanying text).

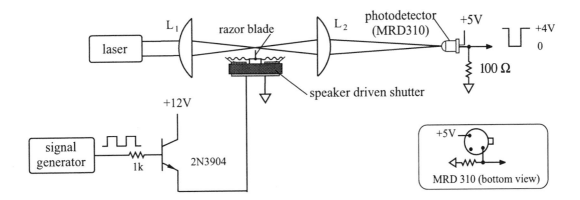

Figure 6-10. A small piece of razor blade epoxied to a loudspeaker movement driven at high speed can be used to shutter a laser beam. Lens L_1 focuses a laser beam to a small diameter (~20μm), decreasing blanking rise and fall times. A second lens L_2 refocuses the beam to a suitable spot size for displaying letters written on a white card.

Inexpensive plastic lenses from hand magnifiers are suitable for L_1 and L_2. Small diameter lenses are best because they usually have shorter focal lengths[4] too, making the apparatus in Figure 6-10 more compact. Plastic lenses from one-time-use cameras are ideal, they have a focal length of about 35mm and are ~1cm diameter. These lenses will focus a low power laser beam to a spot diameter $d{\sim}f\lambda/D$ [$d{\sim}35x10^{-3}x633x10^{-9}/10^{-3}$ ~ 22μm], if a 1mm diameter helium neon laser beam is used for writing. Plano convex lenses shown in Figure 6-10 can be purchased from *Edmund Scientific* C96,008 $5.95 16mm dia. 36mm F.L., C96,011 $6.95 15mm dia. 47mm F.L. Plano convex lenses produce their best focus when oriented as in Figure 6-10, because beam deviations are then shared at each surface (nature seems to love symmetry).

[4] Focal length can be measured by focusing a ceiling fluorescent lamp onto your hand - lens to image distance is the approximate focal length. You can use the sun, or any light source for this test, providing the source is at least 10 focal lengths from the lens

All items in Figure 6-10, except the speaker may be mounted directly on a 3"x½" wooden base board of appropriate length. Lenses and photo detector can be attached to right-angled brackets made from metal or balsa wood. They can then be moved along the base board for focusing.

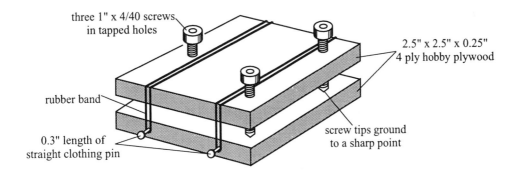

Figure 6-11. Speaker height can be precisely adjusted with a screw adjustable platform shown here. A smooth easy turning action is achieved if screw threads are first tapped into the top platform. Screws are pointed on a fine rotating sanding disk, (a Dremel tool is ideal for this operation). Points bite into the lower board giving a stable assembly held together by two elastic rubber bands.

Shutter height requires adjustment to ~ ±0.001" and is done with a screw adjustable platform illustrated in Figure 6-11. Three screws allow for easy two axis tilting, and sharp points give a no-slip bite and fine control. Tapped holes in the upper board permit screws to be easily adjusted with fingertips. Dressmaking pins are snipped to ~⅜" length and pushed into the wood for ~0.04" using a pair of needle nosed pliers, then hammered in (with plier support) until only ~0.1" protrudes.

Shutter testing procedures

[1]   Make sure speaker drive voltage is polarised so that shutter movement is *UP*, ie. the laser beam should be blocked when a +5V signal is applied.

[2]   Set the laser beam so it runs symmetrically along the baseboard, (a white card supported on modelling clay moved along the baseboard is useful for this operation). Leave this card at the end of the board with a mark on the card signifying the laser spot position.

[3]   Insert L_1 close to the laser and positioned so its enlarged beam is concentric with the spot marked in [2].

[4]   Place shutter and screw-adjustable platform on the baseboard at the approximate focus of L_1.

[5]   Position L_2 so a spot is formed symmetrically at the card mark, from [2].

[6]   Set photodetector to receive light from the laser.

[7]   Apply a 100Hz square wave signal with a signal generator, then adjust shutter height and its location between L_1 and L_2 to get a good square wave photodetector signal. Photodetector output should be ≤4V to avoid signal saturation. If detector level is too high use pieces of white paper as attenuators. A drop of oil on the paper will increase its transmission, if a little more light is needed.

[8]   Galvanometer mirrors can now be attached to the same baseboard making a complete scanning/shutter unit.

[9]   When everything is operating satisfactorily, tape or otherwise secure lens' brackets, speaker shutter assembly and galvos, so they can be transported as a unit.

Galvanometer applications for hobbyists

Mirror galvanometers may be used to scan objects or photos under computer control. This can be accomplished by illuminating the object with an X,Y raster scanned laser beam shown schematically in Figure 6-12. Reflected light from an object is monitored by a photo detector and these signals are processed giving a gray scale image for presentation on a video monitor or printer.

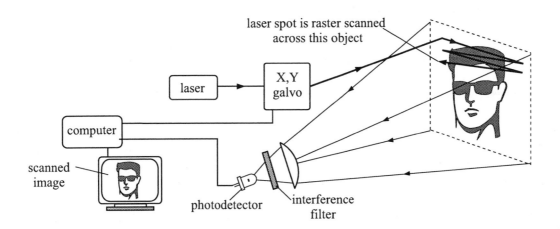

Figure 6-12. Raster illumination of an object or printed page can be done under computer control with a pair of X,Y galvo mirrors. Local variations of an object's reflectivity are detected by a photo sensor and used to construct a two dimensional gray scale computer record. A lens collects light reflected from the illuminated object, improving the detector signal to noise ratio. Spurious background light can be significantly reduced by using an interference filter that allows only light at the laser wavelength to reach the detector. Chopping the laser radiation, and high pass filtering the resulting modulated detector signal also helps. More information on these methods is given in a future book.

Students' optical scanning projects are always crowd pleasers and popular with any aged audience. They are technically challenging from both hardware and software viewpoints and have all the necessary ingredients for a good project. Top groups add extra frills, such as image enhancement, mosaic displays of stored scan records, and hard copies on request. With a 10mW laser and a large collection lens, head and shoulder's portraits take about a minute, (blacked-out glasses should be worn by the subject). A 60" x 30" tabletop setup is ideal for recording objets d'art, or favorite beer bottles using a 5mW 670nm solid state laser, (*Hosfelt* 75-174 $54.95). Laser writing can be done with the same setup.

Optical scanning in total darkness

Optical scanning in total visual darkness can be done with an infrared laser diode coupled to an X,Y galvo, using the same techniques employed for visible light scanning, shown in Figure 6-12. Although conceptually identical to visible light scanning, IR pictures taken in the dark seem more esoteric to the uninitiated. Youngsters are always pleased to take home their picture made this way, and a demonstration evokes oohs! and aahs! from spectators of all ages.

Infrared or total darkness scanning avoids ambient light difficulties but poses alignment problems, because a 780nm IR laser diode is barely visible when reflected from a white card. Procedures for collimating IR laser diode light, detector cards, and using auxiliary beams for alignment are discussed in a future book. Infrared laser diodes can be salvaged from CD players or purchased from Sharp Electronics Corporation. An alternative method of total darkness scanning, using an IR filtered tungsten light source and a Reticon™ linear diode array is also described there.

Home made galvanometers

Because home made galvanometers operate open-loop, satisfactory scanning is only possible when a rastered spot moves smoothly, faithfully following command drive signals. Incorporating position sensing and feedback control into a home made galvo is too difficult for the average amateur, because of mechanical complications. Furthermore, servo positioning is unnecessary for scanning projects.

Two home built designs are described below, but it is pertinent to first mention two 'galvo' techniques that cannot be used successfully for laser writing or optical scanning.

Why dc brush motors cannot be used as optical galvanometers

Although a dc brush motor appears to run smoothly, it really moves in a series of pulsed steps, as commutated power is applied to a rotating armature. Inertia evens out these pulsating movements creating a seemingly smooth action. However, when operated in a bidirectional or back-and-forth mode, a motor shaft will not return to the same position at each reversal, because inertia causes overrun.

Consequently, raster sweeps formed by motor action have jerky discrete motions and these are not repeatable. Armature inertia on most dc brush motors is very large by galvanometer standards and serious positional errors accumulate during fast start/stop sequencing. Gearing down a motor helps, and a model aircraft servo can be used for slow scanning because it contains a positional feedback potentiometer. However even the best model aircraft servos have jitter, and cannot be cycled at high rates.

Why analog current meter movements make unsatisfactory optical galvanometers

Optical galvanometer mirrors are relatively heavy and appreciable energy is needed to overcome inertia during start/stop raster sweep reversals. This energy is provided electrically, and most 8Ω galvos draw an average running current $\geq 0.5A$. This is a huge current to apply directly to any conventional analog meter movement, and most meters will burn out if more than a few hundred microamp is applied directly to their coils. This is because even high current meter movements are very delicate, and most of a measured current is bypassed with a shunt resistor.

A very small mirror mounted on a very large meter movement can be made to oscillate with a sine wave drive signal. However, such motion is deceptive, and insufficient for precise scanning. If adequate power is applied for proper positional control, a meter coil will burn out.

An inexpensive home-built rare earth magnet optical galvanometer

A compact, reliable, galvo design suitable for student and amateur use is sketched in Figure 6-13, and its performance compares very favorably with low end commercial units costing several hundred dollars. A unit can be built in a few hours using basic tools and readily acquired materials. Use only materials specified because these have been shown to work satisfactorily. For example if a steel support blade is substituted instead of brass as recommended, it may be affected by the coil fields. A hobby plywood housing is easy to fabricate, and will not nick the magnet wire's insulation. Wood is nonmagnetic and if these and other parts are made carefully the completed galvo will be rugged and work well.

Motion for this design is based on the motor principle illustrated in Figure 6-5 (b). A flexible hinge on the support blade allows a magnet carrying balsa platform to teeter totter about this silicone rubber joint, as magnetic force is applied. For scanning projects, a galvanometer mirror is driven by an amplified triangle waveform applied to the coil. Scan rates are typically a few hertz or less, allowing an optical detector sufficient time to acquire a good signal from each picture element (pixel) on an illuminated object.

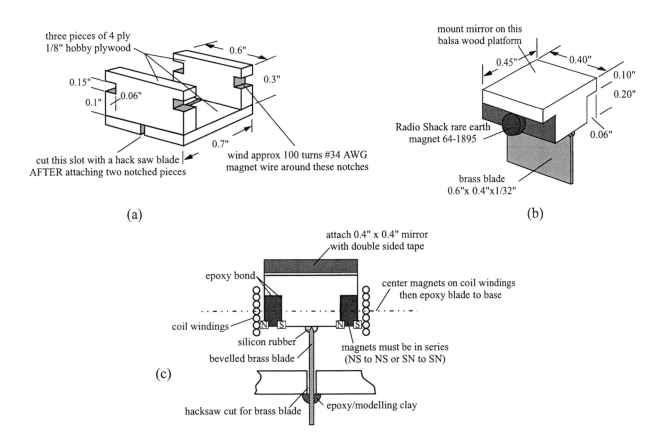

Figure 6-13. These sketches show components required to make a compact and reliable optical galvanometer described in the text. If components in (a) are carefully made, they can be cemented with crazy glue to make a rugged assembly. (b) Silicone rubber is used to attach a nonmagnetic blade to the balsa platform. Keep excess rubber on the sides of the blade to a minimum, since this impedes rocking motion. (c) For clarity, this diagram shows only essential details of items (a) and (b) after assembly. Magnets are centered on coil windings before epoxying blade to the base of (a). Rare earth magnets 0.06"x0.2" dia. are *Radio Shack* 64-1895, two for $1.49.

Radio Shack magnets shown in Figure 6-13 are probably the best buy, and work well in this design. Similar Nd-Fe-B magnets 1/16"x3/16" dia., are sold by *Edmund Scientific* A38,428 $3.75 each, and *McMaster-Carr,* 0.197" dia. x 0.08", 5902K47, $1.46 (1-5), $1.22 (6 up).

However, half inch diameter ceramic button magnets are also satisfactory and are widely obtainable from hardware stores, *Radio Shack* 64-1883 five for $1.49, or in refrigerator magnet clips. These heavier magnets work satisfactorily for scanning but this galvo has a lower natural resonant frequency because of added weight. When using ½" diameter button magnets increase coil support size to 0.5"x0.8" and coil notch to 0.18"x0.1", slotted base becomes 0.8"x0.8" but blade dimensions remain unchanged. Use one piece of balsa ½"x½"x½" for the mirror platform and recess this to leave 0.1" for the platform top while making magnets sit flush to platform edges as in Figure 6-13 (c).

Magnet wire is available from *Allied Radio* #36F 1320 Belden #8057, 4060ft $36.73 or from *Electrosonic* Belden #8057, 4060ft $16.70.This wire is solder strippable with rosin core solder and has a resistivity of ~0.26Ω/ft. Each coil needs about 30 feet of wire and should have a resistance of ~8Ω. Wire gage is not critical providing each coil has a resistance of ≥4Ω, so salvaged wire from a small transformer or ac adapter may be used to save cash.

Rare earth magnet galvanometer construction notes

[1]   Cut the three (3) pieces in (a) and glue them together. Do not cut coil notches or slot in base at this time. Hobby store plywood is a premium product and bonds well with SuperGlue. Panelling and many inferior plywoods splinter badly and must be cemented with white aliphatic glue or epoxy.

[2]   After glue has set, cut coil notches with a jeweller's saw or a small file.

[3]   Use a hacksaw blade to make a slot for the bevelled brass support blade[5] shown in (b) and (c).

[4]   Make mirror platform in (b) from hard balsa wood using a utility knife or razor saw.

[5]   Attach magnets with crazy glue as shown in (b) and (c). Magnets must be arranged in series, ie. inner sides should attract.

[6]   Fasten (b) down on a flat surface so magnets are horizontal. Add a thin seam of clear silicon rubber[6] to blade edges with a toothpick. Now support blade with modelling clay so it touches wood perpendicularly as suggested in Figure 6-13. Allow at least 8 hours for rubber to cure.

[7]   Wind the coil by hand. Windings do not have to be neat, but you will pack more wire in the notches if you're not too sloppy.

[8]   Solder insulated flexible extension wires to the coil, and fasten these so they cannot be pulled away from the module, (#26 AWG wire is ok).

[9]   Insert blade from (b) through base slot of (a), centering magnets on coil windings. Hold blade at base with modelling clay and test assembly as described in 'Galvanometer test procedures' following Figure 6-15. When everything is satisfactory glue blade to the base with 5 minute epoxy.

[10]   Attach a 0.04"x0.4"x0.5" glass mirror to the platform with double sided tape. Edmund Scientific sells 12mmx12mmx1mm mirror, #C31,005 $6 each, but it is more economical to buy their #C41,620 mirror 67mmx67mmx1mm for $11.10 and cut the size you want. These are both first surface mirrors with a protective coating against accidental scratching. Glass mirrors should be cut with a diamond, so check the Yellow Pages under 'glass' for a friendly glazier who will do the job gratis. Thin plastic shaving mirror is easier to cut and works too. Use the back surface of a plastic mirror (which is a first surface mirror), to avoid reflections from other surfaces.

        If you see 'ghost' satellite spots around the main reflected laser beam, these are spurious reflections. They affect image resolution slightly, and disappear when a first surface mirror is used

If stray magnetic fields from the galvanometer coils interfere with other apparatus, they can be substantially reduced by wrapping a ⅛" strip of 0.01" coffee can steel around a coil's exterior. This strip is first covered with tape or heat shrink tubing, making it less likely to nick the magnet wire's insulation.

---

[5] A neat 45 degree bevel on a brass blade is easily made by drawing the blade along a file laid flat on a table.

[6] Silicon rubber is an RTV rubber (Room Temperature Vulcanizing) rubber, and is sold by hardware stores as a caulking compound in large rigid tubes, or in smaller squeeze tubes. Only use a fresh product and always discard a small portion first squeezed from a tube, any time it is used.

A loudspeaker driven galvanometer

Once one becomes familiar with speakers and their wonderful design features, it is tempting to use them in all kinds of transduction applications. The design presented in Figure 6-14 works well and embodies several design characteristics that hobbyists may find useful for other projects.

A 3" or 4" speaker with good compliance works best for this type of galvanometer, and inexpensive speakers are adequate because their suspensions are flexible and have a large travel. Test results in Table 6-1 were measured on 4" speakers salvaged from a portable am/fm radio with dual tape decks.

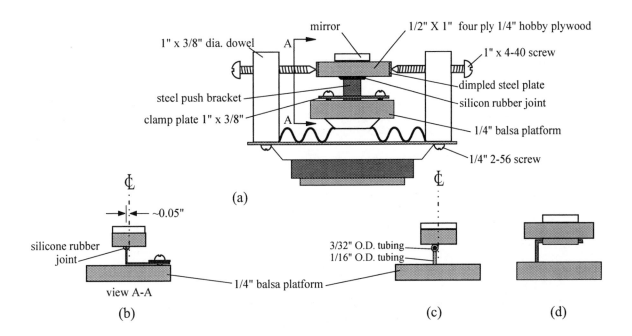

Figure 6-14. Components for this speaker driven galvanometer are sold at good hobby stores, and a 4" speaker from an old radio will give fine performance.

Mirror rotation is achieved by rocking a pin-balanced beam using longitudinal force generated by the speaker coil. This arrangement gives an extremely low friction movement with negligible side play, when screw tips are properly adjusted. A motor driven sanding disk or belt can be used to point screw ends by hand. Each screw is twirled by its head between thumb and forefinger against the abrasive surface, with a little support from a finger of the other hand. An angle of 45 degrees is ok but this is not critical. Use a magnifier to check for surface finish and unwanted concentric grooving, then regrind with a finer grit to get a smoothly tapered surface.

Steel plates from a coffee can (½"x¼"x0.01") are lightly dimpled in their centers with a pointed punch or nail, which should then be rotated to smooth the indentation. Countersink each end of the plywood to provide clearance for dimples, then attach plates to beam ends with crazy glue.

Speaker force cannot be applied to the mirror support platform via a rigidly coupled push bracket, so it is essential that a flexible connection is used as shown in Figure 6-14b,c,d. This is necessary because any point on the plywood beam moves in an arc as it is rocked. However, speaker movement is strictly longitudinal, mandating the use of a flexible coupling. Silicone rubber is a good adhesive for this purpose. It has very good adhesion to most surfaces, and can be flexed repeatedly without exhibiting signs of fatigue.

Figures 6-14c,d outline an alternative pushing method employing small diameter lubricated hobby tubing[7]. This works equally well and probably has a longer life. Recommended tubing sizes in Figure 6-14 should be attached a short distance from the beam's centerline as shown. Tests show galvo drive efficiency and total scan angle both increase when rocking force is applied close to the centerline.

[7] Small size hobby tubing can be cut by first nicking through a tube wall with a file or razor saw, then breaking by flexing the tube with thumbs opposite the notch. Right angle bends in small tubing can be made by supporting the smaller tube in two pieces of the next larger size tube. Slip the larger tubes on so their ends touch at the desired bend point, then bend with fingers.

Speaker galvanometer construction notes

[1] Cut away the speaker's paper cone, taking care not to snip through coil power leads that are also glued to the cone surface. These must be left intact. Extraneous material can be removed from around the power leads.

[2] Use a pair of aviation snips to cut away all unwanted metal framework. Retain as much of the metal base as possible for mounting ⅜" diameter dowels, and for additional attachments needed when two speakers are combined to form an X,Y galvo pair.

[3] Cut two 1¼" lengths of ⅜" diameter wooden dowel with square ends. Drill a 3/32" (0.092") hole by ⅜" deep, then screw in (don't tap first) a 4/40 machine screw. This procedure makes a strong and reliable connection to the speaker's frame. Dowelling splits easily if wood screws are used.

[4] Drill ⅛" dia. clearance holes in the speaker frame then mount both dowels using ¼"x 4-40 machine screws.

[5] Set speaker/dowel assembly on a flat surface then use a supported pencil to mark a dowel ~¼" from its top. Without moving the pencil, rotate the assembly and mark other dowel at the same height. Remove dowels and drill at marked locations with a 3/32" drill, then fit previously pointed 1"x4-40 screws, (screw these in without prior tapping).

[6] Epoxy a 1¼"x1¼"x¼" hard [8] balsa wood platform to the spider's center (cut off speaker dome if necessary [9])

[7] Make clamp plate and push bracket from 0.01" coffee-can steel. Drill holes in clamp plate and make matching tapped holes for 2-56 machine screws in balsa platform from (6). Use a 1/16" drill for the screw holes, then start 2-56 screws by hand in balsa. Tighten screws very gently until clamp plate holds push bracket firmly. If a screw strips the balsa threads, add a sliver of balsa wood to the hole, then refit screw.

[8] Assemble everything except the push bracket and clamp plate. Align and center beam, then adjust pin screws so there is no side play but it spins freely. Level beam so it is horizontal then put a dab of modelling clay on each screw/beam junction to prevent rotation.

[9] Trim push bracket height so it just touches underside of the wooden beam. Now add clamp plate and place push bracket so it sits ~0.05" from centerline of beam rotation axis. Apply a thin seam of silicon rubber with a toothpick to both junction sides of push bracket and wooden beam, and allow this to cure for at least 8 hours.

[10] Add a drop of fine oil to each pivot point.

[11] Galvo is now ready for testing.

[8] Test balsa with a thumbnail and select firm wood. Balsa comes in six densities but not all hobby store proprietors know this.

[9] Do not remove a speaker dome unless you cover the resulting opening immediately. Small magnetic particles are readily attracted to the coil-magnet gap, and these are nearly impossible to remove. You will need a new speaker if this happens.

Magnet galvanometers outlined in Figure 6-13 and a speaker galvanometer depicted in Figure 6-14 meet requirements of hobbyists and students for laser scanning projects. These devices are inexpensive to build and give repeatable and reliable performance. For X,Y scanning systems a pair of galvos should be combined as described in Figure 6-7. A simple wood support structure is adequate, but details have been left so hobbyists can translate their own ideas into suitable hardware. Purchasers of commercial galvos must pay extra for X,Y mounting.

Operating and testing optical galvanometers

A galvanometer driving circuit

Home made and commercial galvanometers can be operated from the circuit shown in Figure 6-15. Two circuits are required for an X,Y galvo system and these will comfortably fit onto one standard 6" solderless breadboard. Install small clip-on heat sinks on all transistors.

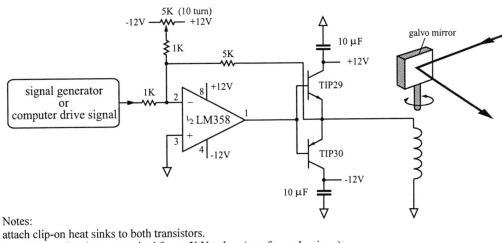

Notes:
attach clip-on heat sinks to both transistors.
two of these circuits are required for an X,Y galvo, (one for each mirror).

Figure 6-15. This driver is suitable for home made or commercial optical galvanometers. It provides ample power for even large galvos, so take care not to overdrive small units. Initial beam position is adjusted with a 5k trimpot, and this should be set to its midpoint when operating a new galvo for the first time.

Galvanometer test procedures

[1]   Arrange laser and galvanometer to give a horizontally swept line projected onto a sheet of quad lined paper.

[2]   Apply a ±0.5V 5Hz sine wave from a signal generator to a driver from Figure 6-15 and use the driver's output to power the galvo coil. This should produce a clean uniform linear sweep, with no local bright spots.

[3]   Reduce sweep rate to ~0.1Hz and the laser spot should move smoothly with no hesitation or jerkiness.

[4]   Switching to a triangle drive waveform, gives instant reversal at the end of each sweep and brightness along the sweep should still be uniform. Local bright spots suggest galvo bounce and may be due to overdriving, or too much silicone rubber at the blade support.

[5]   Total sweep amplitude will remain fairly constant as sweep rate is raised, but increases dramatically at the galvo's natural resonant frequency. Reduce drive amplitude, then increase scan rate to find the resonant frequency, which occurs at maximum sweep amplitude.

[6]   A galvanometer's sensitivity S depends on how much coil current is required to turn it through an angle θ. Sensitivity can be determined by calculating θ from sweep length l over a prescribed projection distance L, for a coil current $i=V/R$. At the low frequencies used with galvanometers, coil impedance is approximately equal to its dc resistance R, and V is the instantaneous applied coil voltage.

For example an 8Ω speaker galvanometer produces a scan length l=8" at a distance L= 22" when its coil is driven by a ±2V 30Hz signal. Total angular deflection is, $\theta = 2\ tan^{-1}\ (l/2L) = 20.6\ degrees$. Peak to peak coil current is $V/R = 2x\ 2/8 = 500mA$ and sensitivity $S=500/20.6\ \sim 24mA/degree$.

[7]   Because home built galvos are operated open-loop, it is important that beam deflection is proportional to coil current in a predictable and repeatable way. Although sensitivity S obtained from dynamic sweeps is useful, it does not guarantee proportionality of movement. A galvanometer can be tested statically by applying various coil currents and measuring the corresponding mirror deflections.

A static calibration curve for a rare earth magnet galvo is plotted in Figure 6-16 using discrete voltage steps obtained by adjusting the 5k offset trim pot in Figure 6-15. Linearity shown in this figure is adequate for projects. It could be improved by sensing the true deflection and using this information to steer coil movement via a servo feedback loop.

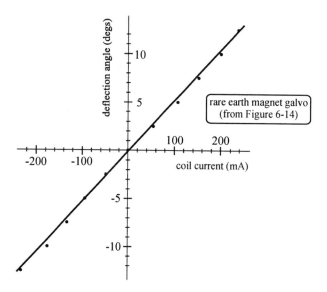

Figure 6-16. This static rare earth magnet galvanometer calibration was obtained using a driver from Figure 6-15, without any additional signal input (ie no signal generator). Discrete measured voltages were applied to the galvanometer coil by adjusting the 5k trimpot, and monitoring mirror rotation by observing deflections of a laser spot on a quad-ruled paper screen. It is not necessary for hobbyists to calibrate their galvos statically in order to use them for scanning, if they are fabricated according to instructions given in the text.

Measurements in Figure 6-16 are best obtained by setting the laser spot at its normal zero position, then quickly applying and simultaneously measuring the voltage necessary to give a prescribed beam deflection. This procedure more closely duplicates dynamic raster behavior and reduces hysteresis effects from the silicon rubber hinge.

Galvanometer performance

Dynamic test results on rare earth and ceramic button magnet galvos are presented in Table 6-1, and these may be compared with those of a commercial galvanometer.

Table 6-1. Operating characteristics of home-made optical galvanometers

Galvanometer	coil resistance R (Ω)	coil inductance L (mH)	total scan angle θ (degrees)	resonant freq. (Hz)	sensitivity (mA/deg)
rare earth, Figure 6-13	7.4	0.42	30	47	28
½" button magnets	10.9	0.89	30	25	19
speaker, Figure 6-14	8	1.30	30	55	8
G320 *General Scanning**	8	30	40	60	30

* *General Scanning Inc. 500 Arsenal St. P.O. Box 307, Watertown, MA 02272 (617) 924-1010 (800) 343-1167.*
The G320 is $295 and can be used for random stepping or raster scanning. A G120 optical scanner at $120 has a total scan angle of 40 degrees, but is designed for continuous sweep scanning only. Home made galvanometers are fine for hobby use and compare favorably with low end commercial units. Extreme precision is unnecessary for student scanning projects and home built modules meet all desirable criteria.

Home built galvanometers have larger scan angles than many commercial units, and are suitable for tabletop scanning. Rare earth and button magnet galvos are very compact, easy to build and can be repaired if necessary. If you blow up a commercial unit, it probably means a costly repair bill.

Electromagnets and computer controlled games

Board games have fascinated and challenged humans intellectually for centuries, and in the past two decades computers have acquired sufficient competitive playing capabilities to challenge even a world champion seriously. An inexpensive chess player, eg. *Radio Shack* #60-2439 $39.99, gives excellent players a good game. Even Gary Kasparov had a few uncomfortable moments when first confronted by IBM's 'Big Blue'. It is conceivable that computer chess machines will eventually beat the best human players, but checkers and many other board games are predicted to give programmers and machines greater difficulties.

At the University of Toronto, early chess machines built by undergraduates employed 6502 microprocessors for logic and interface control. Human opponents' moves were sensed with onboard optical detectors, and computer responses given by mechanically moving the appropriate piece with an electromagnet below the board.

Building computer controlled robotic moving arms, and applying appropriate computer game strategies has real-world payoffs for young engineers. Some former students are now employed by Spar Aerospace in Toronto, where the Space Shuttle's remote manipulator arm is designed and manufactured. Every time a satellite is lifted from the Shuttle's cargo bay and placed in orbit by the 'Canadarm', it is comforting to contemplate that a quiet game of computer chess may have played a minor role.

Moving chess pieces under computer control and playing a winning game is not enough for some dedicated students. In 1992 four 2nd year students, constructed a machine that solved Rubik's Cube mechanically. A standard Cube was installed centrally inside a ¼" Plexiglas™ walled box. Six geared motors, with their shafts penetrating these walls were connected to cube faces for manipulation purposes.

Starting from a random configuration, cube sections were turned by these motors following computer commands. This project used a lot of nifty software, some excellent mechanical design, minor but critical modification to the Cube's moving parts, and just the right amount of lubrication.

It is probably inevitable that in their continuous quest for one-upmanship future students will look to Backgammon, Mah Jong, Peg Solitaire and other piece moving board games for light relief. What game of Cribbage would be complete without a robotic hand sticking marker pegs into the proper holes in response to computer generated commands?

Moving game pieces and other elements in two dimensions

In some board games, pieces can be picked up magnetically then simply slid from one position to another. Checkers is mechanically more complicated because 'crowning' requires a lifting action, then combining two disks that are subsequently moved together as a 'king'. Play at this stage means the electromagnet in Figure 6-17 must be able to move up and down besides traversing the board's surface, (ie. x,y,z or r,θ,z capability is required).

Similar difficulties arise in chess when pieces are moved from their tops, because pieces have differing heights in most sets. Chess can be played with an r,θ arm using flat disks marked with appropriate pictorial denominations. However, this is aesthetically less appealing, and students usually opt for x,y movement control of real pieces from beneath a board.

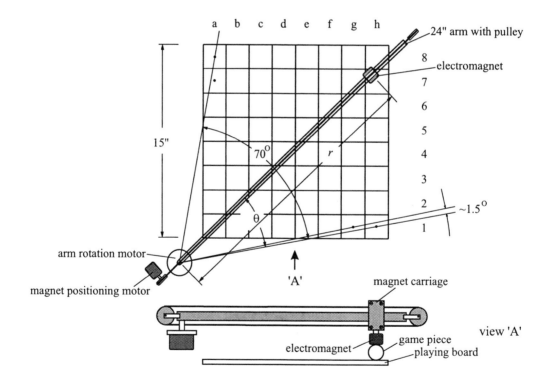

Figure 6-17. An electromagnet can be positioned anywhere along this rotatable arm, and such polar coordinate positioning allows access to every square on the board. Soft-iron pick up elements attached to game pieces permit moving them to new locations, but all pick up elements must be the same height unless the electromagnet can also be moved vertically.

Polar coordinate (r,θ) translation systems

It is instructive to examine a few hardware limitations and possible design tradeoffs, when constructing a polar positioning system shown in Figure 6-17. For example, a heavy magnetic pickup puts more stress on the beam, needs a bigger rotation motor, and possibly extra support bearings. A long arm gives better angular resolution in the corner squares but is more difficult to support, and it may be necessary to use composite materials for added stiffness, lighter weight and vibration damping. Such choices arise in many static structural designs, but a moving arm adds dynamic complexities that also require consideration.

Selecting an appropriate motor for arm rotation and precise positioning requires some deliberation, since an angular precision of about ±0.5° is required when the pickup is on squares g1, h1 or a7, a8. Some pros and cons when using a geared dc motor, stepper motor or a model aircraft servo for arm rotation, are addressed below.

Geared dc brush motor

(a)   This type of motor has adequate torque plus good low speed control, and can operate in both ±θ directions using a bidirectional controller. Typical gear trains have a few degrees of backlash on their output shafts, but even a 'perfect' gearbox has positional overrun due to armature inertia. Such erratic motions produce unacceptable accumulated position errors after a few game moves, making feedback position control essential.

(b)   A potentiometer rigidly attached to the motor's shaft will track arm rotation position accurately. However, no slop can be permitted in the pot/motor connection if the pickup element is to be positioned to ±0.5° on squares g1, h1 etc. A suitable-single turn, low torque (0.25 oz-in) pot, is sold by *Servo Systems* #PS-573 $19.50. Single-turn potentiometers are best for this application, giving better accuracy and precision than an equivalent multi-turn pot. For example, a ten-turn potentiometer has 3600° measurement capability but we need to monitor only 70° of rotation. Therefore, neither pot resolution nor accuracy are fully exploited when a multi-turn potentiometer is employed for position feedback control that uses less than 2 percent of its total capacity.

(c)   A 720 step 360° optical encoder suitable for monitoring arm rotation has 0.5 degree resolution, but such precise encoders are difficult for an amateur to build. *Servo Systems* sells a 4096-step encoder #ADC-245 $68 that is more than adequate for the task, but other options should be considered before following this route.

Stepper motor

(a)   For many applications, steppers operating in their open loop mode are an ideal solution. Stepping motors move in well defined angular steps, need no positional feedback, and have high torque at low rotation rates.

A small electromagnetic pick up weighing 2oz (56.8g) creates a torque of 48 oz-in when moved at the end of a 24" massless arm. This electromagnet can be moved by an ungeared stepper motor having a torque of 50 oz-in (a motor with this torque is about 2¼" dia.x2"). Most steppers in this class move in 1.8° steps (200 steps/rev) and will probably do the job. If a lightweight arm is used, and the electromagnet is retracted to give only a small moment during arm rotation, a stepper motor works well.

(b)   Students have successfully employed a large stepper motor (4" dia.x2.5"), to move a heavy electromagnet on an r, θ system, for playing 'Diplomacy' with ⅜" steel balls.

Model aircraft servo

Model servos are little power houses, developing surprisingly high torque from tiny dc brush motors and a planetary gear train. Standard size servos have output torques of about 40 oz-in, but ¼ scale servos produce 130 oz-in and move at 240 degrees/second. Most servos will swing over ±40° without modification, and when powered from an appropriate circuit, have a resettable resolution of ~±0.5°. Such precise angular positioning is possible because a feedback potentiometer is incorporated into a servo head, giving the necessary rotation position information. The rotating arm of Figure 6-17 can be easily attached to a model aircraft servo, because these units come with a circular mounting platform on their output shafts. *Tower Hobbies* sells ¼ scale servos for $32.99, eg. TA2896.

Beam carriage for an electromagnet pickup arm

A simple balsa wood carriage fitting snugly around the rotating arm in Figure 6-17, may be used to support an electromagnet. If top and bottom beam surfaces are covered with transparent tape this provides a low friction surface, allowing the carriage to be pulled by a string or thin wire.

Small pulley wheels are hard to find, but can be made from hobby plywood. Carefully cut out a circular piece of wood with a jeweller's saw. Support this disk in a drill using a 2-56 bolt through its center, and two nuts tightened against each other. A drill press is ideal for making a groove and edge finishing. A hand drill is also satisfactory if it is clamped to a bench, or gently held in a vise. Grooves can be made with a jeweller's file or home made sandpaper sticks.

A 24"x1/2"x3/4" balsa beam deflects ~1mm when tip loaded by 50g, but a stiffer lower mass beam of the same dimensions can be made from a balsa sheet with composite fiber sides (Figure 8-28).
A 500 amp.turn iron cored solenoid weighing 32g can lift 600g and a twin coil magnet weighing 48g lifts 1550g (Figures 6-19, 6-20).

Some practical properties of magnets

A magnet's lifting capacity is increased when its magnetic flux travels from pole to pole, through high permeability materials. For example, an Alnico horseshoe magnet lifts about 200g from one pole as in Figure 6-18 a, but lifting power increases to over 1000g when the magnetic circuit is completed by a soft iron weight as in (b). As pointed out in Question 36 of Chapter 2, magnetic force lines prefer to travel through high permeability materials. Iron has a permeability about 5000 times greater than air. Consequently it conducts magnetic flux much more efficiently, and accounts for the increased lifting power in (b).

Straight bar magnets behave in a similar fashion because they are really just horseshoe magnets that have been straightened. Figure 6-18c is therefore equivalent to (a), whereas (d) is not quite as efficient as (b) because magnetic force is lost by traversing extra soft iron pieces and additional air gaps.

Scrap metal yards use powerful cylindrical electromagnets for moving heavy metal parts, and a 4" by 8" diameter 110V dc 250W electromagnet can lift 4000lbs, (*McMaster-Carr* #56915K92, $248.76).
A 4¼"x2½" dia. magnet sold by *Edmund Scientific* C60,435 $53.95, operating from one C cell lifts 200lbs. More useful to hobbyists are tiny electromagnets sold by *Edmund Scientific* C52,899, three for $4.95. These little brutes are only 0.35" x 0.55" diameter but operate continuously from 5 volts, and will lift a 2.2lb soft iron load at this voltage.

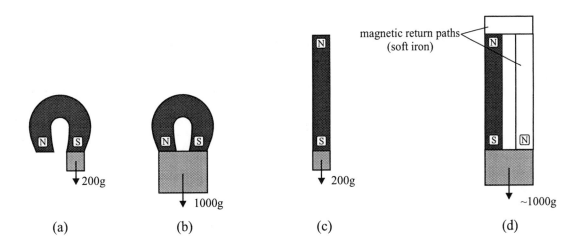

Figure 6-18. Lifting power from a single pole is low for both a horseshoe magnet in (a) and its single bar equivalent in (c). A much larger weight can be lifted by a horseshoe magnet when its magnetic circuit is completed by a soft-iron load as in (b). Lifting capacity for a bar magnet can also be increased, as shown in (d), but this is less efficient than an equivalent horseshoe magnet.

Cylindrical geometry shown in Figure 6-19 (a) is key for high efficiency of all these electromagnets, and reminiscent of the toroidal speaker magnet in Figure 6-2. Cylindrical electromagnets are tops, but difficult to make, because parts must be turned on a lathe. However, electromagnets with almost the same pull as an equivalent cylindrical magnet, can be made using procedures described below.

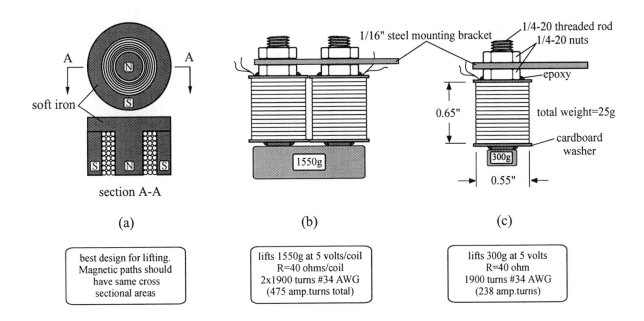

Figure 6-19. (a) Electromagnets with a cylindrical pole pattern shown here give the greatest lifting force, but hobbyists will find the style in (b) has only marginally less pull, and is much easier to make. (b) This double unit has five times the pull of an identical single unit shown in (c), because its magnetic circuit is continuous when a soft iron load is attached. Steel brackets of various thicknesses are sold at hardware stores, and these may be cut to act as both keeper and mounting attachment.

Electromagnet fabrication procedures

Electromagnets in Figure 6-19 (b) and (c) are made from readily available hardware, and magnet wire can be salvaged from ac adapters, old motors or transformers. Always try to make solenoids so they can be powered from standard supply voltages (eg. 5V or 12V).

A single coil magnet in (c) operates from 5V and two such coils in series can be run on 12V. When operated at 12V the double coil magnet shown in (b) increases its lifting capacity to 1900g weight (18.6N).

Construction tips for making an electromagnet

[1]   First screw a nut onto a length of ¼-20 threaded rod, then cut rod to the desired length with a hacksaw. Clean rod ends with a file, then undo the nut to remove burrs. Finish ends of rod square and flat so they will make good contact with a load for maximum lifting power.

[2]   Cut 2 circular cardboard washers of a suitable diameter (0.025" card thickness is ok for coils ≤ 1" diameter).

[3]   Punch a ¼" diameter central hole in each card with a one-hole hand held paper punch.

[4]   Wrap 2 layers of masking tape over rod threads in the coil region.

[5]   Tighten together two ¼-20 nuts at the bracket location.

[6]   Force cardboard washers over the tape, then epoxy cardboard from outside the coil region.

[7]   Poke a needle through cardboard disk at nut location, then pass wire through this hole and tape it to nuts.

[8]   Clamp an electric hand drill to a bench or hold it gently in a vise. Now install the ¼-20 rod/disk assembly in drill chuck.

[9]   Operate the drill from a Variac™ or dimmer switch. You need slow speed control with both hands free.

[10]   If possible, support your wire spool on a rod so the wire can unspool without kinking.

[11]   Count the number of turns, or use your estimated finished coil outside diameter as a winding gage. When winding is completed, twist the wire from [7] together with coil's end wire.

[12]   Thoroughly remove insulation from about ¼" of each wire end. Use a small utility knife and press both wire and blade against a thumb while pulling. Magnet wire *cannot* be soldered unless insulation has been completely removed, so do this job properly.

[13]   Add heavier insulated wire leads. These *must be soldered* to the magnet wire. Use heat shrink tubing or carefully tape each soldered joint.

[14]   Secure heavy wires to coil body with a ¼" wide strip of masking tape. Masking tape can be sliced to an appropriate width with a knife and ruler guide after tape has been stuck to a glass plate. Just peel tape off the glass after cutting. This is neater and faster than using scissors.

N.B.   A two-coil magnet in Figure 6-19 (b) must be wired to give one north and one south pole, just like the horseshoe magnet in Figure 6-18. You can test electromagnet polarities with a compass, or by the lifting ability of a finished electromagnet. Lift will be much greater when coils are wired together correctly.

Moving game pieces with an electromagnet

If a checker board is made from a Plexiglas™ sheet, game pieces can be moved from below using an electromagnet as in Figure 6-20. An edge clamped 15"x15"x⅛" acrylic sheet does not flex appreciably for normal loading by plastic chess pieces, (pawn=10g, king=30g).

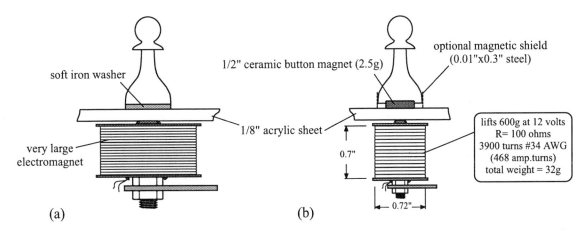

Figure 6-20. (a) Game pieces can be moved to new positions by dragging them from below a smooth acrylic surface with a powerful but heavy traversable magnet.   (b) If a ½" diameter 2.5g ceramic button magnet is attached to each game piece, the electromagnet size and weight can be significantly reduced. An optional steel skirt eliminates repulsive interference between magnets on small pieces, such as pawns.

The magnetic field from a bar magnet decreases inversely as the cube[10] of the distance from a pole, and falls off in a similar way for an iron core solenoid. Consequently, a very powerful magnet is needed to attract a game piece through ⅛" of acrylic, as in Figure 20a.

A much smaller electromagnet can be used to move game pieces if button magnets are attached to their bases, as shown in Figure 6-20b. Interference between base magnets can occur with low mass small diameter pieces, but this only happens when their bases are almost in contact. These magnetic repulsion forces can be eliminated if desired, by adding a 0.3" wide shielding skirt made from 0.01" coffee can steel to pawns, as suggested in Figure 6-20b.

Although a dual solenoid electromagnet from Figure 6-19 is more efficient, a single solenoid is best for moving game pieces, because double pole magnets apply asymmetric attractive forces and tend to move pieces erratically.

[10] It is straightforward to calculate the magnetic field strength B at a large distance r from a magnet with pole strength M. ie $B_r = \mu M \cos\theta/(2\pi r^3)$, $B_\theta = \mu M \sin\theta/(4\pi r^3)$. However, when r is small compared with the magnet's length, things become complicated, and no simple expression is available for determining B. More information on field strength decay for a solenoid shown in Figure 6-20 (b), is given in Figure 6-27.

x, y motion bases

Commercial positioning equipment is expensive, and prices for precision translators escalate rapidly when movements of more than an inch or so are required. For example a 12" single axis linear motion base without motor drive is a bargain at $1000. Most professional translation stages are screw driven, and loads of 400lbs can be moved with micrometer accuracy on a platform of a few square inches. *Edmund Scientific* and *Servo Systems* carry precision units, and *C & H Sales* have inexpensive slides from $32.50 and two axis actuators with 2" movement for $39.95.

One of the simplest techniques for precisely positioning very heavy loads uses screw-driven greased plates. Loads of more than 1000 lbs can be accurately positioned by this method, using ½" thick aluminum greased plates with steel guides. An x,y translation stage using this technique is only 1½" thick, and can be operated from motor driven screws or car jacks, and set to < ±100μm by either a direct reading scale or a linear encoder.

Screw drive mechanisms shown previously in Figures 5-6 and 6-1 are used in commercial translators for good reasons. High precision screws are inexpensive, and excellent micrometer heads with 2.5μm resolution are available for $50. By using two differently pitched screws in series, finer movements are obtainable than when either screw is used independently. This is known as a differential screw, and micrometers based on this principle can achieve resolutions <0.1μm, with an easy turning action and negligible backlash, at a cost of ~$200.

Two longitudinal translation stages must be stacked vertically to give x, y motions, suitable for moving game pieces. Each stage must be capable of about 13" travel if a normal sized chess board is used. A translation unit should not be too heavy because it must be moved by another stage, to produce two-dimensional motion. For this reason it is important to keep motor size, electromagnet and stage weight low. A stage suitable for computer controlled chess moving is shown in Figure 6-21, it weighs only 525g but can move more than ten times its own weight.

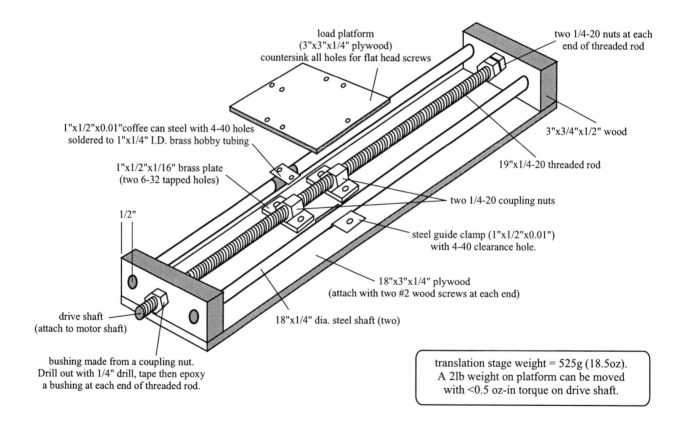

Figure 6-21. Vertical platform loads of several pounds are easily moved with a small geared dc brush motor on this translation stage, and 15lbs can be moved by rotating the drive shaft with only fingertips. Total platform translation is ~14" but this can be extended if a longer and thicker plywood base is used. Larger diameter threaded rod and mating coupling nuts are sold in hardware stores, so this design may be scaled up for both longer travel and much heavier loads.

Construction tips for building a translation stage

[1]   Clamp the two wooden end pieces together. Now drill two ¼" and one ½" diameter holes through both pieces at once, using wood bits. Mark the orientation of these pieces and install them the same way.

[2]   Drill out two coupling nuts so a ¼-20 threaded rod rotates freely but with little side play. Wrap these nuts in masking tape so they are a tight push fit in the wooden end pieces. Epoxy these items to the end blocks *after* the stage has been assembled and tested.

[3]   Cut two brass plates, then drill and tap for 6-32 screws. These plates merely hold the moving coupling nuts to the platform, so attaching them permanently to the coupling nuts is not necessary.

[4]   Make sure the 1"x¼" brass hobby tubing slides freely on the steel rod (¼"diameter music wire). This is a key item and must move smoothly without side play. To solder brass tubing to the steel plate, put a thin nail through tubing then tape both nail ends to a flat surface. Slip plate under brass tubing then solder both parts together. Even a small pencil iron is ok for soldering coffee can steel and ¼" hobby tubing.

[5]   Drill a 4-40 clearance hole in the steel guide clamp before bending this strip. Never hold small metal pieces with fingers when drilling. Grasp the strip firmly with pliers, then support it on a scrap wood block while drilling. Always center punch a dimple on a workpiece before drilling. This prevents a drill bit from wandering. Finger pressure is adequate for bending the guide clamp around a ¼" dia. music wire shaft to make the shape shown in Figure 6-21.

[6]   Use flat head screws and countersink all holes on the wooden platform, leaving a flat mounting surface.

[7]   *Do not* glue or nail baseboard to the end supports. Screws permit disassembly if adjustments are required later.

[8]   Two ¼-20 nuts at each end of the threaded rod are tightened against each other, acting as stops to prevent longitudinal movement of the threaded rod.

[9]   All moving parts, (threaded rod, ¼" guide shafts, end bushings and end nuts), should be lubricated with a light oil.

[10]   A short length of flexible Tygon™ or similar vinyl tubing, makes a handy flexible coupling between threaded rod and a motor shaft. Clear food-grade plastic tubing is sold at hardware stores and ⅛"x¼" O.D. tubing can be warmed with a blower heat gun (or carefully with a lighter), then forced over large diameter shafts giving a tight no-slip fit. Leave about ¼" between both shafts to allow for any misalignment between the two shafts.

Transducers for Braille displays

Students often apply their skills to solving problems for the physically challenged. Controlling a wheelchair from forehead muscle movements, or directing a computer monitor's cursor from head motions, are excellent examples of thoughtful projects constructed by young engineers.

Apparatus that aids the blind always makes a good project, and some useful products now sold in retail stores fall into this category. Voice outputs have been incorporated into talking watches, bathroom scales, thermometers, calculators and many games. These items are well designed, inexpensive and widely available.

Such equipment is useful for most blind people, but the deaf-blind still need Braille devices. Anticipating this need, students built a calculator with keypad input and a 16-character Braille output display. A BCC52 microcontroller was used for logic and interface operations. Many aspects of this project are technically demanding, combining electronic and mechanical problems into a tough challenge.

Braille arrays

A few letter and numbers as represented in the Braille system, are shown in Figure 6-22. Braille is a comprehensive system, and special codes are also available for mathematical characters and other technical symbols. Musical instruments can be controlled from Braille commands, so this is a fertile area for budding roboticists too. Musical notation is handled in Braille using an additional set of character modifiers.

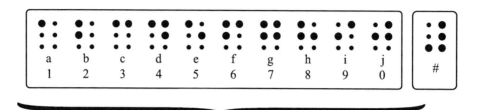

dot combinations shown represent letters,
but signify numbers when preceded by a # symbol

Figure 6-22. An array of six raised dots shown in this diagram should ideally be small enough to be covered and felt by a person's fingertip pad. Larger symbols are often used in public buildings for elevator controls etc.

Solenoids for Braille transducers

A tangible array of raised dots should preferably be sized so all six dots can be felt simultaneously by a single fingertip. This is a tricky transducer problem because magnetic solenoids[11] are probably the only viable option for actuating tactile bumps, if a compact multi-character display is required.

Two possible solenoid options are presented in Figure 6-23. Home made solenoids in (a) provide dot spacings of 0.2", but ready made coils with a slightly larger spacing, shown in (b), are easier to mount.

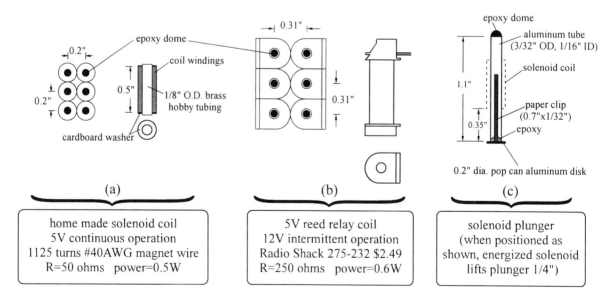

(a)	(b)	(c)
home made solenoid coil 5V continuous operation 1125 turns #40AWG magnet wire R=50 ohms power=0.5W	5V reed relay coil 12V intermittent operation Radio Shack 275-232 $2.49 R=250 ohms power=0.6W	solenoid plunger (when positioned as shown, energized solenoid lifts plunger 1/4")

Figure 6-23. (a) Washers for the coil bobbin can be made by drilling⅛" holes in cardboard, then shaving raised hole edges with a razor blade. A 1½" length of 2-56 threaded rod and four nuts are used to support the bobbin during winding. Coil winding procedures are given in the text following Figure 6-19.   (b) This 5V relay coil must be drilled with a 3/32" drill to accept the aluminum solenoid plunger. Only intermittent operation is possible because this coil is run at 12V to generate adequate pulling force. (c) Aluminum hobby tubing with a steel paper clip insert is used to make this plunger. It may be used with either solenoid shown in this figure.

Assembling a 16-character Braille display composed of 96 small solenoids is labor intensive, and working with close packed arrays of small cylinders can be tiring. Ready-made coils in Figure 6-23b make construction a little easier. These coils are available in Toronto for 50¢ used, and $1 new as a complete reed switch. Make sure your purchases have 5V coils because they must be overrun at 12V to give adequate pulling force. Use a 3/32" drill to enlarge each solenoid's core so a plunger shown in (c) moves freely.

An aluminum plunger with paper clip insert shown in (c), will be attracted into either type of solenoid, when energized. If a solenoid is positioned so it is 0.35" from the plunger's base, it produces a lift of ¼" when power is applied to the coil.

Current consumption escalates rapidly when many coils are energized simultaneously. If half the coils in a 16-character display are energized simultaneously, coils in (a) require almost 5A at 5V and coils in (b) draw 2.3A at 12V. Solenoids are therefore only energized momentarily. They are then held firmly in position by a slotted board that slips under the aluminum disks, but passes around unselected aluminum tubes. This scheme allows bumps on selected solenoids to protrude above a perforated panel for interrogation by a reader. Firm support is provided by this fingered board, and solenoid power is only applied during data updates.

A fingered board can be made from balsa wood and moved by a solenoid or motor/pulley system. Guide rods at the ends of this board can be made from hobby tubing sliding on steel rods made from music wire.

[11] Nitinol (also called muscle wire or Shape Memory Metal SMM), is frequently suggested as a prime mover for Braille display actuators. These shape-memory nickel titanium alloys have found a niche in some commercial systems, but are difficult to use in classroom projects. For most hobby applications Nitinol uses too much current, even when spring bias or innovative mechanical assistance is employed.

Levitation and magnetic forces

Levitation projects always excite viewers, perhaps because of their aura of surrealism, but also because many people cherish a desire to float free from gravity's pull. Therefore it was not surprising that a magnetically-levitated-flywheel received many accolades, when it was displayed during an end of term demonstration. This project was built by three students, and employed a powerful rare earth magnet to support most of the flywheel's weight. Additional auxiliary lifting force was provided by an optical feedback controlled solenoid, which kept the flywheel at a stable levitated height. Pulsed magnetic fields at the flywheel's periphery supplied tangential forces for rotational motion.

Only a few methods [12] are of practical importance for levitating objects. Radiation pressure is very weak compared with gravity or atmospheric pressure, but 0.2mm diameter transparent glass spheres have been supported in a high power laser beam. Spherical bodies are readily suspended in an air jet, and huge loads can be raised with air cushion techniques. Acoustic waves can levitate small objects, and *Dantec* sells an Ultrasonic Levitator that works with solids or liquids from 20μm to several mm diameter. Sound pressure waves from this ultrasonic source give stable non-contact support that resists side forces very well. A samarium-cobalt magnet can be levitated above a disk that becomes superconducting at liquid nitrogen temperature (77K or -196°C) due to the Meissner effect. *Edmund Scientific* sells a kit for this levitation demonstration, A38,169, $39.

A unique pressure gage manufactured by *MKS Instruments Inc.* employs a steel ball that is partially levitated by permanent magnets, then balanced and rotated at 400Hz by an ac powered magnetic field. When spinning power is removed, the sphere's rotation rate decreases because of molecular drag from residual gas surrounding the ball. Absolute pressures can be determined to 1% with this novel device, over a traditionally difficult measurement regime, (5×10^{-7} to 10^{-2} torr).

Using permanent magnets for partially supporting deadweight is common practice for levitating magnetically compatible objects. Nonmagnetic materials can be levitated too, if a soft iron sheet is attached to the article at an appropriate location. Figure 6-24 compares the lifting ability of several permanent magnets with a solenoid coil for moving chess pieces, described earlier in this chapter.

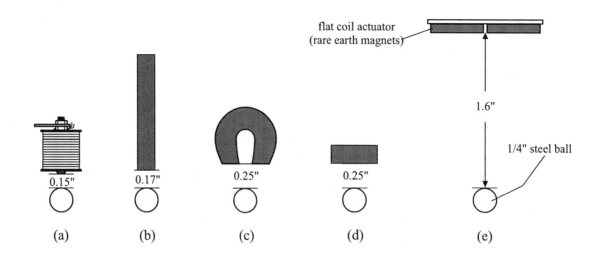

Figure 6-24. Small solenoids and low power permanent magnets give only modest lift, and can attract a ¼" diameter steel ball over a vertical distance of about ¼". A 'chess' solenoid from Figure 6-20b is shown in (a), while (b), (c) are Alnico magnets and (d) is a ½" diameter ceramic button magnet. Impressive pull in (e), is generated by a pair of rare earth flat actuator magnets also depicted in Figure 6-6. Arrangements from (a) to (e) cannot be used for levitating unless they are aided by some additional modulated force.

Magnetic lifting systems (a) through (e) are all unstable, and a steel ball cannot be made to levitate without additional help. In such arrangements only two stable states exist: (i) the ball will be unaffected by the magnet or (ii) the ball will be attracted until it reaches a magnetic pole. Such behavior is annoying but is a reality, and it has been shown theoretically that additional forces are essential for stable levitation.

Many ingenious contraptions have been designed to magnetically levitate an object, without using feedback controlled assistance. Aeromodellers will be familiar with an almost frictionless magnetic support system used by *Top Flite,* in their precision balancer ($19.99 *Tower Hobbies* TA1394). In this remarkable apparatus, two ceramic-8 magnets support a shaft with pointed ends that can be used to test a propeller, a wheel, spinner, tire etc. Top Flite's apparatus does not levitate a test object because one shaft end touches a magnet. Nevertheless, it is an excellent example of good engineering design.

Scientific toys are also sold that can levitate a magnetized top or cylindrical rod above a strong magnetic field. One of the best, called the 'Levitron', keeps a spinning top levitated above a magnetic base for several minutes, when spun by hand. None of these devices violates Earnshaw's principle (derived from Laplace's equation in 1842), which shows stable levitation is impossible using fixed field magnets. Earnshaw's theorem also shows a point charge cannot be supported in a stable manner by a static electric field.

---

[12]   Electrostatic levitation of an 11mm diameter silver sphere weighing 1.3 grams, is mentioned in a very interesting book by R.A.Ford. A reprint of an article originally published in Electrical Experimenter, July 1920 describes how a sphere can be levitated 25cm from an electrode. The author used a Wimshurst generator to lift not only metallic objects but also water globules, cork and wood.
'Homemade Lightning' Creative Experiments in Electricity.   R.A. Ford TAB ISBN 0-07-021528-6 $19.95

If a feedback controlled solenoidal magnet supplements lift from a permanent magnet, true leviation is possible by using position sensing to turn on the solenoid as required. *Maglev's* levitation of a six-inch diameter world atlas globe, is a slick example of such magnetic lifting control. Optical sensing shown in Figure 6-25 is not the only technique for sensing body movement, eddy currents mentioned in Chapter 5, ultrasonics, and other methods have also been used.

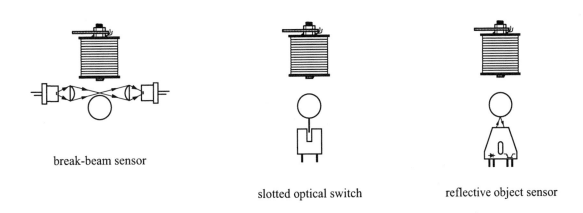

break-beam sensor

slotted optical switch   reflective object sensor

Figure 6-25. A break-beam position sensor is made from discrete optical elements and can be placed at a considerable distance from a levitated object. Infra red slotted optical switches and reflective assemblies are easier to use, and are standard products available from most electronic outlets. Linear Hall effect sensors, positioned at the electromagnet's pole can also be used for feedback controlled levitation of a magnetic body.

Two ways in which permanent magnets may be used in conjunction with lifting solenoids are depicted in Figure 6-26.

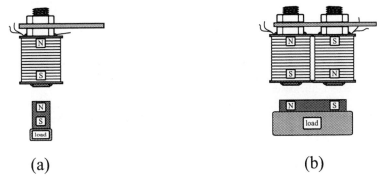

(a)   (b)

both arrangements can be used in either attraction or repulsion mode
(position sensors are not shown)

Figure 6-26. Two lifting techniques employing permanent magnets for supporting deadweight are shown, but (b) is more efficient and has greater lifting capacity.

Magnetic attraction forces

In footnote #10 (following Figure 6-20), it was stated that under certain conditions magnetic field strength decreases as r^{-3}, and iron cored solenoids might behave similarly. From an experimenter's point of view, knowing the exact equation for a magnet's flux field is of limited practical value. Far more useful is information on what attractive force is available at various distances from a magnetic pole. These attraction forces can be easily measured using a soft iron slug, a spring balance, and paper pole spacers. Test results in Figure 6-27 compare magnetic pull forces between a permanent rod magnet and our 'chess' solenoid. Solenoid voltage has been adjusted so its pull force equals that of an equivalent permanent magnet when r is zero.

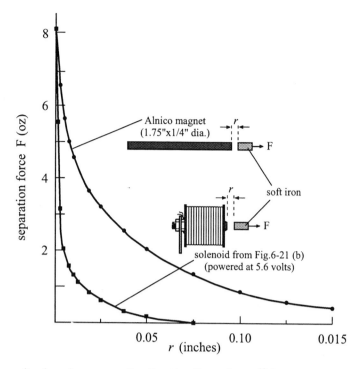

Figure 6-27. These results show how magnetic attraction force decays[13] for a permanent magnet and an equivalent solenoid energised steel core. A spring balance or calibrated elastic threads were used to measure the separation pull force. Paper sheets or cardboard inserted between the magnet pole and soft iron, were used as spacers to give discrete values of *r*.

[13]   The approximate exponential power at which these curves decay, can be found using a mathematical procedure known as 'forward differences'.

A table of 'pull force' values at equal increments of *r* is first made. Differences are then taken between successive values. If these differences are not all zero, the process is repeated using this new set of differences. For r^{-1} decay first forward differences should be zero, for r^{-2} fall-off second differences are zero etc.

At least three forward difference iterations are required when this process is applied to data in Figure 6-27, suggesting magnetic field decay is more complex than a simple r^{-3} relationship. An equation could be derived to fit curves in this figure, but a set of measurements as in Figure 6-27 is of more practical use.

Plunger solenoids for hobby use

A cross sectional view of a powerful commercial solenoid is shown in Figure 6-28. Its housing is made from 0.1" steel plate with outside dimensions 2"x2"x2.5". When energized, both plunger and housing combine to complete the magnetic circuit with only small flux loss. Contrast this design with the simple Braille solenoids in Figure 6-23, where magnetic efficiency was sacrificed for ease of fabrication and overall size.

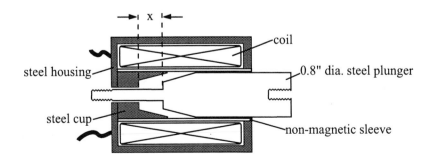

Figure 6-28. Plunger solenoids constructed with the features shown, have useful pulling power over larger distances than a solenoid without a magnetic return circuit. Pulling forces for this solenoid are given in Figures 6-29, and 6-30.

Addition of a steel housing and plunger cup may appear as trivial modifications, but these two elements improve solenoid performance dramatically by completing the magnetic circuit. Not only is total pull increased by a factor of five or more, but useful pulling power is available over a much longer interval. For example, when the plunger in Figure 6-28 is extended so x=1", it still has a pull of 4lb. Pulling force relationships for this solenoid are plotted in Figure 6-29.

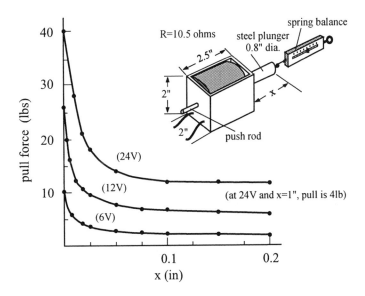

Figure 6-29. Magnetic pulling force in this well designed commercial solenoid decays more slowly when a steel housing and plunger cup assembly are used. Even with a 1" plunger extension, pulling force is 4lb. Not all solenoids are made with both push and pull capabilities, but most can be modified to perform either function. Solenoids are manufactured for either continuous or intermittent operation, and the latter may burn out if operated continuously.

By operating a solenoid above its rated voltage its pulling capacity can be increased, and Figure 6-30 shows how pull improves for the solenoid previously discussed. Care must be taken to reduce the operating duty cycle when running above the specified rating of 24V. Consequently this 40% duty cycle solenoid must be derated to 25% (ie $40 \times 24^2/30^2$), when run at 30V.

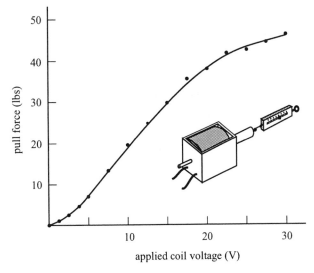

Figure 6-30. Solenoids give enhanced performance at higher voltages, and on many small solenoids pull-force can be increased by 50%. Average total power must remain constant, so operating duty cycle should be decreased accordingly, when running a solenoid above its rated voltage.

Pull force levels for this solenoid do not increase linearly with applied voltage, probably due to flux saturation of the steel housing. Smaller solenoids usually benefit more from higher voltage operation because their housings are heavier in relation to their size, and pull-force gains of 50% or more are often possible.

Measuring pull forces

Pull forces from 0.1oz to 50lbs can be conveniently measured with spring balances and elastic threads. Spring and digital scales are used by people fishing to validate their claims, and are sold at hardware and sporting goods stores. Home made scales are easily fabricated using a spring and a ruler, and are more accurate than most ready-made scales, because spring extension can be measured more precisely with a ruler.

Spring scales and elastic threads are first calibrated by measuring their extension with a known weight. Load on a spring scale is linearly proportional to length extension x in Figure 6-31, when extension is no more than 50% of a spring's unloaded length. If a spring balance is not overloaded, weighing accuracies of ±1% are possible by measuring extension carefully.

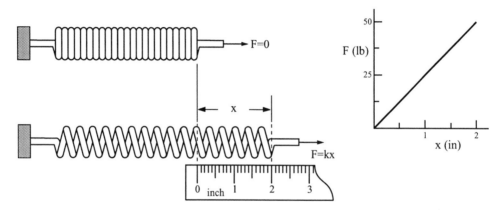

Figure 6-31. Applied force is linearly proportional to spring extension x when a spring is stretched within its elastic limit. Good springs can be extended by 50% of their unloaded length without permanent deformation.

Home made spring scales fabricated using spring data given in Table 6-2, will cover most robotic force measurement requirements.

Table 6-2. Useful data for making load measuring springs

spring type	max load	spring length (unloaded)	coil O.D.	spring material	force constant k= load/extension (x)	
spiral coil	50lb	3.9"	0.6"	0.1" dia. steel	447gm.wt/mm	25lb/in
spiral coil	25lb	2.4"	0.45"	0.061" dia. steel	250gm.wt/mm	14lb/in
spiral coil	5lb	1.6"	0.25"	0.036" dia. steel	110gm.wt/mm	98oz/in
spiral coil	8oz	2.05"	0.242"	0.023" dia. steel	7.5gm.wt/mm	6.7oz/in
spiral coil	2oz	1.25"	0.20"	0.014" dia. steel	3.0gm.wt/mm	2.7oz/in
elastic thread	0.5oz	1.97" (50mm)	single strand	0.020" dia. (0.5mm elastic)	0.60gm.wt/mm	0.54oz/in

Items listed in this table can be extended to 1.5 times their unloaded length, and collectively cover the range from 0.1oz to 50lb. Hardware stores carry a good selection of springs, but *McMaster-Carr* is one of the best sources for all types and sizes of springs. Fine elastic suitable for very light loads is sold at sewing centers, but its fabric covering should be removed for best results. Measurement accuracy of ±1% is possible with springs, and is about ±5% for elastic.

Solenoids and electromagnetic clutches for robotic applications are available from surplus outlets such as *C&H Sales*, or *Servo Systems*, and a few such devices are shown in Figure 6-32.

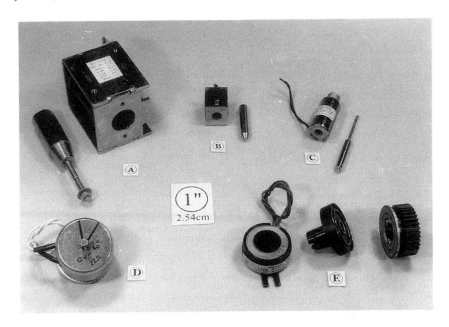

Figure 6-32. Solenoids and clutches are made in many styles and come in a range of sizes. Push/pull forces are specified at plunger shaft extensions, and for the rated operating voltages quoted below.

    [A]   24V (intermittent 40%), R=10.5Ω, 0"=40lb, 0.1"=12lb, 0.5"=12lb, 1"=4lb
    [B]   5V (intermittent), R=5.2Ω, 0"=32oz, 0.06"=12.5oz, 0.25"=3oz
    [C]   12V (continuous), R=41Ω, 0"=16oz, 0.06"=9.6oz, 0.75"=0.8oz
    [D]   12V rotary solenoid, R=22Ω, 40 degree action.
    [E]   Magnetic clutch 24V (continuous), R=120Ω. Static torque=12lb-in, dynamic torque=8lb-in.

Rectangular body solenoids are usually easier to install, because a large hole or bracket is not required on a mounting plate. Rotary solenoids are great for fast flagging operations, eg. *C&H Sales* #SOL9500 $10. Try to select solenoids that operate from 5V, 6V or 12V, and don't forget to install a snubber diode on every solenoid coil, even when testing. Classroom projects can often use 110V ac solenoids salvaged from old washing machines.

Electromagnetic brakes and clutches

Although hydraulic or cable actuated brakes are very effective, electromagnetic brakes are more convenient in some robotic applications. Electric brakes are simple devices, often employing a cylindrical pole magnet design shown previously in Figure 6-19a. A small electric brake made by *WAG* for model aircraft nose wheel braking was originally marketed by *ACE* #25455. With braking torque of 8 in-oz and a 15 ohm coil this unit provided good braking even when operated at 2.4 volts. Currently, aeromodellers seem to favor cable-actuated mechanical brakes, and a servo operated brake is available from Tower Hobbies ROCQ1104 04, $5.79. Electric motors on subway cars are used for both propulsion and braking, and this principle can be applied to model cars and trains too.

    Figure 6-33 outlines the essential features of an electric brake. Although any type of electromagnet can be used to provide some brake effect, a cylindrical pole magnet configuration shown in this figure is most efficient.

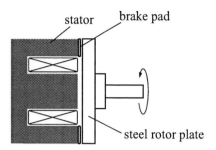

Figure 6-33. A cylindrical pole magnet shown here is the most effective configuration for an electric brake, but most electric magnets can be modified to act as brakes.

An electric brake for stopping a rotating plate is also the basic mechanism of an electromagnetic clutch. The transformation from brake to clutch is straightforward, but requires some thought if optimum torque and efficiency are required. To reduce magnetic flux loss and provide high braking torque, most modern electric clutches incorporate design features outlined in Figure 6-34, and shown in Figure 6-32E.

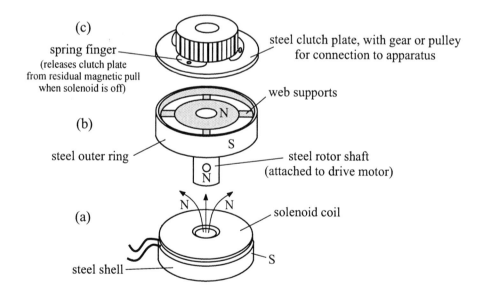

Figure 6-34. Three essential elements of an electromagnetic clutch are shown in this exploded view.
(a) This solenoid coil is always stationary and must be firmly supported. It has central clearance for the rotor shaft and its steel housing completes the magnetic circuit with rotor shaft and outer ring. (b) A webbed support is used to reduce flux loss between rotor shaft and outer ring, because these elements have opposite magnetic polarities. The rotor shaft is directly coupled to a drive motor's shaft. (c) When the coil is energized this clutch plate is attracted to the rotor. A pulley or gear is often attached to this clutch plate and used for coupling to external machinery.

Driving and steering wheeled vehicles

Wheeled vehicles can be driven and steered by applying motive power selectively to chosen wheels. For example, a double-shafted power drive with two electric clutches can be used to drive and steer a full sized wheelchair. Small mobile robots are usually more effectively powered with two motors - one at each wheel. Servo-linked front wheel steering and a single drive motor are used on many four wheeled buggies.

Typical dynamic or slipping torque for a ¼" drive-shaft clutch is about 10lb-in. This is adequate for small vehicles but even on smooth flat terrain, a ½" shaft clutch is the minimum suitable for a full size wheelchair. Surplus electric clutches are available from *C&H Sales*, *Servo Systems* and other surplus supply houses. Prices for a 5/16" shaft clutch with 6lb-in dynamic torque operating from 24V dc at 0.2A, start at $6.50.

Off the shelf clutches range from 2.5in-lb torque for a 1" dia. module with a 3/16" bore, to a 1800 in-lb unit at 8.5" dia. with a 2" bore. These new units operate with drive voltages from 24 to 100V dc. Most surplus clutches operate on 24V dc, so they must be run from a separate battery pack on mobile robots. Bobbins on some 24 volt ready made clutches can be rewound for 5volt or 12 volt operation, simplifying power requirements. If a lathe is available, an electric clutch can also be built from scratch by duplicating features outlined in Figure 6-34. Home made clutches take more effort, but they can be tailored to interface more exactly with other project elements.

Pneumatic, hydraulic and other transducers

Descriptions of vacuum, pneumatic, hydraulic and other techniques useful for hobbyists and students have been reserved for a future book, and include the following:

Powerful moving forces can be produced by bidirectional compressed air cylinders, and these units may be operated under computer control. Air tables, air skids, air tracks and a variety of air film cushion devices can also be used for projects and are described.

Operating principles of explosive bolt cutters, pyrotechnic devices, and hydrostatic switches are outlined. These items may not necessarily be used directly by hobbyists, but their operating features can be applied in many robotic applications.

Heavy spherical objects are easily levitated in a suitable air jet, both directly above the flow, or at an angle, with no apparent support. Techniques for this type of levitation are described

Construction information is provided for an infinite resolution pneumatic displacement sensor capable of measuring to 0.0001".

Details on building a pneumatic gun to launch paper darts at over 400mph (643.7km/h), and techniques for measuring and recording high speed dart velocities with a computer are provided.

Electronic Construction

Equipment requirements

Basic equipment necessary for building robotic projects is listed in Chapter 1, and a good starter kit including a digital multimeter, power supply and tools can be assembled for just over $100. Serious troubleshooting requires an oscilloscope ($325), and a function generator ($239). However a compact generator sold by *C&S Sales* #9600, $29.95 may be used instead, if funds are not available for a more sophisticated instrument.

A hobbyist should keep a small stock of components on hand, and essential items are detailed in Chapter 3. Commonly used resistors, capacitors, diodes, transistors and integrated circuits are required, and these are also listed in Chapter 3. Power rheostats are indispensable for some testing, and a few relays, switches and other items are handy for experimenting.

Special wire, connectors, PCB equipment, perforated breadboards, stand-offs etc. mentioned in this chapter, can be bought as the need arises. Only essential features about these items are included here because better reference works and trade journals are available giving more in-depth coverage.

Chapters 3 and 4 outline some component protection techniques and noise suppression procedures, and these methods will be reviewed here briefly. However, readers will benefit by rereading Chapters 3 and 4 as a memory refresher.

Good technical books often provide answers to difficult problems. Those listed in Tables 3-8 and 3-9 and Appendix B are among the best from a practical viewpoint.

Carbon composition resistors

Most circuits in this book will operate satisfactorily with ¼W 5% (gold band) carbon composition resistors, and these are made in values from 1Ω to 22MΩ. Manufacturing procedures have improved significantly since the introduction of computer controlled production, and many 5% resistors are now within 1% of their designated value.

Carbon resistors have a temperature coefficient (TC) of ~±0.1%/°C, (usually quoted as ±1000ppm/°C). Temperature coefficients for resistors do not always vary linearly with temperature, and this is true for many other physical devices. In fact, over the temperature range from 0°C to +50°C, actual TCs for resistors are usually lower than the specified values. This is fortunate for hobbyists because most of their circuits will be operating around room temperature, so will experience less drift than calculated from TC specifications.

If you want to reduce all unnecessary sources of noise in a special circuit you should refrain from using carbon resistors, because they are inherently noisy. Metal film resistors are a bit more expensive, but worth the extra cost if you wish to attain a really low-noise threshold.

Metal film resistors

Metal film resistors are made by depositing a very thin conducting layer onto a ceramic former. This resistive layer is then trimmed to give the resistance wanted by cutting a groove into the film, either mechanically or with a laser beam. Such procedures not only result in precise resistance values, the technique is extremely flexible and special value resistors can be produced at a reasonable cost. Other benefits also accrue from using metal films, because these resistors have low inductance and may be used in high frequency circuits. By selecting an appropriate combination of metals for the film composition, TCs can be tailored for optimum performance at specific temperatures. Metal film resistors can also be made in the standard cylindrical format, or as Surface Mount Devices (SMD) now used in many commercial printed circuits.

Besides their excellent temperature properties (TC≤±50ppm/°C), metal film resistors have good long term stability and are the best choice when low-noise performance is required. Prices for 1% ¼W metal film resistors start at 20¢ in unit quantities from *Radio Shack*, and ½ watt 2% versions are sold by *Hosfelt* at 6/90¢ in values from 10Ω to 1MΩ. *Hosfelt* also sells NTE metal film resistor kits with 120 resistors in 30 different values. Prices start at $28.69 for their RK02 kit. Several manufacturers have facilities for producing special resistors at reasonable prices, making design of special circuitry easier for professionals.

Most precision metal film resistors are the same size as an equivalent carbon composition resistor, but an extra color band is used to define an additional significant digit on all 1% precision resistors.

DIP, SIP and SMD resistor networks

Single In-line Package (SIP) resistors complement DIP resistor networks, and are available as isolated or bussed networks that plug into solderless breadboards or may be soldered onto PCBs. These items save space when an IC needs a series of pull-up or pull-down resistors, because it is easier and neater to mount and wire one chip than eight discrete resistors.

Good quality SIP and DIP resistors have TCs of ±100ppm/°C and are available in values from 22Ω to $2.2M\Omega$ with each resistor rated at about 0.2W. Individual price for SIPs is about 25¢ and ~50¢ for a DIP resistor. For circuit boards, or whenever space or weight is a factor, these resistor packs are worth the slight increase in cost.

Surface mount resistor networks are widely used in industry but are less useful for the amateur. They have slightly lower power dissipation of ~0.1W/resistor, and TCs of ±100ppm/°C. Cost is higher for SMDs, starting at $1 in unit quantities.

For hobbyists, the biggest drawback with SMDs is their small size. Working with SMD ICs can be a pain in the neck because of their 0.05" pad spacing, and their use is only justified when a particular function is not available in regular size DIP format. For example, NVE's GMR magnetic sensors (mentioned in text following Figure 5-60), and some voice recognition chips are only available as surface mount devices. Two methods for using SMDs in hobby applications are described in Figure 7-17, for those who must work with these devices.

Four, six and eight resistor packs are available in SIP or DIP formats, and eight resistor versions of these chips are shown in Figure 7-1. Some precision resistors and power resistors are also shown in the same figure. Internal network descriptions for DIP and SIP resistor packages are given in Figure 7-2.

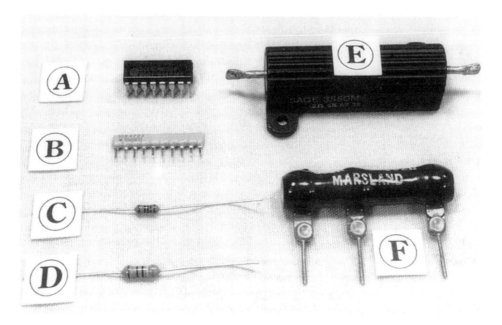

Figure 7-1. Both DIP and SIP resistor networks in [A] and [B] have 0.1" leg spacing, and will plug into solderless breadboards. Precision 1% metal film resistors [C] ¼ watt, [D] ½ watt, can be used at high frequencies. Metal film also gives the best low-noise performance. Both [E] and [F] are wire wound power resistors with good temperature specifications, [E] 50 watt $\pm1\%$, 0.2Ω, ~$5, [F] 5 watt $\pm2\%$, 301Ω, 562Ω, 863Ω, ~$2.

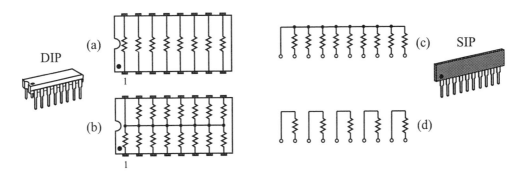

Figure 7-2. [a] DIP with 8 isolated resistors. [b] DIP with 15 common resistors. [c] Bussed SIP networks with 9 common resistors. [d] SIP with 5 isolated resistors. DIP resistors come in 14 or 16 pin, and SIPs are made in 6, 8 or 10 pin packs. Dual line SIP terminators with up to 28 resistors in a package are also available.

Wire wound resistors

Most wire wound resistors have higher inductance than equivalent carbon or metal film resistors, so they should be used with discretion at high frequencies. For example, a 5Ω/5W wire resistor may have an inductance of 0.03mH and this rises to ~0.9mH for a 150Ω/5W resistor. High resistance in a wire wound resistor invariably means it will also have a large inductance. Wire wound resistors with low inductance are manufactured, and these are made using special winding techniques. Wire wound resistors are the most economical high power resistive elements available, and although other power resistors are made, these usually have inferior performance or are more expensive.

Wire wound resistors and rheostats are more difficult to manufacture than most other resistors so their high costs are justified. Precision wire wound resistors are readily available priced from $1 with ±20ppm/°C, and this performance is now duplicated in SMD wire wound components.

Special mixes of metals can give extremely low expansion coefficients and are consequently used to make excellent precision resistors and potentiometers, with TCs as low as ±5ppm/°C. Thin films of these metals are vapor deposited on Kapton™ to fabricate strain and thermal gages discussed in Chapter 5.

Unfortunately power resistors and rheostats are seldom available at surplus prices, and except for homemade power resistors there are no low cost alternatives.

Homemade power resistors

Most high quality wire wound power resistors are fabricated from Nichrome™ (60%Ni, 16%Cr, 24%Fe). This special resistive alloy is made by the *Driver-Harris Co.*, and Nichrome™ is their trademark. Nichrome has about 65 times the resistivity of copper, and for practical purposes its resistive properties are often quoted in Ω/ft for various wire diameters, as in Table 7-1.

Table 7-1. Nichrome wire resistance properties at 20°C.

gage (AWG)	diameter (inch)	(Ω/ft)	gage (AWG)	diameter (inch)	(Ω/ft)
12	0.0808	0.1029	24	0.0201	1.671
16	0.0508	0.2595	28	0.0126	4.251
20	0.0320	0.6592	32	0.00795	10.55

$\rho_{\text{Nichrome}^{\text{TM}}} \sim 1.12 \times 10^{-6}\,\Omega.\text{m}$ $\rho_{\text{copper}} \sim 1.7 \times 10^{-8}\,\Omega.\text{m}$

217

Nickel chromium-wire is very resistant to oxidation at high temperatures and is used for making heaters in hair dryers, or in ribbon form for toasters and clothes irons. Wire salvaged from these appliances can be used to make power resistors, and when heated is useful for making precision cuts in Styrofoam™.

Nickel-chrome alloys cannot be soldered, but homemade resistors can be fabricated by first crimping heater wire in brass hobby tubing. Connectors or flexible leads are then easily soldered to the brass tubing.

Crimping is best done by clamping both wire and tubing in a bench vise, but fine gage wire ($\leq$24AWG) requires a different technique. Thin heater wire is first wound tightly onto a well-tinned copper or brass rod/tube, leaving gaps between the wire turns. After making this good mechanical connection, the joint can be covered with solder. A heater wire connection made in this way can be operated so the wire runs red hot with no difficulty.

Bare Nickel-chrome wire suitable for homemade resistors is sold by *McMaster-Carr* in ¼lb spools, and is available in 16 gages from 14 to 40 AWG. A few examples: 14ga. 8880K11, 20ft/spool, $11.61,   26ga. 8880K24, 315ft/spool, $16.48,   40ga. 8880K39, 8352ft/spool, $44.29.

"I need a 107.3Ω resistor"

Manufacturers can supply any resistor, with a size, power rating, precision, TC etc., compatible with current technology, and the cost is not prohibitive (or cheap!). Yet for most hobby work, the precise value of a resistor is seldom important, and usually a tolerance of ±5% is ok. Using 1% resistors for most hobby circuits is overkill, and a waste of money, try to use the cheapest component that gives satisfactory performance. For those special cases where you need to generate a stable, well-defined voltage reference, just use a 317 or 337 regulator adjusted with a good quality trimpot.

When a resistor gets hot its resistance changes, and this sometimes creates more problems than when it blows completely. Even slight overheating can be a problem if a resistor is a critical element in a low noise circuit, because electronic noise in a resistor increases as a resistor warms up. Thermal or Johnson noise in a resistor scales as $T^{\frac{1}{2}}$, where T is the absolute temperature in K. At room temperature, noise from a 1MΩ resistor is about 4µV for a circuit bandwidth of 1kHz. This is insignificant for most hobby work.

Generally resistor values vary with precision, so if you are looking through a catalog for the first time you may find listings confusing. Intervals between resistor values may seem odd, but they enable resistors to be combined more readily when making intermediate unlisted values.

Because of the wide choice in resistors, it is difficult to find a reference text satisfactorily covering resistor power ratings, precision, size etc. in a useful manner. Catalogs from *Electrosonic* and *Newark* have excellent specification listings on numerous resistor types, and are a good initial source for resistor data.

For hobbyists and professionals, good electronic catalogs are a blessing since they succinctly summarize manufacturers' data in a methodical manner. Their organization of an enormous amount of material in a user friendly way means these volumes are one of the best, (and certainly one of the least expensive) information sources for an experimenter.

Potentiometers

Resistor characteristics also apply to potentiometers, consequently carbon composition pots are the least expensive, and have the highest TCs. Most trimpots use cermet elements with TCs of ±100ppm/°C or better, but single-turn inexpensive round pots are usually carbon composition (TC ~±1000ppm/°C). The best potentiometers have wire-wound elements (~±20ppm/°C), and prices for these items start at $10 even for surplus items.

For most hobby work cermet elements are fine, and a multi-turn trimpot is hard to beat at $1.50. You will find a vertical adjustment trimpot style shown in Figure 3-1 most convenient for use on solderless breadboards.

Questions 7 and 8 in Chapter 2 cover some important points to consider when using potentiometers, and may save a few dollars by preventing some very common mistakes.

Remember 'linear' potentiometers are never perfect, and if you try to use just a small portion of track you may be disappointed with their accuracy. However, when a pot is used over its full scale range, it will meet its specifications.

Wire wound pots can have very high inductance (>1H in some cases). This high inductance coupled with a large stray capacitance (also present in many potentiometers), can cause undesirable circuit oscillations. These comments are not meant to imply wire wound pots are inferior. On the contrary, some wire wound potentiometers are among the most precise and accurate electronic elements manufactured.

High voltage resistors

Is it possible to measure the potential of a 100kV dc source using the circuit of Figure 7-3?

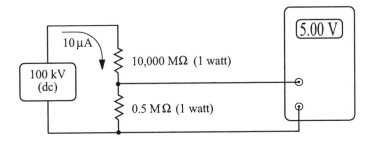

Figure 7-3. High voltage measurements require special techniques described in the text.

Although 1 watt resistors can handle the power, standard resistors cannot be used in Figure 7-3, because these ordinarily have a case length ≤2". In this divider almost 100kV will be applied across the top resistor, and flashover will occur as current almost instantaneously creates a conducting track along the resistor's surface.

Even good insulating surfaces collect dust and grime, so a safe rule of thumb is to allow 10kV/inch in situations where current can creep along the surface of an insulator. For safety, the top resistor in Figure 7-3 should be about 10" long!

Figure 7-4 shows a 15kVdc high voltage probe with a 6" long 300MΩ resistor protected by a clear plastic case. This keeps the operator's hand at a safe distance, and a high value resistor limits current from the source to 50μA (about 100 times below the lethal limit).

High value resistors

A 10,000MΩ resistor encapsulated in glass is shown in Figure 7-4C. Even a fingerprint on its outer envelope can compromise this 1% precision *Victoreen* resistor, by providing an alternate lower resistance path between end electrodes.

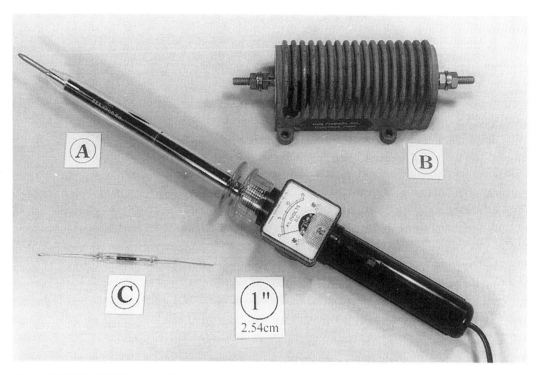

Figure 7-4. [A] This 15-kilovolt dc high voltage probe has a 6" long 300MΩ resistor, encased in an insulating plastic shroud. For operator protection a high value resistor restricts the measured current to 50μA and its length prevents flashover. [B] This 385Ω, 250watt resistor may seem large, but much larger power resistors are made. [C] Very high value resistors are enclosed in a glass envelope to avoid contamination. This resistor is 10,000MΩ (10^{10}Ω), and made by *Victoreen*.

Power resistors

Whenever a power resistor is employed, we should always ask - "is this really necessary?", because useful energy is dissipated as heat. Hot transistors waste energy and power MOSFETs with their low 'ON' resistance can often improve operating efficiency. When a motor's speed is regulated with a series rheostat, we have similar energy waste, and this can sometimes be avoided by using a switching solid state controller. Ballast resistors in older fluorescent lamp fixtures are now being supplanted by electronic switching controls, giving considerable energy savings in large office buildings. Sometimes using a power resistor is unavoidable and often expedient, but it is not always the best solution.

Fixed value wire-wound resistors are stocked in 2,5,7,15,25 watt ratings by *Hosfelt* at reasonable prices. Five watt resistors from 1Ω to 8.5kΩ are 19¢, and 1Ω to 1kΩ are 49¢ for 25W. An adjustable 10Ω 175W slide resistor 81-103 is $8.95 from the same source.

Capacitors

Capacitor kits are sold but (to the author's knowledge), none are available to suit a roboticist's needs. It is more economical and efficient to buy those capacitors you are most likely to use, and some of these are listed in Chapter 3.

Because so many varieties of capacitors are sold, it can be confusing for hobbyists and students who are building their first project. Electrolytics, ceramic, mica, polystyrene, tantalum, polyester, Teflon etc. give a beginner too many choices. If one then looks at size, dissipation factor, insulation resistance, voltage capability, leakage and frequency response, things get worse.

Fortunately things are not as bad as they seem, and just 4 types of capacitors listed in Table7-2 will cover most robotic needs.

Table 7-2. Capacitors for robotic projects

	capacity	tolerance (typical)	temp coeff ($\pm$ppm/$^\circ$C)	breakdown leakage	uses characteristics
ceramic disk [a]	1pF - 0.1μF	$\pm$10%	30 - 5000	50V - 1000V <0.01μA	Decoupling power rails, motors. High freq. signal coupling. Low cost NPO are very stable
electrolytic*	0.47μF - 0.15F	$\pm$20%	~500	10V - 450V 3μA - 4000μA	Power line filters, energy storage. Photo-flash, RC timing. Low cost, high inductance, high capacity per unit volume
tantalum*	0.47μF - 330μF	$\pm$20%	~200	6V - 60V 0.5μA -10μA	Blocking, bypassing, RC timing. Mid price range. High capacity per unit volume
polystyrene	100pF - 0.1F	$\pm$5%	50	25V - 500V <0.01μA	Excellent temperature stability. Low leakage, inexpensive. Low inductance. Low capacity per unit volume

* For safety these capacitors must be connected to the proper polarity or they may explode (Yes even at 12 volts). Non-polarised tantalums and electrolytics are made but these are rarely used for projects.

[a] Ceramic capacitors have been given a bad rap, because their leakage is not as bad as often suggested. Choose mylar or polycarbonate for sample-and-hold circuits, or long period integrators.

Diodes, transistors and ICs

Only a few semiconductor devices are required to make a good start in building robotic control circuits, and these items are covered in Chapter 3. Some guidelines for testing transistors and diodes, and using transistors or power MOSFETs for control functions are also given in Chapter 3.

Additional information on transistors, op amps, diodes, power MOSFETs, relays, resistor and capacitor codes, wire gages, inductance formulas and other useful data, is provided in a future book.

Static sensitive components can be stored in conducting foam or anti-static plastic shipping tubes. However, these methods are ponderous and time consuming for a large and varied stock of parts. The technique shown in Figure 3-16 is superior, and safe when recommended handling procedures are followed.

Inductors

Inductor use in most hobby circuits is restricted to tank circuits, power line filters, and for frequencies above ~1MHz where inductor sizes are small and inexpensive. Traditionally inductors have been underused, because equipment for measuring inductance has been costly and cumbersome. This situation has now changed and a $75 DMM that measures inductances to 20H is described in Chapter 1. This meter from *C&S Sales* LCM-1850, also measures resistance, voltage, capacitance and checks diodes and transistors. Hobbyists and students can now design LCR circuits more easily using this type of meter.

Properties of LC filters can be predicted from reliable, well-tested design equations, given in many electronic texts. Most public libraries have copies of the *'ARRL Handbook for Radio Amateurs'*, and this book gives a feel for the potential uses of LC filters. For those who love to fiddle, *Radio Shack* sells a 30 piece inductor kit for $2.99, #273-1601. *Hosfelt* stocks fixed inductors, chokes, ferrite beads, rods, toroids, and ferrite cored variable inductors, all starting at 20¢.

Making fixed and tunable inductors using ferrite slugs is straightforward, and practical guidelines for making these items and conventional or toroidal transformers are given in the *ARRL Handbook*.

    Air cored inductors are extremely stable, and have many uses. Stiff paper or cardboard makes a good coil former, and PVC pipe is acceptable. Professional high Q radio frequency coils are sometimes wound on slim polystyrene support rods, but this method needs special winding equipment.

    An air core inductor can be designed using Equation 7-1. For a 1000 turn coil with a 1sq.in. cross section, $\mu_{air} \sim 4\pi \times 10^{-7}$ Wb (amp.meter)$^{-1}$, N=1000, A=6.45x10^{-4}m^2, l=2" (5.08x10^{-2} m), and L=16mH.

$$(7\text{-}1) \quad L = \frac{\mu N^2 A}{l} \quad = \quad \frac{4\pi x 10^{-7} x 10^6 x 6.45 x 10^{-4}}{5.08 x 10^{-2}} \quad = \quad 16mH$$

Equation 7-1 is valid only if $l/r \gg 1$, so coil inductance will not be exactly 16mH because $l/r \sim 3.5$ in this example. Inductance can be increased substantially for air coils by using a ferrite or powdered iron core.

In classrooms, students usually favor active filters because of their versatility and wide operating range.

An excellent book covering all essential features of active filters is:

        Active Filter Cookbook,   Don Lancaster, $29.95 Sams, 21168

Wire and Cable

"Why do I need both solid and stranded wire?"

Because student projects are frequently transported between homes and classroom, faults due to wire breaks are not uncommon. Most of these breaks occur where insulation is stripped from solid core wire, and wire has been nicked by a stripping tool. Subsequent flexing causes a break in the wire, often with surprising ease, and intermittent contacts are sometimes formed at these breaks. These 'intermittents' are hard to find, and difficult to fix when buried in a mass of wires that may develop more breaks if moved.

    Solid core insulated hook-up wire has definite merits, but is grossly overused by beginners. Solid wire definitely has the edge when beautiful bends or artistic wire routing is required, but it should never be used on connectors or anywhere flexing is likely to occur. Repeated bending of copper wire work-hardens the metal, making it more susceptible to breakage.

When to use solid wire

(1)   Use insulated solid copper hook-up wire for interhole connections on solderless breadboards.

Precut wire kits for solderless breadboards are useful, but your circuits will be neater if connections are made with wires cut to the exact length from bulk color coded #22 or #24 solid wire. For example *Newark* sells #24 AWG solid telephone hook-up wire (red, black, yellow and green wires in a sleeve) #89F868 500ft, $29.07. This quantity will last a lifetime, or you can share it with friends. Telephone wire can also be obtained free from a friendly telephone repair person, because a great deal of old wire is discarded during new installations.

Strip and bend one end of the wire and insert this into a breadboard socket. Then indicate the spot where the other end should have its insulation removed, by gently marking it with the tips of wire cutters. Always use proper wire strippers when removing insulation. This does a neat job, and with practice you will learn how to remove insulation without marking a wire's core, (an important skill).

(2)   Solid wire can be used for supporting heavy components.

Heavy gage solid wire can be used to support high power transistors and other large devices on plain or solderable perf. board (eg. 3"x4¼" *Hosfelt* JALPC-4 $1.75). For these tasks use tinned solid copper bus wire, or insulated solid wire in sizes from #20 (0.032" dia.) or larger. If very large currents are involved use solid copper residential house wire, sold by the foot at hardware stores. Tinned #24 copper wire is sold in 50ft rolls by *Radio Shack* 278-1341 $1.79, and all sizes are stocked by *Newark*.

When to use flexible or stranded wire

(1)   Use stranded wire for interconnecting individual solderless breadboard circuits.

Flexible hook-up wire or ribbon cable should be used to connect solderless breadboards or other circuits together. Solid wire should only be used for interconnecting breadboards when they are securely held on a common baseboard.

An excellent flexible wire that strips readily and solders easily, is a #24 gage wire composed of seven strands of #32 tinned copper wire (Belden 8525). This wire is sold by *Newark* 37F641WA, 100ft, $19.93, but any similar wire is suitable.

(2)   Use stranded wire for connectors, power leads, motor leads, battery connections, relays, solenoids, and other devices that may be moved frequently, or subjected to vibration.

A small #26 wire may be composed of 7x34 gage wires but if high flex duty is a required 10x36 gage gives longer service. For high current heavy flexing duty a #12 wire may be constructed from 250 strands of #36 wire. A #8 133x29 wire is sold by *Hosfelt* #60-352 at 45¢/ft. 'Monster'speaker wire has excellent flexibility and costs about $1/foot for twin #12 cable that can carry 20 amps.

Where continuous duty demands the ultimate in conductor flexibility, flat plastic tapes coated with parallel thin film conductive strips are employed. Such tapes can be salvaged from dot matrix printers or photocopy machines.

Although solid wire use must be restricted to certain tasks, flexible wire can be used in any application. Flexible wires can be twisted together by holding one end of a bundle in a drill chuck. The twisted wires can be prevented from unravelling with a heat shrink tubing ligature, as shown in Figure 7-5 (B)

Choosing the correct wire size

If you select a larger diameter wire than necessary your project will work, but heavy wire can be difficult to handle and may place unnecessary mechanical stress on small parts. On the other hand a wire that cannot carry the required current will drop excessive voltage, and may overheat or cause a fire. For most robotics wiring #20 gage wire is fine, but if large currents are involved then choose a wire according to Table 7-3.

Table 7-3. Current carrying capacities of insulated copper wire in free air at 25°C/77°F

gage (AWG)	diameter (inch)	Ω/1000ft	capacity (amps)	gage (AWG)	diameter (inch)	Ω/1000ft	capacity (amps)
#4	0.204	0.25	70	#20	0.032	10	2
#8	0.129	0.63	35	#24	0.020	26	0.5
#12	0.081	1.6	20	#28	0.013	65	0.15
#16	0.051	4	6	#32	0.008	164	0.08

Ribbon cable

Flat ribbon cable is ideal for multiple low current leads since it is readily traced, and makes for neater assemblies. Surplus stores sometimes sell old ribbon cables with female header connectors still attached, as shown in Figure 7-5 [C]. These connectors have 0.1" spacing and can be attached to solderless breadboards with a double-ended male header shown in the figure. Extending these cables is sometimes easier, neater, more reliable, and less expensive than attaching new connectors to a ribbon cable without the proper tool. Hosfelt sells 2½", 34 pin headers #91-169 $1.79. A Radio Shack connector tool 276-1596 at $12.99, works better than clamping header connector parts in bench vise jaws.

Most ribbon cable is 28 AWG and can only handle about 200mA a line, because it is designed for low current digital signals. Rainbow ribbon cable is not only colorful, it makes wire tracing a breeze. It is more expensive than gray ribbon cable, but like gray cable is also sold in 10 to 64 conductor sizes, with 0.05" spacing. Try to use colored ribbon cable in a logical manner, by assigning the numerical value for each color to appropriate pin numbers or functions. This makes it easier for a stranger to fault-find a circuit. *Hosfelt* stocks 20 to 60 conductor gray ribbon cable, $2.90/10ft to $8.60/10ft, it is also available in 100ft lengths.

Wire wrapping wire

Wire-wrap is an excellent technique for bread boarding prototype digital circuitry. As a rule it should be reserved for digital work because 30 gage wire (the most popular size) is a bit too small to carry analog currents without excessive voltage drops. Fifty foot spools of #30 wrapping wire are sold by *Radio Shack* for $2.79, red 278-501, white 278-502, blue 278-503.

Wire wrap IC sockets with long pins are inserted into 0.1" spacing perforated board, and point to point connections made using a wire wrapping tool. One of the best wire-wrap hand tools is sold by *Radio Shack* 276-1570, $7.49, shown in Figure 7-5 (D). Insulation is easily removed using a stripper (item E in the figure), normally concealed in the tool's handle. Wire can be wrapped or unwrapped with this versatile tool.

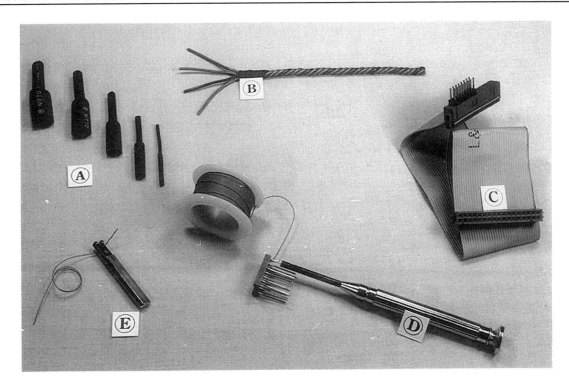

Figure 7-5. [A] Heat shrinkable tubing contracts to about half its original diameter when heated with a heat gun, or a match flame. [B] Flexible wire leads can be twisted together in a drill chuck, and ends secured with heat shrink tubing. [C] Ribbon cables with connectors can be salvaged from computer drives, and are also sold at surplus outlets. These connectors can be attached to a solderless breadboard by using a double male header (shown inserted in the left part of the upper connector). [D] Wire wrapping wire stripper sold by *Radio Shack* 276-1570, $7.49, can be used to wrap or unwrap wire. [E] This wire stripper is normally stored in the handle of the tool in [D].

Safe operating temperatures for common wire insulations

Maximum recommended temperatures for common wire insulations are listed in Table 7-4. Other materials such as polyethylene, silicone rubber, Kapton ™ and rubber are used for jacketing wire, but those listed in this table are most often used by hobbyists for robotics work.

Table 7-4. Maximum operating temperatures for wire jacket insulations

	maximum operating temperatures
black PVC	-40°C to +100°C
Teflon™ (TFE type)	-55°C to +260°C
fiberglass	-55°C to +250°C
magnet wire	-40°C to +200°C

Hobbyists should stick to tabled recommendations when working at 115 volts or higher, but wire can be safely used at higher temperatures with lower voltages. For example Kynar covered 30 gage wrapping wire was used for oven experiments at 250°C in Figure 5-34, well above its rated limit of 135°C. Similarly, magnet wire and Teflon insulation can be used above 250°C at low voltages. A test is the best method for establishing safe operating temperature limits for wire insulations.

Magnet wire

Magnet wire is used in the manufacture of transformers, electric motors, solenoids, relays, speaker coils and a host of other devices. Consequently, magnet wire can be obtained in a variety of diameters by selectively salvaging these items. *Radio Shack* sells three spools of magnet wire 22ga. (40ft), 26ga. (75ft), 30ga. (200ft), 278-1345, $3.99.

Insulation on magnet wire has some impressive properties. It is a rugged protective varnish that resists most solvents, and can be removed only by abrasion or with special chemicals (eg. *G.C.* Strip-X). Some magnet wire insulation is ultra high vacuum compatible, can withstand 750°C, and may be repeatedly flexed without cracking. Powerful electromagnets are sometimes constructed with a rectangular cross section wire, because this can be closely wound without leaving gaps between adjacent turns or layers. Coils wound with this wire provide higher current density, and can be cooled more effectively than those made with circular cross section wire.

Solder-strippable magnet wire can be tinned directly with a soldering iron if rosin flux solder is used. Acid flux or acid core solder are both unsuitable for solder-strippable magnet wire, since neither removes the protective coating.

Insulation on all types of magnet wire can be removed with a utility knife. Use either the sharp edge, or back of a blade to scrape insulation until the underlying copper is clearly visible. It may then be tinned with solder, or mechanically connected as required.

Project wiring tips

For some hobby projects wiring layout is unimportant. This means you can have a rat's nest of wires that crisscross haphazardly, yet everything will work just fine. Naturally a great deal of time is wasted fault finding such circuits, and it is better in the long run to have a neat layout using different colors for certain wire functions. A choice of red and black for positive and negative power leads makes sense, and since green and ground both start with the letter 'G', this will be easy to remember.

As operating frequency in a circuit rises greater care must be taken with wiring, or things will not work properly. You might like to take another look at Question 21 in Chapter 2, which illustrates how capacitive coupling can cause difficulties. If you are using motors, solenoids or relays, these must be decoupled or clamped (Questions 24, 25), because it is possible to destroy valuable equipment if appropriate safety measures are not employed. To avoid some common difficulties, power rails and individual ICs need special attention shown in Figure 3-14. In this chapter we will see that line drivers and line terminations may also be required at times.

Supply leads for motors and other inductive devices will radiate less noise at radio frequencies if they are twisted together to reduce their inductance, (use a hand drill for long leads). For severe pickup problems, leads can be placed inside a braided shield or sheathed in aluminum foil. Shields should always be connected to power ground for maximum effect.

It might be some consolation to know that even experts cannot build a complex high frequency circuit for the first time without difficulties. They may have to move a few wires around, add a capacitor here and there, shield a line or several components, add extra ground wires or a ground plane, twist a pair of leads, use a ferrite spike grabber, install an input line filter, try an ac line isolation transformer, freeze circuit components, heat circuit components, switch components . . .

This is not intended to be an intimidation list, it is given to show that no one is perfect. Successful robotics designs are not divinely inspired. They develop asymptotically to an acceptable level of 'perfection' by diligent hard work and a large dose of 'Suck it and see'. So if you find yourself fiddling, it may be comforting to remember that every practical robotics engineer does this too sometimes.

"Why do I need a larger gage wire for a common ground line on my power supply?"

A power supply's common/ground line must carry the sum of currents from all other polarities. For example, on a supply with +12V 3A, -12V 2A, +5V 8A, -5V 2A a common line carries 15 amps, or nearly twice the capacity of any other line.

It is tempting to use a light wire gage for power leads, especially when this is the only wire on hand. This can cause serious problems if a circuit draws large currents in either dc or pulse mode. An example in Figure 4-5 illustrates one difficulty that can arise if #24 AWG wire is used to carry 10 amps. This situation can happen quite unwittingly as explained in the text which accompanies that figure. Information given in Table 7-3 is useful for selecting the correct wire gage, and forestalling such problems.

Power leads are usually sized for average power requirements. However, current pulses develop larger voltage spikes on small diameter wire, than on leads designed to carry higher currents.

Coaxial cable and twisted pairs

If your project fits on a normal tabletop then no special wiring protocols are needed for running under computer control, provided recommended procedures regarding wire gage and decoupling are followed. Buffers should be installed in all signal lines longer than about 2ft if these go to a computer or microprocessor. This will clear up anomalous behaviour as described in text following Figure 3-18. When very long wires are used to transmit high frequency or timing signals, some attention must be given to the type of wire used and how it is terminated.

Students and hobbyists should have some understanding of how high frequency signals behave when they are sent from one point to another over wires or a transmission line. Although signal propagation in transmission lines is a fascinating topic it is outside the scope of this book, and we will only deal with simple practical considerations. For those interested in transmission line effects and their interaction with microprocessor busses, a very good account is given in the following book:

Analog Electronics for Microcomputer Systems, Goldbrough, Lund and Rayner. Sams 21821 (1983), ~ $40.

Transmission line signal distortion at high frequencies

When a 1MHz square wave is sent along a long length of wire it may arrive distorted and attenuated as shown in Figure 7-6. Only two cases in that figure show the original signal transmitted with negligible distortion, and this is achieved using coaxial cable terminated by its characteristic impedance.

Long wires always present difficulties for an experimenter working with high frequencies. A circuit that works fine with 12" leads may exhibit erratic behavior if these leads are extended to 30". This is because signals do not travel instantly from one end of a wire to the other. They move at about 65% of light velocity, or about 8" a nanosecond.

Upper traces in Figure 7-6 show how rise and fall times are degraded as wire length increases. For clarity only a 15ft example has been shown for the twisted pair. High frequency signal fidelity in all lower traces is improved by installing a 50Ω resistor across the wire ends. Improvement is most noticeable with coaxial cables (a) and (b), but a twisted pair in (c) benefits too. Note that when a terminating resistor is added signal amplitude drops due to loading, and post amplification may be necessary if CMOS or TTL triggering levels are required.

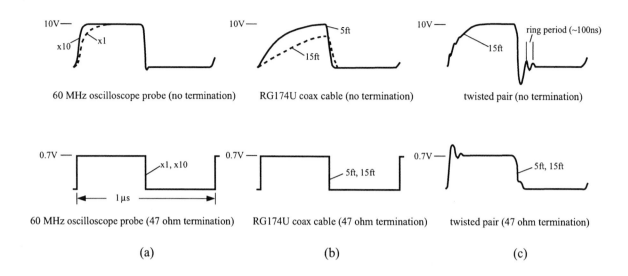

Figure 7-6. High frequency signals are not always transmitted faithfully over long wires, and only the lower traces of (a) and (b) depict true renditions of the original input waveform. Twisted wires in (c) are inferior to a coaxial cable but do an acceptable job if a suitable termination resistor is employed. Quarter watt 47Ω resistors are used across wire ends as parallel terminators in Figure 7-7. Input signals for these tests were derived from an SN7426 high voltage NAND gate sourcing 10mA into 1.5kΩ.

Rise and fall times are degraded in the upper three traces due to signal integration by distributed capacity in the wires or cable (see Question 16 in Chapter 2). Modulation or ringing in (c) results from signal reflections or echos, created as wave energy bounces back and forth along the cable. Wave velocity in a copper cable is about 65% of light velocity (c~1ft/ns). Therefore, echo transit time is about 100ns over a 15ft line, as indicated by the ring period in (c). Ringing is negligible in a properly terminated line because a wave dumps all its energy into a matched load resistor, and very little energy is reflected.

Installing line terminating resistors

Switchable 50Ω terminators are provided on some oscilloscopes, and are convenient for checking signal fidelity when using 50Ω coax cables. Quarter watt 47Ω or 51Ω resistors are suitable for terminating signal lines, and these can be installed on a solderless breadboard as shown in Figure 7-7. Solderless breadboards can only be used to about 10MHz because of capacity between adjacent sockets (~5pF).

When a terminating resistor must be spliced into a line (eg. at a scope's input), a BNC Tee or an RCA jack may be used, as shown in the figure.

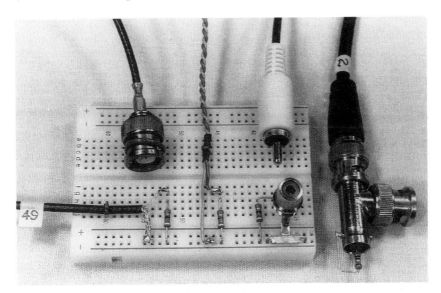

Figure 7-7. Terminating resistors can be installed on a breadboard, or attached to a BNC Tee connector. Left to right: RG174U coax with a 47Ω resistor installed at the cable's end. A twisted pair and an RCA socket with onboard terminations. A resistor soldered directly to a BNC Tee connector for terminating a coax cable. Put a dab of solder on the resistor's tip so it makes good contact with the Tee's center conductor.

Coaxial cable characteristics

A few useful characteristics for some common coax cables are presented in Table 7-5. Capacitance for a twisted pair depends on wire size, insulation diameters and also twist pitch, but is about 30pF/ft for 20AWG Belden 8502 stranded wire.

Table 7-5. Coaxial cable and twisted pair characteristics.

	O.D.	capacity	max volts (RMS)	impedance	propagation velocity*	cost (approx)
RG58U	0.195"	28.5pF/ft	1400	53.5Ω	0.66c	15¢/ft
RG174U	0.100"	30.8pF/ft	1500	50Ω	0.66c	25¢/ft
21-597 *Amphenol*	0.150"	20pF/ft	2500	75Ω	0.66c	18¢/ft
twisted pair		~30pF/ft			~0.6c	

* RG62U foamed polyethylene core coax has a velocity of ~0.84c (c=light velocity ~1ft/ns)

Male and female RCA connectors (*Hosfelt* MS14B 25¢, MS15B 25¢), are easily installed on coax cable for high frequency projects. RG174U is a small diameter coaxial cable (*Hosfelt* #60-238 19¢/ft), and ideal for hobby projects or making video cables. Ready-made video cables are available from *Hosfelt* starting at $1.25 #894-RG, for a 6ft cable made with RG59U.

Attaching BNC connectors is not difficult, but is rarely done successfully at the first attempt, so it is easier and sometimes cheaper to buy ready-made coax cable assemblies. *Hosfelt* sells ready-made RG58 cables with male BNC fittings at both ends in 3,6,12,25 ft lengths from $3.50, to $8.95, #60-232 to #60-235. They also carry a large selection of BNC and UHF connectors and other fittings at bargain prices.

For those well-heeled hobbyists demanding high frequency performance in the smallest sized package, miniature coaxial cable and compatible connectors are manufactured and sold by *Microdot*.

Cable ties and clamps for tidy wiring

Wiring always looks better when it is neatly routed and tied, but this does nothing to improve performance, and minimal marks should be assigned for such neatness. Functionality must always be the preeminent concern, because if a project is neat but does not work it has not achieved its goal. Top projects not only work well, they are usually neat too.

Radio Shack, Hosfelt and other suppliers carry a variety of wiring fasteners, cable ties, clips etc. Black or white 4" self locking cable ties, 44-132 or 44-133 are 100/$1 from *Hosfelt*. Wires can be held down with clips or tied to supports with Velcro™ strips, cable ties or tape, and these items are available from Radio Shack or Hosfelt.

Bands of ¾" black insulating tape wound around wire bundles every 6" or so, also do an acceptable job. Black tape becomes tacky after a few months, but will look good for 'show day'.

Good looking home-made cable straps can be cut from plastic bottles with scissors. They can be pierced with an awl or a nail for screw mounting, or attached with sticky foam tape. These easily made clamps can be cut to fit snugly around any sized wire bundle and cost nothing.

Wire markers

Some projects have literally hundreds of wires, and may need tagging unless you have a photographic memory. Power leads should always be identified, just in case your buddy decides to hook things up when you are not around.

Special wire identification markers are sold, and these are excellent for professional use, but amateurs can get by with masking tape identifiers, rolled around a cable or as a tag, shown in Figure 7-7. If you are fussy, laser-print special labels on white paper, then attach these with transparent tape.

Skin effect and Litz wire

Students are sometime puzzled by fine multi-strand wire windings used to make high frequency inductors. Is this black magic, or is there really a difference between a solid wire, and multiple wires with the same current carrying capacity?

At high frequencies current flow generates self-induced emfs, forcing most current to flow through the outer surface of a solid wire. This phenomenon significantly increases a wire's resistance at 100kHz, but for a ¾" diameter conductor, resistance at 50Hz is only 2% higher than its dc value. At 1MHz all current is forced to travel through an outer layer or 'skin' about 0.066mm thick, and this decreases to ~0.0025mm at gigahertz frequencies.

'Skin depth' (δ) is that level below a conductor's surface where current density has fallen to e^{-1} (~0.368) of its surface value. For example, skin depth of a copper conductor at 60Hz is $\delta_{skin} = (2\rho/\mu\omega)^{1/2}$, or about 8.5mm. For copper, δ_{skin} can be calculated using the following parameters:

$$\text{copper resistivity } \rho_{Cu} = 1.7 \times 10^{-8} \ \Omega.\text{m}, \qquad \mu = 4\pi \times 10^{-7} \ \text{H.m}^{-1}, \qquad \omega = 2\pi f$$

Litz wire is made from many strands of insulated wire that are carefully interwoven, then connected in parallel at their ends. Multiple wires have a larger total surface area than a single wire with the same overall cross section, so Litz wire neatly solves high frequency resistance loss up to a few MHz. Litz wire is also flexible, and manufactured in a wide range of wire sizes and strand combinations.

Skin effect is also responsible for the use of circular or rectangular waveguide tubes seen in gigahertz equipment. These guides are often silver plated, because conductivity for silver is about 6.5% better than copper, and only a very thin coating is necessary.

Heat shrinkable tubing

Heat shrink tubing is very useful stuff and is used to make tamper-proof seals on food produce jars, and medicine containers. Shrink tubing is readily available from 3/64" to 1" diameter and much larger sizes are made and widely used in food and liquor industries. Heat shrink tubing forms a tight but semi-flexible seal over wire, metal, wood etc. It has a shrink ratio of about 2:1 and can be heated with a heat gun (an electric paint stripper ~$25 is ideal), or even with a cigarette lighter or match flame.

Light duty flexible shaft couplings for small motors can be formed with shrink tubing, which is especially handy for this purpose when two shafts of different diameters must be joined. Heat shrink tubing also makes a neat protector around joints, where leads are soldered to connectors. A shrinkable aluminum/PVC laminate tubing is sold for shielding purposes, and shrink 'boots' are used on very large connectors to prevent wire breakage during rough handling.

Shrink wrap is widely used for covering crimp lugs, terminals and splices. Hard shrink collars on these items are rated to 180°C, and are very rigid. Flexibility in shrink tubing is often overlooked, but shrink materials are made in ten grades - from very flexible to most rigid. Some properties of heat shrinkable tubing are summarized in Table 7-6.

Table 7-6. Heat-shrinkable tubing properties

	Teflon™	**Polyolefin**	**PVC**	**Kynar**	**Neoprene**
dielectric strength	1500v/mil	≤1000V/mil	≤1000V/mil	≤900V/mil	≤900V/mil
shrink ratio [a]	2:1	2:1	2:1	2:1	2:1
shrink temperature	~300°C	~125°C	~100°C	~175°C	~350°C
flexibility [b]	2 to 6	2 to 7	5	10	1
price (approx) [c]	$3.67/ft	60¢/ft	80¢/ft	$1.90/ft	$4/ft

[a] Shrink ratios to 6:1 are possible   [b] 1= maximum flexibility   [c] *Newark* price per 100ft

Top quality heat shrinkable tubing is about 20¢/ft at surplus prices (eg. 10ft of ⅛" diameter #33-132, *Hosfelt*, $2.25). Lower grade tubing is satisfactory for hobby projects, and is discounted as low as 4¢/ft in Toronto. An assortment of shrink tubing sizes in different colors is available from *Radio Shack* $2.19 for seven pieces, 278-1610.

Adhesive lined shrink tubing is available, but a dab of epoxy or crazy glue also works well for sticking shrink tubing to motor shafts and other tasks. Although hot or cold plastic fusion tape is sold, its cost is rarely justified for hobby work. Regular ¾" black electrical tape can be stuck to a glass plate, then sliced to convenient widths for covering oddly shaped connections.

Soldering techniques for electronics

Every roboticist should be capable of making good electrical connections by soldering. But good soldering technique goes far beyond this simple task. By developing your soldering skills you will be able to fabricate many mechanical items that are often constructed with glue or nuts and bolts. Good soldering technique for roboticists is so important it is also covered in Chapter 8, where heavy soldering methods are discussed. Learning to solder well is not difficult. Like anything worthwhile it takes time and practice to become really proficient, but dividends are worth the effort.

Equipment for soldering

Beginners will be astounded at the sophistication employed for electronics soldering. Temperature controlled hot air/soldering iron stations with mouth-watering features have prices that top $2000, and some surface mount stations cost more than $5000. Desoldering stations meeting military specifications, can handle surface mount devices or plated through holes on PCBs. These often have their own continuous vacuum pumps. Cordless and butane irons can be used when an ac plug is not available, while solders, creams, pastes, and wicks need several pages for complete listings in a good catalog. In mass production facilities, manufacturing an eight-layer circuit board using silk screening, wave soldering, ultrasonic cleaning etc. is an impressive sight.

Fancy equipment is essential for professionals where time is money. However, with care almost any soldering task can be done by an experienced worker, using inexpensive basic equipment described below.

An iron, stand and sponge recommended in Table1-1 at $5 are adequate for all small soldering jobs encountered when building projects. Moderately heavy objects such as ¼x20 screws and nuts need more heat, and a 100watt soldering gun *Radio Shack* 64-2193 $12.99 handles such items. Brackets, slides, holders and many mechanical gadgets can be fabricated from pieces of galvanized sheet steel, joined together with solder. A solder-tip propane torch is best for such work, and details are given in Chapter 8.

Only special temperature controlled soldering irons can be operated continuously from 115V outlets. Tips of inexpensive irons will be ruined if they are not run at a reduced voltage, which is about 65V for an iron recommended in the robotics kit. Students at the University of Toronto use a controller shown in Figures 1-1 and 7-8 with their soldering irons. This unit can also be used for other ac control functions up to10 amps. Radio Shack now has a similar soldering iron controller, #64-2054 $12, (not listed in 1997 US catalog).

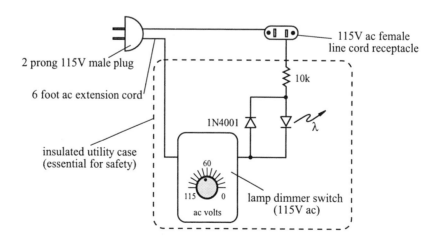

Figure 7-8. This ac voltage controller can be built for about $10 in parts and is suitable for controlling soldering iron temperature, ac brush motor speed, lamp brightness etc. A soldering iron controller is also available from Radio Shack.

Soldering iron tip's are plated to prevent the base material (usually copper), from oxidizing. Once this plating is removed, a tip is worthless. This underscores the importance of protecting a tip by using either the dimmer switch iron controller in Figure 7-8, or a Variac™ to reduce ac line voltage to the iron.

A wet sponge should be used to wipe an iron's tip before soldering, and special sponges are available for this purpose, ordinary bathroom sponges are unsuitable because they melt. Many irons do not have heat resistant power cords, so care must be taken to ensure a hot tip does not burn the cord insulation. A proper stand for an iron reduces the risk of tabletop burns and a possible fire, since paper can reach its ignition point when held against a hot iron's tip.

It is easy to solder brass, copper, silver, gold, tin and nickel with standard 60/40 rosin core solder, and a 1lb roll of this solder can be purchased from *Hosfelt* 0.031" #46-148, $6.95. This size solder can be used for heavier work by twisting several strands into a bundle, or may be flattened with a hammer then sliced to make very thin solder strips.

Rosin core solder cannot be used to solder iron or stainless steel wire or the plating on some connectors. These metals and coatings require acid core solder, or plain solder plus solder paste or liquid flux. Fluxes just mentioned are all corrosive, often containing zinc chloride and hydrochloric acid, so they should be used with care. Try not to inhale fumes when soldering, and be sure to clean any items thoroughly with water and detergent if they have been soldered with acid fluxes.

Solder should always coat a soldering iron tip readily, flowing freely over the complete tip surface. This is called 'wetting'. If a soldering iron tip gets a non-wettable skin, just scrape the heated tip with a sharp edge of soft metal such as brass, copper or aluminum. Don't use any harder metals or you will destroy the tip plating. Once this skin has been removed, a tip can be rejuvenated by dipping the hot tip in solder paste, then immediately coating it with rosin core solder. Be a pro and always leave an iron's tip wetted with solder when putting it aside, or before turning the power off. Special tip cleaners are sold (eg Radio Shack #64-020, $5.99), but solder paste is more generally useful.

Student's are eternally inquisitive and always want to know why aluminum cannot be soldered (usually after they have attempted this unsuccessfully). In fact aluminum can be soldered and a technique is described in Chapter 8, but for practical purposes it is seldom worth the extra effort.

Cold joints

This has nothing to do with arthritis, but is a commonly used term to describe a joint between metals which is covered with solder that has not wetted the metal surfaces. Beginners sometimes camouflage their lack of soldering expertise by making these poorly soldered 'mechanical' connections, and later suffer the consequences. These consequences can appear subtly, because cold joints are often thermally and vibrationally sensitive. For instance, a robot may stop working when it is inclined at a certain angle or when it reaches a particular temperature. Such faults can be difficult to trace, especially if many such connections need checking.

"Should I twist wires together before soldering them?"

Old timers sometimes say *"make a good mechanical connection before soldering."* This may be good advice for high current heavy wire connections, but is definitely bad advice for most robotics electronic work. In general, wires should **never** be twisted together or bent through solder lugs, before they are soldered. Trying to undo a soldered connection where wires have been intimately twisted, can be frustrating. In addition hot solder can spray over several feet, if wires coated with hot solder are suddenly pulled apart.

On a properly soldered connection a wire can be pulled away from the joint with a shearing action, without breaking the wire. This is because wire has a higher tensile strength than solder, and the latter should give way first, but some solder should remain attached to all parts that were soldered initially.

Soldering techniques

Good soldering technique depends far more on the individual wielding the iron than equipment being used, although it is hard to convince some of this fact. For instance, a fancy temperature controlled soldering station costing $500, heats items just like a $5 iron from the kit in Table 1-1. Neither iron will eliminate cold joints if used improperly, and work quality from either iron is indistinguishable when done by a skilled worker.

Basic steps for successful soldering are simple, and if followed carefully will yield impressive results you will be proud to show others.

[1]   For proper bonding, all items must be clean and free from grease or oxidation.

[2]   Pre-tin (cover with a wetted solder layer), all items before they are joined.

[3]   Keep an iron's tip clean by wiping it frequently on a wet sponge, then wet the iron's tip with solder.

[4]   After pre-tinning, solder pieces together by applying the *iron tip* and *solder simultaneously* to the *pieces* you are soldering.

[5]   Remember heat can only be efficiently transferred from an iron's tip if molten solder is present for conducting heat to the workpieces.

An experienced person can make look soldering look simple (which it is), but it is a lot harder than it looks when soldering either very large or very small items for the first time.

A pro can support two wires together (end to end) in one hand, so their opposing ends overlap. This junction is then moved until it touches solder sticking up from a spool. A quick dab with the iron - presto! A perfect solder joint. A beginner will need to hold one wire in a mini vise to do the same job. However, even pros use a mini-vise as a third hand for holding connectors and other tricky objects, as shown in Figure 1-6.

Heat sinks for soldering use

You will seldom see an experienced solderer use a heat sink, because they know intuitively just how hot the transistor etc. is getting while it is being soldered - a deft touch and the job is done! Not so for a beginner unfortunately, who is still trying to coordinate both hand movements - one holding an iron, and the other some solder.

If you are apprehensive about overheating a delicate electronic part, use either a heat sink shown in Figure 7-9, or a small pair of pliers with an elastic band wrapped around its tips in Figure 8-32E.

Desoldering pumps, salvaging and replacing ICs on a PCB

With a little care, components soldered to PCBs (Printed Circuit Boards) can be removed intact, by using a desoldering pump or bulb solder sucker. Apply a well-tinned iron tip to the part, making sure the pump's orifice is very close to the solder. Activate the pump a split second after solder starts to melt. This is easier to write than to do, so practice on some old parts first until you develop the correct timing, (something like playing a triplet with the right hand in the same period as two notes with the left). Suck solder from both sides of the board if necessary, then wiggle each pin with a probe to make sure it moves freely.

Transistor legs on some boards are clipped so short it is nearly impossible to get an iron's tip to the right location. Similar difficulties arise when trying to remove integrated circuits for reuse, especially if they are attached to a PCB with plated-through holes. Usually it is a waste of time attempting to remove parts under these circumstances, because you may end up overheating and destroying the part you wish to save.

An alternate and safer method for removing parts from a PCB is simply to cut all IC pins, remove the chip, suck out solder, then remove any pieces of pins from the hole. If the purpose is to reuse the original chip, just solder the chip's pins to a suitable socket. Augat's machined-pin style sockets illustrated in Figure 7-17a are best for this job.

Desoldering braid

Desoldering pumps do not leave a PCB surface in a pristine state, and the last vestiges of solder are best removed with copper soldering braid or wick. Braid is placed on solder that must be removed, and then heated with a hot iron tip. *Hosfelt* sells 5ft spools of Chem-Wik® desoldering braid in six widths from 0.030" to 0.190", from $1.45 to $1.99, #2-5L to 19-5L. Figure 7-9 shows a $5 pump (*Hosfelt* #46-131), bulb-type solder sucker (Radio Shack #64-2086, $2.79), solder wick and heat sink.

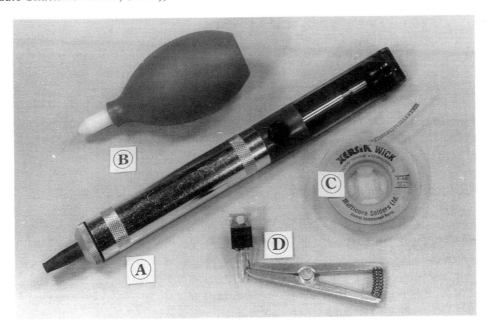

Figure 7-9. Desoldering accessories shown here are only needed for repairing printed circuit boards, and their use is described in the text. [A] Desoldering pump. [B] Bulb solder sucker. [C] Desoldering braid. [D] Clip-on heat sink is part of a four-piece soldering kit sold by *Radio Shack* 64-2227, $3.75.

Heat sinks are rarely used by experienced solderers, but are handy when a critical part must not be overheated. A makeshift heat sink is easily improvised, by wrapping a rubber band around tips of a pair of needle-nosed pliers. Plier tips can then be clamped onto legs or pins of heat sensitive components, see Figure 8-32E.

Repairing a broken IC pin

A broken pin on a valuable IC can be repaired by cutting a pin from a defunct IC. Mount the damaged IC in a mini-vise, then tin both damaged pin and the new pin. Hold the new pin in pliers and solder it quickly to the old pin stub, overlapping as much as possible. Naturally this should be done with an old chip first, unless you have your soldering degree.

Electronic hardware and miscellaneous parts

Connectors

Building a project without using a single connector is possible, but this is seldom a good idea. Top projects usually have a modular design, so individual sections can be quickly removed, tested then replaced. Easy disassembly also means a large project can be moved in sections with less chance of damage. Installing connectors may take a little longer initially, but is a more professional approach and saves time in the end. Modular design is widely used by industry for sophisticated products from the space shuttle to television sets, and represents the finest in engineering design. Therefore, it behooves a neophyte to emulate these standards.

Connectors provide breakpoints between each module, allowing individual members of a team to concentrate their efforts on a particular project element. Each element can be tested separately to ensure it meets predetermined specifications, then combined with other elements in a logical checkout sequence.

Connectors often take more space in an electronic catalog than any other component. Hundreds of connector types are available, from Microdot miniature connectors to special military grade connectors used on aircraft or anywhere reliability is of primary concern. These latter connectors sometimes carry a mix of coaxial feeds and high current lines, while some even have filter networks built into a single pin's socket. High quality connector pins are often gold-plated to give lower contact resistance, and deter corrosion. Cadmium plating (used on many common connectors), is shunned for space experiments because it can migrate to other parts of a payload, and form shorting whiskers on compatible surfaces. Fortunately excellent inexpensive connectors are available for hobby work that will handle most amateur needs.

D subminiature connectors

D subminiature connectors shown in Figure 7-10 [A] are readily available in 9, 15, 25 and 50 pin versions. *Radio Shack* carries 9 and 25 pin D connectors from 99¢, and *Hosfelt* sells 9, 15, 25 and 50 pin solder cup versions from 39¢. When wires are soldered to a connector's pins, it is good practice to slip a piece of heat shrink tubing over the exposed wire section. Metallized plastic hoods to protect exposed wires are made for all sizes of D connectors, and these are sold by *Hosfelt* from 39¢.

Because D sub-mini connectors are so widely employed on computer peripheral equipment, a great variety of ready-made cables, gender changers, extension cables and adapters, are available at reasonable cost from computer outlets. Radio Shack's catalog lists 30 stock items employing a D type connector on at least one end of a cable, all retailing for less than $13.

Filtered D connectors are available starting at about $10 for 9 pins, and microminiature D type connectors with 0.05" pin spacing (about half that of a regular D sub-mini), start at $27. Regular D connectors handle 5 amp per pin, but decreases to 3 amps/pin on microminiature D type connectors.

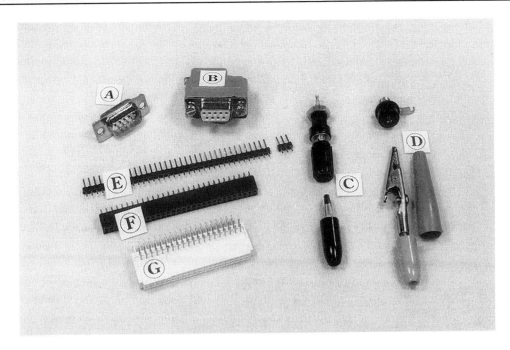

Figure 7-10. [A, B] D subminiature connectors are made in 9, 15, 25, 37 and 50 pin styles and all except 37 pin types are widely available from about 50¢. [C, D] Banana plugs can carry 15amps, and fit into mating jacks, binding posts or special alligator clips as shown. [E, F, G] Header strips are the most useful connectors for hobbyists. They are inexpensive, with 0.1" pin spacing and come with straight or right-angle pins.

Leak tight connectors for submarine, vacuum and high pressure projects

Hermetically sealed connectors are necessary for underwater, vacuum or high pressure projects, but sealing connectors at home is tricky. Although D sub miniature hermetic connectors are manufactured, they are expensive (>$100 for 25 pins). The best solution for amateurs is to pass solid wires through holes drilled in an acrylic sheet, then seal the high pressure side with epoxy or silicone rubber. Make feedthrough leads sufficiently long, so connectors can be soldered onto these wires without overheating the encapsulant. Air can easily leak between any wire and its insulation jacket. Therefore, insulating sleeves on all wires must be removed where they cross a pressure interface to get a leak-tight seal. Stranded wire may only be used for home made hermetic connectors if it is first saturated with solder where it passes through a pressure interface.

Header strip connectors

Strip connectors shown in Figure 7-10 [E, F, G], are the most useful connectors for roboticists. They can be bundled together to form very large connectors, or snipped to short lengths, making mini connectors. Pin spacing is 0.1", so these strips can be stuck into solderless breadboards or easily attached to ribbon cable. Each 0.025" square pin is rated for 3 amps, and pins or sockets can be removed with a pair of needle-nosed pliers. To avoid shorting between adjacent pins, it is a good policy to slip a heat shrink sleeve over each wire connection made to a header strip.

Both male and female headers are made in single and double row formats, and are also available in a variety of straight or right angle versions. Single and double row male or female headers are usually notched, so they can be snapped apart at a convenient point. Some dual female headers are not notched. These can be sectioned by removing two opposing pins, then cut apart with a razorsaw or hacksaw.

Headers are sold by *Radio Shack* and most good electronics stores. *Hosfelt* stocks almost every style used by roboticists, and a double 40 row male straight connector #21-278 is 95¢. A snappable female dual 18 row header #21-287 is 65¢. Buying long strips is most economical when purchasing headers, since these can always be cut or snapped to make smaller connectors.

DIP headers made for cable connectors and wire wrap purposes, are very useful for supporting components. Active or passive filters and other small circuits can be built on a suitable DIP header and used as plug-in modules for solderless breadboards. Such modules permit quick circuit modifications, without affecting normal circuit board layout. Good quality machine-pin headers are sold by *Hosfelt* 28 pin, #21-312, 20¢.

Banana plugs and alligator clips

A banana plug and jack is the simplest and least expensive high current, high voltage connector available. Many types of banana plugs and jacks are made but styles shown in Figure 7-10 are most useful for hobby work. A standard binding post jack and plug is shown in [C] and a panel jack and a mating alligator clip with its rubber shroud in [D]. Wires can be either soldered to a plug or slipped through a machined groove, then bent so a short length of wire is compressed by the cap. With appropriate wire, banana plugs and jacks are rated to 15A and up to 5kV. A #16 stranded wire is satisfactory up to 15A.

For classroom and professional use, special banana leads, clips, posts, dual plugs, dual jacks, adapters, jumpers, test prods etc. are available. These well-made accessories are ideal for laboratory use, allowing students to make circuit attachments and modifications without soldering, when running formal experiments.

Telephone cable and connectors

When long, low current lines complete with connectors are needed, it's hard to beat telephone line cords. They are inexpensive, widely available, easily extended, have low connector resistance, and can be obtained in 4, 6 or 8 conductor types. Quick-disconnect telephone jacks are very handy when many elements must be operated from a common host. Several devices can be operated in parallel from three way jacks, and 50ft extensions are sold in many stores.

A stranded-wire telephone conductor is very flexible, but is ordinarily only 28 gage and therefore too small for low current analog signals. It is also more difficult to solder stranded telephone wire because this is composed of very fine wires twisted around a fabric cord. Solid telephone wire is available up to 22 gage, and this can be used for solderless breadboard connections.

Telephone connectors and solid wire cables are used to transfer strain gage signals to monitoring circuits, for a third year laboratory experiment at the University of Toronto. Ten different materials, including aluminum, Kevlar, GRE, magnesium, Plexiglas™, bronze and balsa wood, can be tested in under an hour. A telephone jack is installed on each test beam for monitoring gage resistance with remote equipment. Longitudinal and lateral gages are installed on each beam, and Young's modulus, Poisson's ratio, material damping and resonant frequency are automatically calculated using a computer interface. This experiment evolved from a third year student project, and has been in regular use for the past eight years. Principal features of the measurement system for this experiment are outlined in the first part of Chapter 5.

Sockets for integrated circuits, relays. DIP, SIP and ZIF sockets

Top quality machine-contact sockets shown in Figure 7-11[B,C,D,] are sometimes worth the extra cost. They can be stacked as shown in [D], inserted into solderless breadboards, used to support small circuits or even SMDs as shown in Figure 7-17. Machine-contact sockets hold IC pins better than any other socket type, and are the best choice for high vibration environments. If test boards made for student use are constructed with machine-pin sockets, they will survive thousands of insertions/removals without problems. *Hosfelt* stocks 8 to 40 pin quality sockets in standard widths, for wire-wrap or circuit board mounting. Prices start at 45¢ for an 8 pin machine-contact socket, and economy sockets are available from seven cents.

ZIF or Zero Insertion Force sockets (Figure 7-11 E), are used in quality EPROM burners and for testing VLSI or similar devices with large numbers of pins. By moving an over-center handle a ZIF socket applies only side pressure to pins as a chip is clamped securely to the socket.

Students may insert/remove a chip from a solderless breadboard dozens of times during testing, but only rarely do they break a pin. Broken pins on an integrated circuit can be repaired as described in text following Figure 7-9, so hobbyists will rarely need to install a ZIF socket, (prices begin at ~$10 for 14 leads).

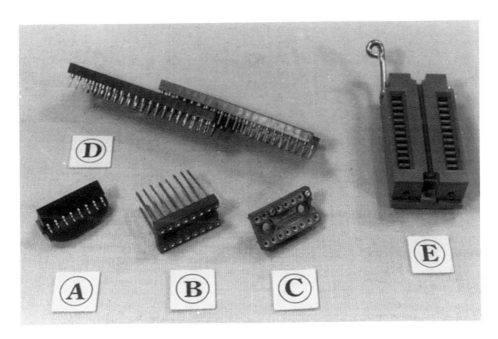

Figure 7-11. [A] Solder-tail sockets are made for circuit boards, and the item shown can be bought for seven cents. [B, C, D] Quality machine-pin sockets are more expensive (45¢ up), but can be inserted into solderless breadboards, stacked together as shown in [D], or soldered to PCBs. They have excellent pin retention and are tops for high vibration conditions. [E] ZIF or zero insertion force sockets clamp an IC's pins securely but very gently. They are used in EPROM programmers, or when many chips must run through a standard set of tests.

Switches and their quirks

More than 100 pages are devoted to switches in Newark's catalog, so it is difficult to do justice here to the large variety of switches available. Switch types such as DIP, Hall, illuminated, keyboard, knife, magnetic, membrane, mercury, micro, reed, rocker, rotary, slide, thumb wheel, toggle can be subdivided into further classifications.

We would have to include various mounting styles, and could not in good conscience omit photoelectric, surface mount and even solid state switches. The last item includes the 4051 (8 channel mux/demux), and 74150, 4067 (both 16 channel mux/demux), providing multiplexing and demultiplexing capabilities for roboticists. Multiplexing and demultiplexing can also be done at slower rates with mechanical switches, and solid state mux/demux devices are covered in a future book.

Basic mechanical switch functions are summarized in Figure 7-12, and these are all useful at times in projects. A DPDT reversing switch is not normally available in stores, but is easy to make by adding just two wires to a standard DPDT switch.

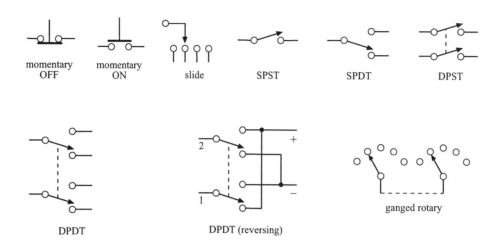

Figure 7-12. These commonly used switch symbols apply to various switch designs and shapes, and some typical switches are shown in Figure 7-13. A reversing DPDT switch is made by adding two external wires to a regular DPDT. Small rotary switches have only one section, but subminiature and regular size ganged-rotary-switches are widely available with up to four sections.

Voltage and current ratings must be observed when using a switch. Many 5 volt switches look the same as a 125V switch, but cannot be used at this higher voltage. Ratings are always stamped on switches, so check these parameters first, before use.

Some switches have special contact actions, such as 'shorting' or 'non-shorting', 'make before break' etc. These features are mainly found in rotary switches, but if you find anomalous behavior check resistance at switch contacts with a DMM, while moving the switch actuating lever/knob.

Sometimes students use household light switches for high current applications. These switches are inexpensive and sold in every hardware store, so overall this is a good idea. However, just because a 115v ac switch is rated for 20 amp does not mean it can be used at 12V and 192 amps. Contact current handling does not scale with switch power capability. In fact most ac switches have a lower current capacity when operated on dc. For example a 115V, 20A ac switch can only handle 20 amps when switching 12 volts. A compact and more versatile solution for this latter task would be to use an automobile switch, such as *Radio Shack* #275-710 DPDT toggle, 12V 20A, center off, #2.99.

Don't forget that relays are also switches, and a SPDT 12V 40amp relay from *Hosfelt* is only $3.50 and can handle many high current project needs. This relay is shown in Figure 3-5, and wires can be soldered directly to its pins, if a female socket is not available.

Roboticist's favorite switches

A few of the most commonly used switches in robotic projects are shown in Figure 7-13. Reed switches should be included in this figure, but do not reproduce well in a composite photograph, their essential features are shown in Figure 5-52.

Mercury switches can be used as level detectors or tilt indicators, *Hosfelt* #51-334, 75¢. They are strictly on/off devices and can only be used to sense a discrete tilt. A series of mercury switches must be installed at defined angles if these switches are required to sense different angles. Mercury switches are simple, reliable, inexpensive, and can be used as angle limit devices in conjunction with analog tilt sensors.

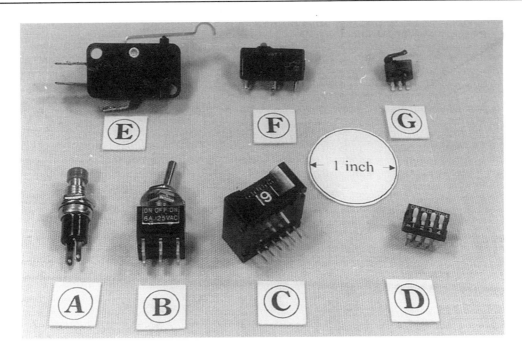

Figure 7-13. [A] Functions such as momentary ON, momentary OFF, push ON or push OFF are available in this type of switch, priced from 25¢ with a typical rating of 3A at 125V. Other push-button type switches with lower profiles offer DPDT, momentary and latching functions.   [B] Toggle switches come in momentary, latching, shorting or non-shorting actions and in many sizes, styles and ratings from 45¢. Toggle switches with multiple poles are also available as SPST or DPDT.   [C] Thumb wheel switches can be plugged into solderless breadboards, and some thumb wheels interlock to form modular switch banks.   [D] SPST DIP switches have 0.1" pin spacing, with two to 10 individual switches. *Hosfelt* DIP switch prices start at 39¢.   [E,F,G] Micro switches are very useful as limit switches, and for obstacle sensing. Operating force can be as low as 0.03oz if a 2" lever extension is added to [G]. A 125V 0.1A switch similar to [F] but in a 0.5"x0.25"x0.31" package, is 51-197, *Hosfelt,* 45¢.

Micro switches work well as limit detectors and if placed in series with motor leads can cut off power to prevent overdriving a finger joint etc. They are good insurance if an overrun might damage valuable mechanical components

Operating force and current ratings vary with micro switch dimensions. A large switch [E] may have a rating of 125Vac and 15 amps, and an operating force anywhere from 3oz to 20 oz. Some small micro switches can handle 20A at 125Vac, but it is wise to check ratings of surplus items before using them at high currents. Item [G] in Figure 7-13 is the smallest micro switch made, with an operating force of 0.2oz, and is rated for 2amps at 30V with an inductive load.

Operating force on a levered micro switch can be reduced by extending the lever, and item [G] functions reliably at 0.03oz when a 2" lever extension is attached.

Mechanical switch bounce

It is appropriate to remind readers that no mechanical switch can be turned on or off without some contact bounce. As contacts meet, mechanical vibrations create transient voltage fluctuations for a brief period before solid contact is established, (see Question 39, Chapter 2). Therefore, mechanical switches cannot be used by themselves to give reliable on/off logic signals for IC triggering purposes. Debounce circuits are used to clean up noisy switch signals, and are provided in a future book.

Terminal strips, tie boards and terminal blocks

If your project has many external wires, you will inevitably reach a stage where several wire ends must be joined together. They can be twisted in a bunch and then soldered, but this is a poor method if they must be unfastened later, and this is often the case.

Professionals solve this problem with the help of some components shown in Figure 7-14, and these are used in many commercial instruments. Tie boards and terminal blocks can simplify fault finding too, if wiring or circuit diagrams are annotated with wire numbers and their locations. For example TB1(23) may indicate that a particular signal can be found on Tie Board #1 at terminal #23. If you record normal operating voltages at such test points, it will be easier to check these later, when things go awry.

Figure 7-14. [A] Double row screw barrier blocks are made in various lengths, and with different screw head sizes. Several wires can be clamped with a single screw as shown, or wires can be inserted in a crimp lug first. Assigning a number or letter to each terminal helps when searching for a particular wire. [B,C] Solder tie boards can hold many wires, and should also be numbered. [D] Wire gages up to 16 AWG can be screw clamped into these pin barrier terminal strips, and may be used with solderless breadboards or printed circuit boards. Three #20 wires will easily fit into each socket, as shown in this figure.

Double row barrier blocks [A] in Figure 5-14, are sold by *Hosfelt* in 1.4" to 11.9" lengths. Because barrier blocks are easily cut with a hacksaw, it is more economical to buy the longest item, which has 30 sections, #3000-30, $4.95. Each terminal is rated at 20 amps, 250Vac, and contact jumpers priced from 11¢ can be used to join individual sections for higher current capacity. Pin barrier terminal strips [D], are made to handle from 4 to 38 connections, *Hosfelt* 99¢ to $2.99.

PBX and telephone systems use nifty connection blocks for interconnecting many lines, by just inserting wires. A quick-connect block holding wires from 20 to 26 AWG is available from *Hosfelt* 56-314, $7.95. Wires do not need stripping, and one block handles 31 wire sets (up to six wires/set).

Radio Shack stocks terminal strips, PCB wire terminal tie boards, and terminal barrier blocks. Their 'Euro' style barrier strip #274-677, $4.19 has 12 positions, and is similar to that used for power supply output leads in Figure 1-1. *C&H Sales* has a very good selection of barrier strips and tie boards at excellent prices.

Fuses - are they necessary for projects?

Fuses are seldom needed in amateur robotics, because most projects use relatively low currents. Some simple examples may help you decide when fuses should be installed.

(i)   Obviously if any part of a project has valuable components that can be protected by a fuse, then use one. However this is easier said than done, because most solid state devices will blow long before a fuse even gets warm. Motors, lamps, power control circuitry etc. sometimes benefit from fusing. In those cases a fuse should be installed right at the device for optimum protection.

(ii)   If a short circuit in an apparatus might overheat wires or other components and cause a fire, then install a fuse. A fuse placed at an auto battery terminal will provide overall protection, but this may not protect individual elements from being destroyed.

Generally it is best to install only fuses that are absolutely essential. Give each fuse a headroom of about 50% above normal operating current, (use rms current when fusing ac lines). Well designed commercial equipment invariably has only an ac line fuse, because it is difficult and uneconomical to protect individual components against all contingencies.

Select a fuse according to its current rating. For example, a 250V 3A fuse can be used to protect apparatus operating at voltages lower than 250V, and will blow at 'about' 3 amps. Although they are just resistive devices, fuse action can be complex. Consequently 'about' is the level of certainty, when a fuse is used at other than its rated voltage. A 250V fuse may blow in 5 seconds at 200% of rated load current, but at 32V and 200% load, blow time will be 10 seconds, (ICs sometimes blow in milliseconds). Except for some automobile fuses, few low voltage fuses are readily available.

Always use 'standard' type fuses. This means a style sold by the local hardware store, a nearby electronics outlet etc. A 1¼"x¼" diameter 3AG fuse probably comes closest to being a 'standard' fuse, and these are widely available in fast-blow, normal or slow-blow versions. *Hosfelt* carries fast-blow 3AG in 42 current ratings, from 1/32A to 30A, $2.65 to $6.49 for a box of five. Normal-blow are available from the same source in 32 current capacities from 1/32A to 10A. Most are $1.19 to $3.75 for a box of five.

Uncommon fuses can be difficult to get in emergencies, and holders for such fuses are always more expensive. Many DMMs use odd sized fuses, and this is annoying when a replacement is required.

Rectangular clip-type fuse holder blocks are best for low voltage dc work. They can be attached with screws, or even stuck down with double sided foam tape. In-line or panel-mount cylindrical fuse holders are safer for higher voltages, or when using ac lines. Remember to slip heat shrink tubing over connections for safety, if high voltage leads might be touched accidentally. Clip-type 3AG fuse block holders 43-168 are 35¢, and cylindrical panel mount 3AG holders, FH66 are 49¢ from *Hosfelt*.

Battery holders

Flat-pack battery holders are best for holding small cells and batteries. They can be fastened to decks with sticky foam tape. Batteries may either be charged in situ, or removed and placed in a regular battery charger. *Radio Shack* carries 14 types of holders for AA, AAA, N, D and C cells, prices are from 79¢ to $1.99. They also stock heavy duty 9Vsnap connectors, 270-324, five for $1.89.

Rectangular or circular style gel cells (SLAs), and shrink wrapped NiCd packs can be secured with metal or plastic straps. Hardware stores sell perforated nylon coated steel strapping in 10 foot rolls for $2.50. This material is ½" wide with 3/16" holes at ¾" intervals. It may be cut to convenient lengths for holding heavy batteries. Slip some foam under a strap to cushion clamping forces, when using steel straps. Smaller battery packs may be supported with mounting cable ties, or homemade plastic straps.

Large stainless steel screw-type hose clamps are excellent for holding heavy objects. A 5" diameter hose clamp costs $1, and can be joined to similar clamps to make a longer adjustable band, suitable for holding even large auto batteries, (see Figure 8-23A).

Nylon screws, washers, nuts and spacers

Nylon screws, washers, nuts and spacers are invaluable when components must be 'floated' from ground or another potential. Nylon fasteners are non-corrosive, so are often used in submersibles and similar projects. Hobby stores sell nylon bolts, and *Hosfelt* stocks nylon pan-head screws in 4/40 to 8/32 sizes and various lengths, at 10¢/screw. Matching nuts, washers and spacers cost from 5¢ to 25¢, also from *Hosfelt*.

Using solderless breadboards

Building and testing circuits can be one of the most satisfying parts of robotics. Translating a circuit from this book into working hardware on a solderless breadboard is straightforward, and circuits built on these boards can be used directly in most projects. There is no need to make printed circuit boards except in very special circumstances. Even if you plan to make printed circuit boards for your apparatus, it is a good idea to test a circuit first, by building it on a solderless breadboard.

Use stranded wire for all power leads, and solder these leads to U jumpers as described below. Label power leads, because it is so easy to make mistakes when you are tired or keyed up just before a demonstration. Different colored wires not only make a finished board more attractive, they are very helpful when tracing a complicated layout. Try to use red and black for plus and minus power leads, and green for ground/common. If everyone in a classroom uses the same scheme, boards can be swapped more easily without making mistakes.

Try to lay out components on a breadboard in a logical manner. This means major items such as amplifiers, comparators, tone decoders etc. should be located on a breadboard in the same left-to-right sequence as they appear on a circuit diagram. As an example, transmitter and receiver boards for the ultrasonic proximity detector of Figure 5-68 are shown in Figure 7-15.

Figure 7-15. An ultrasonic transmitter and receiver built from circuits in Figure 5-68 are shown. Connecting wires have been cut to size from 24 gage solid telephone wires, and all component leads trimmed. A tidy board laid out in a logical manner is easier to check and takes less effort to make, (if one includes time spent tracing and checking a disorganized circuit). Miniclips are attached to a U ground jumper and a straight wire probe. An oscilloscope probe can be connected in the same manner.

Both units are made from one standard size 6.5" breadboard, cut with a hacksaw. Parts have been laid out as suggested above, with 567 tone decoder and lock detector LED to the right of the LM386.

Double-row power distribution strips run parallel to board length at the top and bottom of most solderless breadboards. On some boards these strips are not continuous and users must be alert for this anomaly. Always check distribution strips, adding jumpers if continuity is required. Jumpers are visible on the receiver board about ¾" in from left board edge. They ordinarily are at the center of a board, but this board has been cut.

When a unipolar power source is used, spare distribution rails can be used as additional common/ground rails. This has been done on both boards, and long ground jumpers are visible next to transducers on each board. Extra ground rails provide shorter paths for decoupling capacitors, extending useful high frequency operation for a solderless breadboard, (not a factor with low frequencies employed in this circuit). With careful layout, solderless breadboards can be used with many circuits up to 10MHz.

U clips made from #20 gage solid hookup up wire may be used to attach power lines to a solderless breadboard, by soldering leads to U clips as shown in Figure 1-3. For most work 20 AWG stranded wire is fine for power leads, but this should be increased to #16 gage or two #20 wires if current exceeds four amps.

It is easier to check voltages and signals, if a short U shaped jumper bent from bare hookup wire is inserted in a ground rail. An insulated probe can be made from a short length of wire for examining signals at various board locations. Two miniclip leads are shown hooked onto these wires in Figure 7-15. Oscilloscope probes may be connected to a board in the same manner.

"How much current can a solderless breadboard take?"

Because breadboards are so convenient, students are tempted to use them for every circuit. Charred board areas are not uncommon in the classroom, but with care solderless breadboards may be used at 8A.
Listed dc ratings for a breadboard are:

Maximum current:	4 amps.
Maximum wire size:	20 AWG (.032" diameter)
Contact resistance*:	0.0025Ω (two adjacent contacts)
Capacitance*:	3.5pF (between two adjacent 5 pin rails)

* varies with manufacturer

Power dissipation is the limiting factor in deciding how much current a board can handle. It is much easier to pass 10A along a 50-pin distribution strip which has a resistance of 0.055Ω, than to pass the same current through two adjacent pins with only 0.0025Ω between them. This is because 5.5W just warms a 6" rail, but things get a bit hotter when 0.25W must be dissipated in 0.1" (the distance between two pins).

When using a breadboard above 4 amps, solder all leads to #20 U jumpers. If you try to apply high current through a single pin connection, things will get hot and may destroy a socket. Double U jumpers are also necessary for bypassing breaks on distribution strips, since single jumpers overheat and increase board socket resistance. Only use high currents on small portions of a board so that heat generated has a chance to escape. Feeling with fingers will let you know if things are getting too warm.

Current between adjacent pins should be restricted to 8A for continuous operation. Keep an eye on any high current apparatus, especially during an initial test run.

"Which type of solderless breadboard shall I buy?"

A standard 6.5"x2.2"x0.4" board, *Hosfelt* 42-103 $5.49 is recommended, because it is inexpensive and can be used to make either larger or smaller boards. This type of board shown in Figures 1-3 and 7-15, is available from many other sources eg. *Radio Shack* 276-174, $12.49. Larger boards are sold, and a 7.4"x7.5" board with banana jacks is $29.95 from *Hosfelt* 42-107.

Double distribution strips on breadboards make things easier when building analog circuits requiring bipolar power supplies, so check for their presence when buying. When constructing large logic circuits or microprocessors, breadboards can be interconnected using dovetail slides or pins provided at board edges.

Perf boards

Perf boards are available as plain sheets of perforated phenolic, or as printed circuit prototyping boards usually made from epoxy glass composite. Specimens of both types are shown in Figure 7-16. Hole spacing varies, but more board is sold with 0.1" spacing than in any other category. Board [A] in the figure has 15/64" spacing and is the thickest sample at 3/32". Boards [C] and [D] are 1/16" thick and [B] has 3/32" hole spacing.

A variety of pins, pegs, sockets, wrapping pots etc. are sold for use with perf board. Special insertion tools, and pad cutters for making isolated areas around IC pins are sold to complement perf board technology.

Vector® prototyping boards are favored by many electronics workers, because they function well at high frequencies. Wide soldered traces on these boards carry heavy currents with only small voltage drops. Modifying a perf board circuit is easier and causes less heartache than making changes to a brand new PCB. Many hobbyists use perf boards to gain experience in using both sides of a plane surface, so perf board is a stepping stone toward making printed circuit boards.

Plain phenolic perf board is also useful for supporting high voltage components, and is easily mounted on nylon spacers with nylon screws, nuts and washers, giving extra isolation. Always support a perf board on spacers to prevent pins, studs or other circuitry beneath the board from shorting, when placed on a conducting surface.

Figure 7-16. Perforated circuit board is used for wire wrap applications, high frequency circuitry and any time high density population is required for regular size DIP and SIP components. Most perf board sold has 0.1" spacing and Vector® board in [D] has solder coated bus, circuit and ground patterns. Vector® boards are great for getting PCB performance in 'one-off' applications. Perf board can be used to support power transistors or high voltage components as shown at right. Nylon spacers can be used to mount perf board, but flexible push rod sheath shown in this figure works well too. One variety of push rod suitable for spacers, is marketed as 'Gold-N-Rod' (Figure 8-21D), and available from hobby stores,

Radio Shack stocks seven types of perf board with 0.1" spacing at prices from $1.19. Solderable perf board with 0.1" spacing is sold by *Hosfelt* JALPC-4, 3"x4¼", $1.75. Buying large size perf board is more economical because it cuts perfectly on a foot operated metal shear. A plastic cutting tool or a hacksaw can be used for cutting perf board too.

Working with surface mount devices

Many new integrated circuits are now only available in SMD format, and this will become more prevalent as time progresses. Hobbyists will find this a retrograde step, but for commercial ventures it is a wave of the future. Right now more than 50% of components on video cameras boards are SMDs, and over half of all PCBs presently manufactured have at least one SMD.

Machine placement, solder resist, hot air levelled PCBs, are current buzz words that have yet to filter down to many hobby roboticists. In fact for hobby use, SMDs are a real pain. As these components get even smaller it will eventually become impossible for amateurs to mount such devices by hand.

Fortunately many SMDs are still made with 0.050" leg spacing, and these can be placed on an intermediate larger platform, then handled as a standard DIP. Two methods for using 0.05" leg spacing SMDs are shown in Figure 7-17. SMDs with 0.05" spacing have been successfully mounted using the technique in Figure 7-17a, by students with no previous soldering experience. A simple homemade circuit board method is shown in (b).

These diagrams make the procedure look disarmingly simple - it is not. You will need a steady hand and plenty of patience. Always practice on an inexpensive chip first and test your finished work to make sure you have not overheated or short-circuited any pins. *Hosfelt* sells surface mount 74LS04, 14-pin hex inverters for 59¢, and this is one of the least expensive SMD chips available in unit quantities. It is also easy to check an inverter's performance, after it has been soldered as recommended in Figure 7-17.

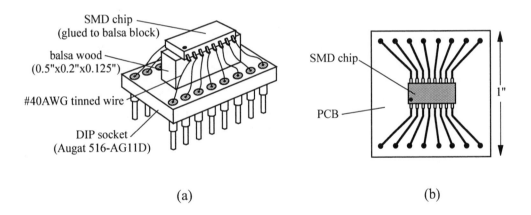

(a)   (b)

Figure 7-17. Surface mount devices with 0.050" leg spacing can be attached to either a DIP socket in (a) or a hand-drawn, etched PCB (b), so they may be handled more easily. Both procedures require some patience, and the method in (a) has been used successfully by students with no previous soldering experience.

Tips for remounting SMD chips

Attaching an SMD chip to a DIP socket using method (a) of Figure 7-17 is straightforward, and not too difficult using the method outlined below.

A machine pin socket DIP IC is recommended and a suitable 16 pin Augat 516-AG11D is sold by *Newark* at $2.48, or *Electrosonic* for $1.75. The same style socket is also available from *Hosfelt* 21-118 for 45¢.

After remounting, valuable SMDs can be shared by many users, because they can be plugged into solderless breadboards. Augat sockets are also stackable, so remember to solder an empty Augat socket onto a printed circuit board if SMD chip-sharing capability is required.

[1]   Plug an Augat style socket into a solderless breadboard. This keeps it stable for soldering.

[2]   Attach a balsa block to the socket, using glue or double sided carpet tape. Make sure the block is narrow enough so it does not interfere with SMD legs.

[3]   Stick SMD to balsa block with glue or double sided tape.

[4]   40 AWG (0.003" dia.) is ideal for making connections to the SMD. Wire wrap wire is normally 0.010" and is too stiff. A stranded wire composed of #40 wires is best, and Belden 8599-100, 7x40, 32AWG, works well. Six inches of this wire is sufficient for several SMDs.

[5]   Tin each upper cup on the Augat socket, and each chip pin. Either use a microtip soldering iron or add a wire extension tip to a regular iron, as described below. Only use a little solder - too much increases the likelihood of getting solder bridges between SMD pins. Narrow solder can be made from larger diameter solder, using a method outlined below.

[6]   Strip about ¾" of insulation from one end of the wire, then bend all but one wire back 180 degrees. This leaves only one wire protruding, so you can work unimpeded by the others.

[7]   Tin the wire's tip. It is easier to do this if you set a reel of solder down, then pull a length of solder so it sticks up. Now touch both wire end and iron to the solder simultaneously.

[8]   Rest the wire-holding hand on the table. You need a steady hand and this support will reduce trembling.

[9]   Solder a wire end to an SMD leg. Don't rush but don't take too long either - a few seconds is ok but more than five seconds is near the limit.

[10]   Bring the free wire end down to a socket cup, then snip it off at that point with side cutters.

[11]   Solder wire to cup, then repeat for the remaining pins

A no-cost micro soldering iron tip

A short length of #22 AWG tinned solid copper wire, wrapped tightly around the tip of a pencil soldering iron is great for soldering SMDs. Wire is more easily attached to straight solder tips, but only a little extra effort is required to make a solid thermal connection on a conical tip. Use pliers to twist the wire tight but go easy, because copper wire has low tensile strength and snaps when overstretched. Leave about ¼" of wire extending from the iron's tip, and tin this before soldering any components.

If you are using a soldering iron controller similar to the one shown in Figure 7-8, there will be no problems with overheating. If you are not using a controller or a temperature controlled iron you will have difficulty in completing steps outlined above, or in using an alternate method described below.

Making very thin solder

Pros can solder to 0.050" SMD legs using a standard iron tip and 0.030" diameter solder. But even these people find it easier to use a microtip iron and thin solder.

You can make 0.10"x0.10" solder from regular 0.030" or larger diameter solder, by first hammering solder flat on a steel block. Just do an inch or so at a time, which will be sufficient for many connections. Flattened solder can be trimmed to narrow widths with manicure scissors, or by slicing with a utility knife and a ruler guide.

A circuit board mount for SMDs

Figure 7-17 (b) shows a possible layout for a surface mount device circuit board. A one-inch square board allows sufficient room for attaching wires, and the board may then be connected to other circuit elements.

Kits for making small circuit boards are sold through major electronic parts outlets. *Radio Shack's* kit #276-1576 at $13.99 includes two 3"x4½" copper clad boards, tank, drill bit, cleaning pad, resist pen, tank and etching solution. An extra fine point etch-resist pen is essential for working with SMDs, and GC Electronics 1/64" pen #22-222 is satisfactory. This pen is available from *Electrosonic* 22-222 for $3. Electrosonic stocks a complete line of professional hardware and chemicals for making printed circuit boards of commercial quality. These materials can also be purchased from *Electrosonic* in small quantities used by many amateurs.

Directions for drawing and etching the circuit board in Figure 7-17 are given below, and if followed will yield good results.

[1]   Cut the size of board needed with a hacksaw. Edges can be finished with a file.

[2]   Thoroughly clean the board with an abrasive cleaner such as Ajax™. Use a water dampened cloth with Ajax, to remove all traces of oxidation from the copper board. Rinse with water, and dry at once with a clean paper towel.

[3]   Tape your board to a flat surface (about 6"x6"), using masking tape across the board's corners.

[4]   Fasten the SMD onto the board with a ⅛" x2" strip of masking tape.

[5]   Using a magnifying glass, mark board with the pen's tip, placing a tiny dot exactly at the end of each SMD leg. If you make a mistake use rubbing alcohol to remove ink, then reclean board as in [2].

[6]   Remove chip. Then using a ruler guide, trace out the pattern you would like, making a circular pad at the end of each line as shown in the figure. No lines should be drawn beneath the chip.

[7]   Your board can be etched in ferric chloride (32 fl.oz. 22-238 *Electrosonic* $5.50) or other etchant. If you use ferric chloride, place your board ink-side up in a Pyrex tray, then cover with ferric chloride. A depth of ¼" is adequate in a 2"x2" dish. Etching is greatly accelerated if the solution is warmed by placing the tray on a hot plate or stove-top element, (make sure your tray is really Pyrex because ferric chloride leaves obstinate stains on fingers too). Rock the tray periodically. This brings fresh solution to the copper surface, shortening etching time.

[8]   Board is ready when copper has been removed from around ink lines. Do not leave it etching for too long or solution will undercut inked lines. Remove board, rinse in water, then dry. If your board is unsatisfactory, the same solution can be used for another board or two.
        Small quantities of ferric chloride can be flushed down the sink after appropriate dilution with water.

[9]   Remove ink lines with rubbing alcohol, then solder each line and pad with rosin core solder.

[10]   Tape SMD to board, and align using a magnifier. Using thin solder and a micro iron tip described above, tack ONE leg of SMD to the board with solder. Check with a magnifier - if everything looks good then solder one

249

leg in the opposing corner. Recheck alignment, then solder remaining legs.

Glass epoxy boards are normally drilled with carbide tip drills. *Hosfelt* sells an assortment of 10 drills 80-241 $4.95. A regular drill bit can only be used for a few holes on glass composite boards before it needs resharpening.

Troubleshooting electronic circuits

Fault finding in electronics is both art and science, and it is impossible to do more than scratch the surface here. Fixing a complicated circuit requires a thorough knowledge of electronics, wide experience, intuitive skills and the ability to test likely solutions quickly, and in an effective way. Even a synthesis of all these elements is sometimes helped with a little luck, or a pertinent comment from a co-worker.

Because these disparate skills must be melded together in just the right combinations, good troubleshooters are rare, and cherished by their employers. Even rarer are good troubleshooters who write books about their art, and that is a pity because this information is sorely needed, in a practical, readable form. The author knows of no book that hobbyists can turn to for assistance in this area. Paperbacks abound covering specific topics, such as TV or VCR repair, but usually contain only minimal information for an amateur roboticist.

What follows is an attempt to cover some typical problems arising in robotic circuitry. No effort is made to deal with point by point circuit analysis or signal tracing.

If you are constructing your very first circuit, practice on something simple at the start. Begin by building all circuits in Chapter 2. This will boost your confidence, and you will gain a better understanding of practical electronics. Circuits in Figures 3-7, 3-15, 3-19, 3-22, 3-23, 3-24, 3-25, employ some basic principles, and you will benefit by breadboarding these too.

'Obvious' things to try on a dysfunctional circuit

[A]   *"My circuit doesn't work"*

Do not turn on power until you have gone through all checklists that follow.

Check the circuit layout. If it looks ok, then get someone else to check your circuit. Like proofreading, one often overlooks mistakes when checking personal work.

If you have no reason to suspect the circuit diagram is in error, then completely rebuild the circuit on a solderless breadboard. If you are already using a solderless breadboard then commence rebuilding at a different place on the board.

Work slowly and methodically to make your circuit neat, rechecking as you go. A neat circuit is easier to trace.

Make sure power and ground are present along all distribution strips on a solderless breadboard. Rail jumpers are required at the center of most boards if continuity is a requirement.

Check Figure 3-14 to ensure ICs are installed with their proper orientation.

Does any chip require pull-up or pull-down resistors?   Look at Figure 3-14

Are your diodes installed correctly?

Measure each resistor with your DMM. Some color combinations are confusing and a 1MΩ resistor will not give the same performance as a 510Ω.

Test every component if you can. Try ICs in another circuit, test transistors with a meter or use information in Figure 3-10.

Have you connected power leads to the correct outputs on the power supply?   If leads are accidentally reversed then chips, diodes and polarised capacitors may be destroyed.

Proceed with the following tests *after* you have familiarized yourself with the remaining checklists.

Remove all ICs from the circuit board.

Install components one at a time and measure all potentials to see if they seem reasonable. Make sure the correct voltages appear at each pin on every IC.

When power is applied, do any components get hot? If a resistor looks discolored check its value with a DMM.

[B]   *"Everything was fine, but it's not working now"*

Are any components discolored, hot or too warm?

Remove all ICs. Go through list [A] step by step.

Disconnect all auxiliary equipment.

Chip-swapping is your best approach. Start with all new chips then put the old chips back one at a time.

Feel each component frequently during the swapping procedure. After a faulty component has been found, you should attempt to find the reason for its demise.

Are you using a PCB that has just stopped working?   Make sure no traces are broken, give the board a twist in all directions to test for intermittents on traces.

Are power line wires sufficiently heavy to carry total current when all items are connected?   Remember that a power supply's common/ground line must carry the sum of currents from all other polarities. This means that on a supply with +12V 3A, -12V 2A, +5V 8A, -5V 2A a ground line may carry 15 amps, (about twice the capacity of any other line).

Do power rail potentials sag when a circuit is fully loaded?   Turn auxiliary equipment on sequentially, while monitoring rail voltages with an oscilloscope, and a DMM.

[C]   *"It's working fine, but my outputs are noisy"*

A major part of any solution is correctly identifying the problem. Is it thermal noise, dark current, power line fluctuations, etc.?

Is noise present on the power lines?   Check each rail with an ac coupled oscilloscope on 10mV sensitivity. If noise and ripple exceeds about 100mV you may need extra filtering, a line filter or an extra power supply

Have you already decoupled your power supply lines with tantalum capacitors?   If so, try adding an 0.1μF ceramic capacitor across each tantalum.

Try moving your circuit to another ac outlet. Borrow an alternate power supply and see if things improve.

If noise comes from the power supply, then substitute an ac adapter, batteries or a dc-dc converter. Try a linear power supply if you are currently using a switch-mode power supply.

Power line filters are very effective for attenuating EMI/RFI travelling through an ac power line. Corcom, Delta, Sprague etc. manufacture these filters. They are easy to install and available from *Newark* and *Electrosonic*. A 10 amp Delta male

115/250V ac outlet, EMI/RFI filter #27-135 sells for $3.99 from *Hosfelt*.

Is other equipment interfering with your apparatus? This is rare in a classroom but can happen when rf experiments, or highly inductive switching circuits are close by.

Try heating sensors gently. Measure RMS noise with and without light on the optical detector. Monitor all power rails with an ac coupled scope probe.

If you can tolerate a drop in frequency response, add an RC filter to your output as in Question 17, Chapter 2. This is the quickest brute-force fix.

[D]   *"Whenever I touch this wire my output goes haywire"*

When you wave your hand near the apparatus, does pickup or noise increase? If so noise may be due to radiative pick up, and your circuit may benefit from shielding. Try a quick and simple Faraday cage shield, described in Answer 14, Chapter 2.

Does noise or pick up increase when you touch the ground rail, or components on a circuit board?
If it does, you probably have a ground loop and should read text preceding Figure 4-5, covering some common problems relating to indiscriminate use of signal and power grounds.

Try grounding your circuitry at different points, or rerouting ground cables so they can be connected at a single point.

Sometimes a persistent grounding problem can only been cured by breaking the circuit. This is done at a convenient point, usually between a transducer/amplifer/output device, and a receiver (which may be a special meter, computer, multiplexer, buffer stage). This means signals must be transmitted between these elements without using any wires. Digital signals are easy to transmit through an opto-isolator. Linear signals are a bit trickier, but an opto-isolator circuit described in a future book handles input signals from 1.5V to 11V with a transfer linearity r > .9999. Pre-amplification with a summing amplifier is used so signals down to zero or below, can be monitored

Isolation transformers capable of better than 1% linear transfer resolution at 100kHz are manufactured, and rf or ultrasonic links are both viable alternatives. However, for hobbyists an opto-isolated link is the least expensive and most practical option for breaking a persistent ground loop. Opto-coupling is also a good technique for providing high voltage isolation.

[E]   *"Why does this signal keep drifting?"*

All components are sensitive to temperatures, and some are more finicky than others.

Try heating components gently with a heat gun, or freezing them with a canned refrigerant, but go easy with these thermal tests. Circuits in this book have been designed for room temperature operation (20°C). Temperature compensation adds extra components and more complications.

Occasionally a faulty component may be overly temperature sensitive, and this is easy to confirm by substitution.

If a circuit cannot be used fruitfully without some temperature compensation, give some consideration to placing critical circuitry in a heated box. This container can be thermostatically controlled at a few degrees above the maximum anticipated ambient temperature. If this is too difficult to implement, it will be necessary to add a diode, transistor, positive or negative TC thermistor etc. to provide a bucking signal to compensate for the initial drift. It is easy to say this, but with some sensors this can turn into a project by itself.

Some electronic anomalies

When a DMM selector switch is set to 'ohms' mode the positive probe will be at a positive potential. Probe voltage can be 3V on some meters, and may affect sensitive circuitry.

Digital multimeters use high frequency timing signals for clocking A/D converters etc. These can promote oscillations in some high impedance amplifiers, causing errors in a DMM's output.

Current measurements must always be made carefully, since blowing up a DMM is easy if appropriate care is not taken. If the meter switch and probes are both set to measure current, and then probes are connected across a voltage source . . . Poof! Hopefully only a fuse will blow. But if you have bypassed the fuse because you had no spare, then your meter will be rendered useless.

Measuring current by inserting an appropriate resistor in series with current flow is always safer. A voltage measurement across this resistor plus Ohm''s law gives the current. A resistor should be selected so it causes minimal perturbation of normal current flow, yet produces a voltage drop that is easily measured. Don't forget to use a power resistor for high currents.

Always discharge capacitors before measuring them with a DMM. They can easily store enough energy to ruin a meter, so short the leads together for a second or two before measuring with a meter. Incidentally, large capacitors will often develop a significant charge even after they have been discharged, so short them every time their capacity is monitored. Extremely large high voltage capacitors can develop a lethal stored charge after being temporarily shorted. Capacitors of this type should always be stored with a shorting strap.

When using a logic probe make sure it is connected to the ground/common of the circuit you are testing otherwise your measurements will be invalid. This is easy to overlook if you are using two or more supplies without a common ground.

Mechanical Construction - tools, materials and techniques

Workshop equipment and accessories for classrooms or hobbyists

Lathes, milling machines, jig borers, drill press, router and similar tools are essential for professional robotics engineering. However, hobbyists with limited funds can build sophisticated devices using simple inexpensive hand tools. For example, one student built a complete robotic hand, wrist and arm (shown on the front cover of this book), using only a hacksaw, pliers, files and an electric hand drill. Three fingers and an opposed thumb had passive spring restraints and were cable operated by geared dc motors, using feedback control circuitry. A tilt sensor described in Figure 5-12 was used to monitor wrist rotation. Ingenuity and perseverance go a long way when fueled by a desire to succeed.

When items must be turned on a lathe or milled but these machines are not available, ask a local machine shop for an estimate. Hand drawn sketches with dimensions and tolerances, are adequate for them to prepare a quotation and they will appreciate the business. For metal parts specify aluminum if possible because it is relatively inexpensive, and can be worked at high speed reducing labor costs. Aluminum does not tarnish easily, can be anodized for a hard insulating finish in many colors, or alodyned to leave all surfaces electrically conducting.

An equipment list given below is intended for classrooms, but hobbyists can buy those items fitting their budgets. Operating instructions and tips following this list, are given in the same sequence as the listing, and beginners will benefit from reading this information.

Workshop equipment

Fire extinguishers
Smoke detector
First aid kit
Safety glasses
Dust masks
Ear protectors
Flammable liquids cabinet
Storage lockers for projects
Storage racks for materials
Cabinets and drawers
Sink with hot and cold water
Compressed air
Electrical power outlets
Work benches
Bench vise
Drill press
Drill vise
Metal shear/bender/roller
Grinder
Tool storage facilities
Hole punch

Hand tools

Hacksaw
Hand woodsaw
Variable speed jigsaw
Hand drill
Drills, center punch
Wood bits
Tap and die set
Screwdrivers
Vise grips®, pliers
Wrenches (socket, spanner)
Hex head/Allen keys
Clamps
Tape measure
Calipers
Squares
Hammers
Tinners metal hand shears
Aviation snips
Files
Utility knives and scrapers
Plastic sheet cutters
Hole punches
Glass cutters
Diamond hones
Punches
Levels
Propane soldering equipment
Heat gun
Glue gun, glue pot

Auxiliary items

Solvents
Oils, greases
Polishes
Adhesives
String
Tape (masking, insulating)
Construction materials
Fasteners
Pipe, tubing
Threaded rod

Circular saws and similar tools have been omitted from this listing but may be added if supervision is available. Power saws, sanders, routers etc. generate fine dust, often making the atmosphere in a small workshop hazardous. Painting fouls air in a confined space and is especially undesirable in buildings with communal airconditioning. Many exterior caulking compounds and other materials are health risks when used indoors.

First aid and emergency procedures: Phone numbers for police, fire, and hospital departments should be posted in a workshop and close to nearby telephones. Hospital locations and directions can be posted too, since these are helpful when an injury does not warrant an ambulance. Most safety items necessary for a small workshop are stocked at local hardware stores.

[1]   Emergency guide charts are sold by safety supply houses, they provide simple guidelines for untrained people who wish to give assistance.

[2]   A first aid kit and manual are usually best placed near a workshop sink, where wounds can be washed and dressed.

[3]   First aid kits must always have an eye bath. Eye splashes need flushing with frequent changes of water for at *least a minute*.

[4]   Burns and scalds should be held under cold running water for *several minutes* to reduce long term skin damage. Do not apply salves or greases to burns before seeing a physician.

[5]   Move electrocution victims with an insulator, making sure power is turned off before touching a victim. Persons may stop breathing after receiving a severe shock, so check for breathing, giving artificial respiration if necessary.

[6]   Fight a fire if it can be controlled with one fire extinguisher. Only use ABC fire extinguishers in workshops, since this type of extinguisher can be used on electrical fires too.

[7]   Always apply firm pressure with a sterile pad to staunch any profuse bleeding. Do not use tourniquets as they may cause gangrene, if blood flow is restricted for a protracted period.

[8]   Never move a fall victim or anyone who may have injured their spine, such action may turn a treatable victim into a paraplegic.

[9]   Call for professional help before things get out of hand.

Safety glasses, dust masks, ear protectors: Consumer grade versions of these items are adequate for laboratory or hobby use. Disposable dust masks, ear plugs and safety glasses/goggles are sold at hardware stores.

Metal cabinet for flammables: Flammable solvents, oils, adhesives and paints may be kept in a locked metal cabinet, satisfying many building safety codes. Students should be reminded that despite the apparent inconsistency, *flammable* and *inflammable* mean the same thing, ie 'easy to ignite'.

Classroom or laboratory workshop: A classroom workshop need not be large if ample table space is provided in a lab area. However, it should be attached to the robotics classroom/laboratory if possible. Total workshop floor area of 24ft x 20ft is sufficient for a class of 50 students, because not all students use a workshop simultaneously. While some are designing, others may be testing, building electronic circuits etc.

Storage lockers: If space permits, storage lockers for projects are very handy. Otherwise, boxes may be used to keep each group's project parts neatly stored under tables or benches.

Compressed air: A laboratory air line with a regulator can be used for pneumatic tools, reed playing instruments, air motors, air cylinders, and various pneumatic experiments. Without an air line these projects must use a compressor, gas bottles or an inflated automobile tire.

Power outlets: Six-plug electrical bars with surge protection cost only a few dollars, and can be used to supply power for every laboratory table or workshop bench. When combined with three pin extension cords, power bars are the most flexible and economical way to distribute ac power from permanent wall sockets.

Project tables and workbenches: Sturdy flattop wooden tables about 6ft x 3ft are most useful in a project classroom, larger tables are more difficult to move and must usually be shared (an additional inconvenience). These tables can be used for electronic workstations shown in Figure 1-8, mechanical assembly, or butted together for large projects

Workshop benches take a beating so they should be heavy, rugged and able to take abuse. Economics affects any choice, but hobby quality bench prices start at $50. Even an old flat-faced door or two sheets of ¾" plywood glued together can be used for a tabletop. Ideally a workbench should not move when it is used, so fasten it to a wall or floor if necessary.

Bench vise: Swivel-base vises are most versatile, but not essential for a workbench, and a fixed base vise with anvil and 4" jaws is adequate for most work. Large heavy vises are more stable when fastened to a workbench, with bolts passing through a bench top.

Drill press: Either a tabletop or free standing drill press is satisfactory. Variable speed is essential with at least ½" drill capacity and an ability to hold a 0.010" diameter drill. Ideally a drill chuck should close to zero gap, so that very small drills may be held securely.

A spring loaded chuck key removes the hazard posed by leaving a key in the chuck, then turning on the drill - a very dangerous practice. Alternatively, a chuck key may be secured by an elastic shock cord that must be stretched slightly to reach the chuck.

A good quality 8" five speed bench-mount drill press can be purchased for less than $200. Floor models starting at $400, usually have more features and accept larger workpieces. Prices for an eight inch, five speed, economy-quality drill press with a ½" chuck, begin at $75.

Drill press vise: A portable tilting drill press vise is essential when holding small work pieces for drilling holes at odd angles. Small items should always be held in a vise rather than with fingers, because serious injuries can occur when a drill snags a workpiece, causing work to rotate.

A single-angle drill press vise with 3" jaws and 6" base that can be tilted to 90 degrees is shown in Figure 8-1. This is a good vise for both hobbyists and classroom use, and usually has Vee grooves across both jaw width and height. Such grooves are necessary for securely holding circular cross section materials. A single-angle good quality vise costs about $75, and is adequate for most robotics drilling. Utility grade single-angle vises sell for $20, and compound angle, or swivel base drill press vise prices start at $200.

Figure 8-1. This single-angle tilting drill press vise will hold rectangular and circular cross section workpieces at any angle for drilling. A vise should be clamped to the press when using drills larger than about ⅛" diameter.

Cutting and forming sheet metal: A metal shear suitable for a small workshop is beyond the financial capabilities of most hobbyists, but almost indispensable for large classes. A 30" combination shear, bender, roller shown in Figure 8-2 is sold by *McMaster-Carr* 2384A72, $635.37. An optional floor stand can be used to support the shear at a proper working height, *McMaster-Carr* 2384A77, $293.69. Brackets, boxes, cylinders and many other items can be fabricated from 1/32" mild steel or 1/16" aluminum sheet using this machine.

Figure 8-2. Three of the most useful sheet metal operations can be performed with this versatile combination brake, slip roll and shear. It is ideal for a small workshop and if casters are attached to an optional floor stand, the machine can be moved for working, then returned to its stored location.

A long blade secured with seven bolts is pressed into a lower Vee grooved channel for bending metal, and movable 'fingers' permit tricky bending when making boxes. Acrylic sheet up to ¼" thickness is easily broken by using the bender blade on the reverse side of a plastic sheet, scored with a plastic sheet cutter, Figure 8-11.

Curve shapes in sheet metal are adjusted with three rollers at the top of the machine, and material thickness is set by two wing screws. Roll diameter may be changed to make smooth curves from a minimum diameter of 1½", in steel, aluminum or other soft metals.

A double handle lever arm on the right of the photograph is used for shearing and when rolling material. Shearing action is good, and material placed on the lower platform is automatically clamped while shearing

Heavy gage aluminum (to ~0.25"), can be cut with a router or carbide tipped circular saw blade, (whenever possible use a guide with these tools). A power jigsaw will cut metal sheets thicker than 1/16" with an appropriate pitch blade. Very thin metal sheets can also be cut with a jigsaw, if material is first securely attached to a wooden backing, using double sided sticky tape on the cutting line.

Nibblers for metal or plastic sheet: Nibblers are sometimes handy for notching holes, or edging thin metal or plastic sheet. However, using a file or jigsaw for these jobs is often better, since nibblers are easily broken, cannot be repaired, and have a short classroom life. Other inexpensive hand tools may be used to cut metal or pliable plastic sheet, and some are shown in Figure 8-10.

Bench grinders - wheel loading and dressing: Drills and hardened cutting tools can be sharpened to new condition on a two-wheel bench grinder with safety shields. After grinding, a small abrasive hone is used to remove any burrs on edges of drills or tool bits.

Hardened steel shaft is sometimes used for axles but is extremely difficult to cut with a hacksaw, and always ruins a blade. If shafting is only case hardened, its thin tough exterior skin can be removed by grinding a narrow notch around the shaft. A hardened rod can then be snapped apart at this notch, and its ends subsequently ground to give a smooth finish.

Inexpensive grinders are usually sold with bonded aluminum oxide wheels, primarily intended for use on hard tool steels. Soft metals will load (fill the pores), of most grinding wheels, but rubber bonded aluminum oxide wheels fabricated for all purpose grinding can be used on steels, cast iron, bronze, copper, aluminum, and stainless steel.

If a grinding wheel's surface becomes loaded with soft metals it presents a hazard, because this increases the likelihood of a tool digging into the wheel causing operator injury. Surfaces on a plugged wheel, or one that has developed a rounded profile, can be restored by reshaping with an abrasive wheel dressing stick, or a grinding wheel dresser.

Drill sharpening guidelines are given on page 263.

Tool Storage facilities: Inexpensive tools may be stored on wall-mounted pegboards, using standard wire brackets, grips, holders and hooks. If each tool's shape is outlined on the pegboard with shelving paper or paint, it is easier to check for missing tools. More expensive tools and delicate instruments are best kept in a steel drawer cabinet.

Sheet metal multi-hole turret punch: It is difficult to drill a hole to an exact diameter in a metal sheet, because all drills wander to some extent increasing hole size. Swarf carried by a drill also enlarges a hole, so it is good policy to remove shavings continuously, if on-size quality is required. Accurate holes in sheet metal or thicker stock can always be made by drilling slightly undersize, then finishing the hole with a reamer.

Hole punches produce nearly 'perfect' holes in a metal sheet, with clean perpendicular edges and no burrs. A turret punch is every sheet metal worker's dream machine, making light work of one of the trickier fabrication problems.

It is hard to beat a turret punch for quickly making holes from 1/16" to 2" diameter in metal sheet. However, spending $7600 for the hand operated turret punch in Figure 8-3 may be difficult to justify, even for large institutions. Square, rectangular, hexagonal and other special punch shapes are available for this excellent machine, which is rated for 1/16" steel, or 5/32" aluminum plate.

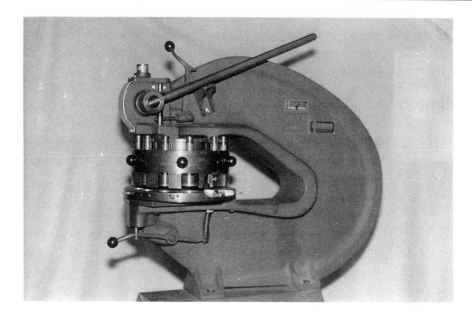

Figure 8-3. Hole size is easily changed on multi-hole turret punches by rotating top and bottom plates that hold punch elements. Rectangular, hexagonal, special punch shapes and knife blades are also available for these machines.

A lever punch for making holes in sheet metal:

No inexpensive metal hole punches are available for hobbyists, and even simple chassis knockout punches cost more than $100 for each size. However, for small classrooms a more affordable alternative to a turret punch is the lever punch shown in Figure 8-4.

The main disadvantage of a lever punch is its small 'throat', or distance from punch to support yoke. With a turret punch shown in Figure 8-3, holes can be made up to 24" from the edge of a metal sheet, but this shrinks to 2.25" for the lever punch in Figure 8-4. For many applications this throat restriction is still acceptable, especially when comparing a lever punch price of $275 with that for a turret punch. Most lever punches accept punches from 1/16" to ½" diameter, and handle the same metal gages as a turret punch.

A lever punch similar to that illustrated in Figure 8-4 is sold by *McMaster-Carr* 3433A21 at $273.97, punch-die sets 3434A21 are $18.11. It takes about a minute to change punch sizes on a lever punch, but hole quality is identical to a turret punch.

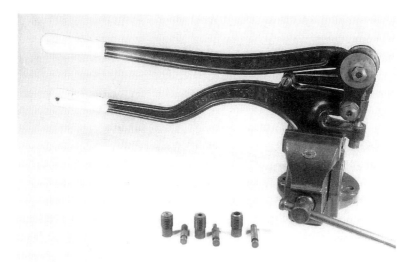

Figure 8-4. Excellent quality holes can be made with this 25" lever punch. At $275 this machine is a less costly alternative to a turret punch. Throat limitations and time required to change a punch, are discussed in the text. Punch and die inserts shown at left, are available in 1/32" increments from 1/16" to 9/16".

Metal hacksaws: A metal hacksaw is one of the most useful basic hand tools, so it pays to spend more for a sturdy frame model that holds a variety of blade lengths. The model shown in Figure 8-5B allows blade installation at 0, ±45 and ±90 degrees. An additional blade can be installed in the upper frame as shown, which is handy for cutting pipe close to a wall etc.

Ordinarily a hacksaw blade is installed so it cuts on a forward stroke, but reversing a blade so teeth cut on a back stroke allows better blade control on tricky cuts. This technique is frequently used with a jewellers saw when following small radii contours. Novices often saw too rapidly, which quickly dulls a blade, especially on hard materials. Slow deliberate strokes with even pressure extend blade life and give a faster cutting rate. Metal hacksaws also make excellent cuts in wood, their fine teeth leaving a surface finish that is easily glued.

'Tight spot' saws shown in Figure 8-5A are used where their name implies, but a short length of blade, wound with a tape, or cloth handle works well too.

Razor saws: Saws with very thin blades are called razor saws. They are invaluable for cutting thin walled brass hobby tubing, soft steels, and a variety of tasks demanding narrow slot cutting ability. Small razor saws such as shown in Figure 8-5C are sold by hobby stores in three blade widths, 0.75", 1" and 1.25". Razor saw blades are commonly available in 0.008" thickness with 42 teeth per inch, or 0.015" and 24 teeth/inch.

Jewellers saws: Jewellers saw blades are made in 20 steps from #8/0 with a thickness of 0.0063" thick and 89 teeth per inch, to a #14 at 0.0236", 18 teeth/inch. These blades are designed to fit a saw shown in Figure 8-5D, and for most work a blade is inserted with teeth pointing toward the hand, giving finer control. Always allow a blade to do the work using only the saw's weight. Do not apply extra pressure with a hand because blades break easily. Even the finest jewellers saw blade cuts through brass or soft steel, and has a very long life on wood and plastics. Hobby stores sell jewellers saws for $10. Blades are $3 for a pack of twelve, and are commonly available in five grades from #2/0 to #3.

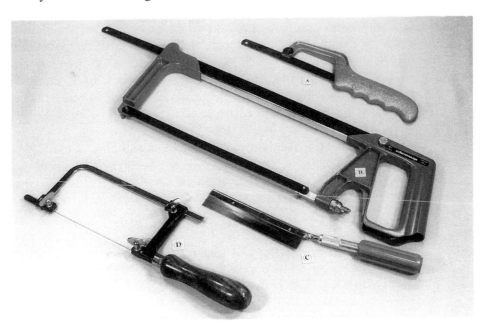

Figure 8-5. Saws are used for making many mechanical robotic parts, and this selection when supplemented with a power jig saw will handle most robotics jobs. [A] A homemade 'tight spot' saw can be made by taping a hacksaw blade, but this blade holder costs only $2. [B] A good hacksaw will accept all length blades. Blades can be tilted at ±45 or ±90 degrees in this model. [C] Razor saw blades may be purchased in hobby stores for $2, and are held in a reusable five dollar handle. [D] Jewellers saws are used for the finest cutting and slitting tasks.

Crosscut wood saw: A hand saw makes less noise, creates less dust and is a lot safer than a circular power saw. Use a short backward stroke to start the main cut, pressing a thumb closest to the work against the side of the blade as a support, while making this starter stroke. Cutting continues with firm steady pressure only on *forward* strokes. Medium paced motion gives a faster cut and better directional control.

Heavy wood pieces can be cut with a 26" crosscut saw. Hole saws, miter saws, tenon saws, bench and radial arm saws are convenient but unessential for hobbyists. A power jigsaw is a more versatile and better investment than a collection of special purpose saws.

Wood saws should not be used for cutting metal, although this sometimes happens in a classroom. If a woodsaw is clamped in a bench vise, its teeth can be sharpened with a fine triangular file, by copying original tooth angles and offsets.

Buying and using a variable speed jigsaw: This important tool is the second power tool to purchase, (after a hand drill). It can be used for wood, metal, plastics, and is perhaps the most versatile saw manufactured. A good jigsaw with blade tilting capability can be used for edge beveling. When used with a straight edge guide, a jigsaw and the proper blade makes very good cuts.

Most work done on a table saw, scroll saw or bandsaw, can be accomplished with an ac power jigsaw. Scroll and bandsaws are restricted by their throat dimensions from handling large work, but make short work of small intricate jobs.

Good variable speed control it most important in a power jigsaw. Some models have jerky controls that cannot be set for a suitable cutting speed, so if possible test a jigsaw before buying. Speed should vary smoothly and continuously, preferably from a finger controlled trigger lever, that can also be mechanically locked at any speed setting. Speed should change minimally under load, and may be checked using a coarse blade on a 1.5" cut depth in pine or other softwood.

Jigsaw blades are made with fine or coarse teeth, double cutting edges, long or short blades, wide or narrow blades, and rectangular or triangular blade profiles. Special purpose blades make short work of wood, plastic, metal or ceramics, and even leather, sponge, rubber, cardboard and other materials can be handled with a knife cutting blade. Carbide grit and diamond coated jigsaw blades are made for working on hard materials.

Blades can often be reused if they break, by grinding off some teeth to make a new shank. A blade modified in this way will be shorter, but still usable. Dulled steel blades clamped in a vise can be resharpened with a fine file, using the original blade contour and teeth offsets as guides.

Drilling a pilot hole for jigsaw blade clearance is not necessary when cutting a surrounded hole in wood. Support a jigsaw so its blade tip is positioned on a part of the proposed hole. The blade should touch the wood at a shallow angle (~30 degrees). Turn the saw on slowly and the blade will cut into wood, making its own hole. It is then used normally to finish the hole.

Circular power saw: This versatile piece of equipment must be used with extreme care, especially in a classroom. Beginners can easily acquire a bad or dangerous habit, such as supporting a two-by-four on a knee while cutting - a common practice on some building construction sites.

Used with a straight guide, a hand-held circular saw produces cuts rivaling those made on a table saw. Splintering or tear-out can be reduced when cutting across wood grain by using masking tape on the cutting line. Shallow blade-width cuts, made at the cutting line location with a utility knife before sawing, are also effective in preventing tear-out.

Circular chain saw blades, rip, crosscut, combination, plywood, veneer, panel, carbide tip, dado, masonry, and diamond tip blades are available for circular saws. Reinforced cutoff wheels can be used in a circular saw to cut through steel beams or hardened stock. Diamond saw blades will cut through ceramics, fiberglass and other brittle materials like butter. Used with discretion, a circular saw is a wonderful tool but must be used with caution by a novice.

Buying a hand drill for robotic use: A variable speed ac line cord hand drill with ⅜" chuck capacity can be purchased for as little as $25, and is the first power tool to buy. A reversing power drill is more useful, since it can be used for inserting and removing screws. Line cord ac drills have greater torque, higher speed and never suffer from low batteries, so they are more practical for a classroom than cordless drills.

Keyless chucks are great time savers. They hold drills as tightly as a keyed chuck, and drills can be removed or replaced in a much shorter time. Some hand drills have a variable clutch to prevent over-torquing when driving screws etc.

Cordless drills: A cordless variable speed reversing drill is ideal for an individual experimenter. Good quality cordless drills start at $50 but prices are falling as more of these tools are used. Look for a drill with a top speed of at least 600 rpm. Limited battery capacity of cordless drills is an impediment when shared by many users, as in a class environment.

Don't buy this type of chuck: Always check a drill's chuck before buying, because many chucks are inadequate for hobby work since they do not close completely. A chuck must be able to hold very small drills (~0.010" diameter) or such drills must be held by a pin vise (Figure 8-6B), which is then held in a drill's chuck.

Dremel™ tools: High speed drilling with tiny drills, grinding, cutting, polishing and other small precision jobs are often simplified with a Dremel™ or similar rotary tool. This type of tool is not essential for either classroom or hobby use, but is handy for sharpening tiny drills and cutting small ceramic or tough metal parts.

Twist drills for hobby work

Complete sets of fractional, numbered, and letter drills from 0.015" to 0.5" are required for professional work, but are big investments for amateurs. Drills are always less expensive when purchased as a set, and a set of drills from 1/16" to ¼" sufficient for small jobs, can be purchased for under $10. Larger, smaller, and intermediate sizes are necessary for some tapped holes, making couplings for motor shafts, and many other jobs.

Metric drills are not essential, because a hole tolerance of 0.002" is adequate on most hobby work, and an appropriate American gage drill can be found that is close enough to satisfy such needs.

Sharpening regular size twist drills: Sharp tools make any cutting or drilling job go faster, with less chance of an accident. Twist drills are easily dulled on stainless steels and case hardened metals, so occasional drill sharpening is necessary.

A toolmaker can explain how to sharpen a drill by hand on a bench grinder. With a little guidance and some practice you will be able to sharpen all size drills from 0.012" to over ½" diameter. Not only is this a wider range than covered by any affordable sharpening jig, you can probably surpass the performance of most drill jigs. Practice sharpening dull drills at the start, then once this has been mastered, try to refinish a broken drill.

Using a fine grinding wheel, rotate a drill half turn from each cutting edge, while progressively slightly increasing the amount of material removed. This technique creates a necessary back clearance, since cutting should only occur at the sharp radial edge on a drill. Any contact other than at the sharp leading edge merely rubs a workpiece, heating the drill. Follow the original drill's contour and alternately grind one side of the tip then the other, removing only a little material each time. A line forms where both sides of a drill's tip meet, and this should be symmetrically disposed when viewed head-on along the flutes. Use a new drill as a guide for checking a sharpened drill's contours, since cutting is faster with the correct tip angle, (typically 118°).

After grinding is complete, remove all burrs with a fine hone, as these impede drilling and prevent swarf from peeling away cleanly.

Sharpening tiny drills: Very small diameter drills ≤ 0.04" are most easily sharpened using a clamped Dremel type tool with a fine abrasive disk. No attempt should be made to use the roll-off procedure previously described, when hand sharpening very small drills. This is too tricky even for an artisan. Merely grind two sharp edged flats that are symmetrically disposed, and at the same angle as those on a new drill. A loupe or magnifier is essential for checking finished tips of tiny drills for burrs, which should be removed with a hone.

Drilling procedures: A pilot hole must be provided before drilling any large diameter hole. This is necessary because pointed ends of twist drills are flat at their tips. Clearance must be provided for this flat, otherwise considerable wasted effort is required to force large drills through a workpiece.

Some common drills and drilling accessories are displayed in Figure 8-6, showing a flat tip on a large drill tip in [A], and a pin vise in [B] holding a #80 drill (0.0135" diameter).

Large drills are ordinarily run at slower speeds, because there is an optimum linear surface cutting velocity V for any material. As drill radius r increases, a drill's rotation frequency f must decrease to keep its linear velocity constant, since $V = 2\pi f r$ at the outer tip edge. For high speed drills, linear cutting speed V is about 250feet/min on aluminum, and decreases to 30fpm for stainless steel. Cutting speed is slower with cheaper carbon steel drills, and faster for carbide tipped drills.

Drills work more effectively when they are run at the correct speed. This is especially important with plastics that can melt, causing drill seizure - a potentially dangerous situation. Most ordinary twist drills have 118° tips, but special drills for plastics and laminates use 60° points without a sharp leading edge. Drilling small holes in plastic with regular drills is usually not a problem, but large holes are more easily drilled if sharp cutting edges on 118° drills are smoothed very slightly, so they cannot dig into plastic. This is most easily done with a light touch on a grindstone, and can be undone to restore a drill to its initial state.

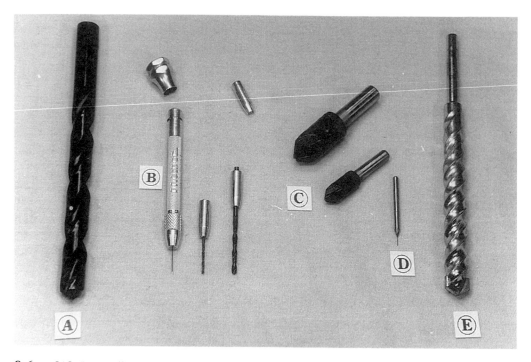

Figure 8-6.   [A] Large diameter twist drills have a flat at their tip, but if a pilot hole is first drilled to provide clearance for this flat, drilling is easier with less drill wander.   [B] This pin vise can be used to hold small diameter drills. Four different size collets are included with the pin vise which can also be used as a hand tool, when a spinner cap at top left is attached. Available from *Hosfelt* #58111 $5.95.   [C] Six-flute countersink bits are ideal for deburring holes, or countersinking for flathead screw clearance.   [D] Short length tiny drills may be purchased with a larger diameter shank, and are widely used for circuit board drilling.   [E] Standard ½" diameter carbide tipped masonry drill with ¼"shank.

Using a center punch: Before drilling a hole in any material, a small dimple should be made to guide the drill and prevent initial wander. Spring loaded 'automatic' center punches are more difficult to use for precise work, so most professional toolmakers prefer a basic punch. A center punch should be hand pressed to precisely mark the point of interest, then tapped with a hammer to dimple a surface for drilling.

Countersink bits, deburring holes, using flathead screws: In some robotic assemblies, clearance problems demand the use of flathead screws. A countersink bit shown in Figure 8-6C is the correct tool for making an appropriate depression, so a screw head can lie flush with the work surface. However, a drill with its diameter a little larger than the screw head may also be used for countersinking, if a countersink bit is unavailable. Seating angles made with a drill will be in error by a few degrees, but this is usually inconsequential for hobby work.

A very slow speed must be used when using a drill for countersinking, alternatively the drill may be slowly rotated by hand. A hand held drill is also very effective for removing burrs from rims of drilled holes.

Making large holes in metal, wood or plastic: A hand held jigsaw is one of the best and safest tools for making holes in thick metal, plastic or wooden plates. If a circle or other shape is carefully marked, and a fine blade at slow speed is used for cutting, good quality holes are easily made.

When 'perfect' circular holes are essential, these must be bored with a jig borer, milling machine or lathe. A fly cutter shown in Figure 8-7A can be used in a drill press, if a workpiece is securely clamped and cutting speed matches the material. Most fly cutters are adjustable and drill press models use a central drill for stability. This means a hole is automatically drilled in any disk cut from a plate. It is usually unwise to operate a flycutter in a drill press without its drill, because this places undue stress on the drill press shank and housing. Even on a large drill press, only light gage material can be cut with a fly cutter. Typical thickness limitations are 1" for softwoods, ½" in Plexiglas®, ¼" with aluminum and 1/16" for steel sheet.

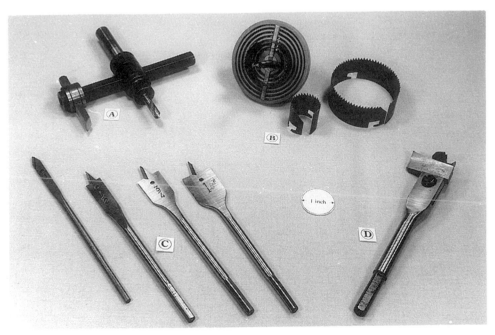

Figure 8-7. Circular holes can be made with tools shown here, but a hand held power jigsaw is more generally useful. [A] An adjustable flycutter may be used on wood, plastic and metals.   [B] Hole saws designed for cutting holes in wood can be used in a drill press. Heavy duty hole saws are made for cutting up to 6" diameter holes through 1" of hard steel, and carbide tipped hole saws give the same performance on ceramic tile, brass and other nonferrous materials. [C] Spade wood boring bits from ¼" to 1½" are one of the fastest ways to get good clean holes in wood. [D] Adjustable spade wood bits are used to drill holes from ⅝" to 5" in diameter. Large diameter holes with an expansive wood bit should be made on a drill press, but holes less than 1.5" diameter can be drilled with care by hand.

An economy fly cutter similar to that shown is sold by *McMaster-Carr* 3082A2, $22.36 and can be adjusted to make 1.75" to 7.875" diameter holes. When using a flycutter always provide a solid disposal backing behind the workpiece, to protect a drill press platform and cutter.

Cutting fluid should be used on metals and plastics for drilling large diameter holes, or when making holes with a flycutter. Multipurpose fluids are available for all kinds of cutting or tapping jobs, and these improve hole quality and cutting capability.

Small holes are easily made in wood with metal twist drills but holes >½" in wood are best made with a spade or auger bit. Spade bits with ¼" shanks are the best buys, these are inexpensive and can be used with both hand drill or a drill press. A set of spade drills from ¼" to 1½" will handle most requirements. Expansive spade bits, fly cutters or hole saws also shown in Figure 8-7 make excellent holes, but can be dangerous in inexperienced hands.

Neater hole edges, free from splintering can be drilled in wood using the following technique. Drill into wood until the tip of a drill just shows on the other side. Stop at this point and commence drilling from the reverse side, using the exit witness as a guide for drill centering

Tapping holes and threading rod: Nuts and bolts are the most commonly used fasteners in robotics, and come in a vast variety of types sizes and materials. Hobbyists should be capable of tapping holes to accept screws, and cutting threads on a rod. These skills are not difficult to learn and greatly enhance construction capabilities.

Some basic equipment for tapping and dieing is shown in Figure 8-8. Utility grade SAE or metric size tap and die kits are sold at auto or hardware stores, starting at $15. Handles and wrenches from such kits are useful, but generally hobbyists need smaller size taps and dies than supplied in such kits. Hardware stores and hobby shops sell handles, wrenches, taps and dies starting at 2-56 or 2mm size. Very small taps such as 1-72 and 0-80 are sold by machine tool suppliers.

A light oil may be satisfactory as a lubricant when tapping or dieing large threads. However, tapping a ¼" deep 0-80 thread in stainless steel is a tougher challenge and needs the correct lubricant. Proper tapping fluid always makes the job easier, it may also be used for many drilling and cutting tasks. It is definitely superior to WD40 oil, and significantly reduces tap breakage.

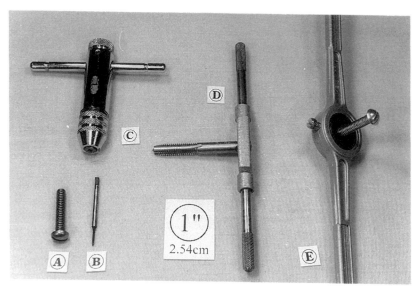

Figure 8-8.   [A] ¼-20 screws are strong enough for even large robotic structures.   [B] Tiny 0-80 holes need this tap, which is the smallest conventional thread size. Only screws used in watches and some spectacle frames have threads smaller than 0-80.   [C] A ratchet tap wrench is handy for tight spots and holds taps up to ¼-20, *McMaster-Carr* 2544A1, $22.50.   [D] Straight handle tap wrenches provide more torque and hold larger taps.   [E] Die wrench is shown with a ¼-20 die installed

Most bolts used by amateur roboticists are less than ¼" in diameter, and Table 8-1 gives useful information for commonly used sizes. When repairing equipment that uses 3-48 or 5-40 screws, it is usually best to re-tap these holes to accept 4-40 or 6-32 screws respectively. Don't waste time looking for metric taps and dies either, although 2 to 5mm sizes are available at hobby stores. They are only rarely needed and a metric threaded hole can be re-tapped to accept the closest American gage screw.

Table 8-1. Most commonly used tap and die sizes for amateur robotics

screw size	tap drill diameter [2]		body drill diameter [3]	
0-80	3/64"	0.047"	#52	0.063"
1-72	#53	0.060"	#48	0.076"
2-56	#50	0.070"	#43	0.089"
4-40	#43	0.089"	#33	0.113"
6-32[1]	#36	0.107"	#28	0.140"
8-32	#29	0.136"	#19	0.166"
10-24	#25	0.150"	#8	0.199"
10-32	#21	0.159"	#8	0.199"
¼-20	#10	0.194	letter 'E'	0.250"

1   Examples: (a) Drill a 0.107" diameter hole for tapping a 6-32 thread (#6 screw with 32 threads/inch)
                        (b) A 0.140" diameter hole allows a 6-32 screw to pass through the material.
2   A quick method for choosing a tapping drill is to align a drill along the tap. Look at both drill and tap when they
     are juxtaposed. If a drill's body just reaches the bottom of a tap's threads, then you have the right drill for tapping.
3   Clearance holes for a screw are made with drill sizes selected from this column. Alternatively, if a drill's body is
     aligned with a screw and just covers the peaks of a tap's threads, that is the correct drill to use.

Tapping holes without breaking a tap: Taps are very hard, extremely brittle and very easily broken, so it is important that an amateur only taps threads by hand. An appropriate tap wrench must be used to hold a tap, and the hand is used to sense a tap's progress through the material. Apply tapping fluid to the hole and gently screw the tap into the hole a fraction of a turn, then back it out, then screw it in again a little further next time. Repeat this procedure until a hole is completely threaded, always sensing for possible tap seizure with the hand.

To minimize tap breakage, drill the correct sized hole using data from Table 8-1 or the table's footnotes. Never make a tapping hole undersize, since this is most common cause of broken taps. However, if a tap hole is too large, tapping will be easy, but screws and bolts will be a sloppy fit. An oversized hole has less material to support a bolt, consequently holding strength is reduced, which can be a serious problem with shallow holes. In such circumstances a shallow hole can be threaded over its complete depth by employing a bottoming tap. A bottoming tap can be made from a regular taper tap, by grinding off its tapered portion.

A broken tap can seldom be removed easily, and spark erosion is too costly for amateurs, so it is wise to tap cautiously, since even pros occasionally break a tap. Dies are more rugged than taps, but caution must still be exercised when cutting threads on thin metal rod or plastics, because these may snap if excessive torque is used.

Screwdrivers: Good quality screwdrivers are worth the extra cost. A blade tip that bends or distorts when handling a tough job is not only annoying, it may also ruin a screw head at the same time. Multi component screwdriver sets are excellent and inexpensive. They have the advantage that their hex shanks can be used in a power drill, and the disadvantage that small bits are easily misplaced in a classroom setting.

Phillips screws are widely used, but screwdrivers often slip in their cross slotted heads, sometimes mangling a head or distorting a screwdriver's tip. In robotics designs where screws must be heavily torqued, amateurs will find hex head, Torx, square head or even slot head screws are a better choice.

Vise grips® and pliers: Vise grips® deserve special mention as one on the most widely used applications of the over-center principle. This nifty tool is now manufactured in many copies, but the original is still best (sometimes by a wide margin). Some genuine Vise grip models have a small wire cutter that easily severs a 0.2" diameter stainless steel aircraft cable, or the same diameter in a mild steel rod. Do not attempt to cut hardened rod or piano wire with Vise grips, this will ruin the cutting blades. As mentioned earlier, hardened shafts can be snapped after notching with a grinder, or parted with an abrasive cutoff wheel.

Vise grips are great for removing seized nuts or bolts with mangled heads, but be careful not to break the bolt or stud. If a nut does not move, heat it gently with a propane flame then apply Vise grips again. This procedure is usually adequate for the most stubborn cases. Vise grips are also the best tool for removing nails and screws that cannot be budged with other techniques. Lock onto a nail then bend it over so the Vise grip forms a fulcrum close to the nail, levering action will now be irresistible and better than any other nail remover. If a good grip can be obtained on a screw head, it can be unscrewed or even torn from wood with Vise grips.

Long nosed pliers and a universal pick-up tool: Multipurpose flat-nose pliers, and bent long-nose electrical pliers supplement pliers recommended for the robotics kit in Chapter 1. Flat-nose pliers are useful for bending small metal brackets but are not essential. Long-nose pliers and magnetic wands are sometimes used for retrieving inaccessible items, but brass and aluminum are not attracted by magnets and often pliers cannot reach through small apertures. A small ball of modelling clay, putty or Blu-Tak® attached to a length of a steel coat hanger, makes a versatile universal pickup tool.

Wrenches, spanners: Only rarely are special wrenches or spanners required when working with robotics related equipment. For those circumstances a little improvisation can usually save the day. Socket sets, open-ended spanners and ring wrenches are inexpensive, and can be stocked as funds permit. More generally useful, are adjustable crescent wrenches and slipjoint pliers.

Allen or hex head wrenches: Because Allen socket screws are probably the best screw fasteners, they are widely used by commercial manufacturers, so hex head Allen wrenches are needed when working on many surplus items. Hexagonal Allen key kits are sold by hardware and auto stores but never contain all sizes, some of which are essential for dismantling or servicing equipment. A list of hex key sizes up to ¾" is given in Table 8-2 so hobbyists can check for missing keys. Obtaining spare keys, especially in the smaller sizes can be difficult, because these are only carried by machine-tool suppliers. A set of 15 short-arm L type hex keys with all sizes up to ⅜" is sold in a vinyl pouch by *McMaster-Carr* 7158A4, $6.72.

Table 8-2. Complete list of hex key sizes up to ¾" (measured across key flats)

0.028"	0.035"	0.050"	1/16"	5/64"	3/32"	7/64"	1/8"	9/64"	5/32"
3/16"	7/32"	1/4"	5/16"	3/8"	7/16"	1/2"	9/16"	5/8"	3/4"

Hex ball wrenches: Ball wrenches are socket hex-head wrenches with a hex profiled ball shaped head, (difficult to describe in words). They are invaluable when working with socket screws in hard to reach places, since the wrench shaft does not have to be in line with a screw's axis. Larger size ball wrenches are sold at hardware and auto stores, and hobby stores normally stock five smaller sizes: 5/64", 3/32", 7/64", 9/64", 5/32" for 2-56, 4-40, 6-32, 8-32, 10-32 socket screw heads. Hobby stores also sell 1.5mm, 2mm, 2.5mm, 3mm and 4mm ball wrenches for 2, 2.5, 3, 4, 5mm socket head screws.

Ball drivers are a pleasure to use, giving a good grip even at large offset angles. Screws can be inserted in awkward spots, by sticking a screw to a ball wrench with a small piece of modelling clay (this works with other screw heads too). Do not apply excessive torque, because ball wrench heads are weaker than a regular Allen screwdriver tip.

C clamps and bar clamps: A range of clamps from 2" to 12" capacity is required for classroom use. These are essential for holding work on a drill press, supporting materials during glue setting time, and for prototype assemblies. Movable clamp fixtures attached to standard metal piping, are the most flexible and least expensive method for clamping large structures.

Tape measures, rules: Ten foot tape rules are necessary in a workshop, and 25ft tapes are useful for ranging experiments. Locking steel tapes with an automatic return feature save time, and also self supporting over reasonable extensions.

American and Canadian scientists use SI units for all their calculations and paper measurements. However, they must often revert to inches, feet and fractional measures such as $\frac{3}{8}$, $\frac{5}{8}$ etc. when designing and building apparatus. Hobbyists and engineering students face the same dilemma, and must be familiar with both measurement systems.

Six and twelve inch steel rules graduated in eighths, thirty seconds etc. are widely available and are inexpensive. Less common, but more useful, are similar rules having additional graduation marks in tenths or even hundredths of an inch, *McMaster-Carr* 6" 2042A85, $4.91 or 12" 2042A86, $8.83. These satin chrome machinists' quality rules make layout and fine design as easy as working in metric.

Calipers, micrometers: Plastic vernier calipers (*Hosfelt* 54-115, $1.95), are adequate for student use in a workshop, they permit inside and outside measurements and are sufficiently accurate for most work. Dial calipers are a better investment than a micrometer for robotic applications. Inexpensive dial calipers (~$30) measure to 6" with an accuracy of 0.001", and are easier to read than a micrometer.

Figure 8-9A shows a 6" metal dial caliper ($29.95), and a utility grade device with a precision of 0.002" costing $15. Prices for an outside micrometer with 0 to 5" capacity, start at about $250.

Squares, angle and center finder: An eight-inch 90 degree carpenters square is fine for most hobby work, but if possible buy a machinists combination square/angle/center finder in Figure 8-9C, *McMaster-Carr* 2007A8, $23.98. Both right angle and protractor attachments on this tool have bubble levels, permitting measurement of tilt angles to better than one degree.

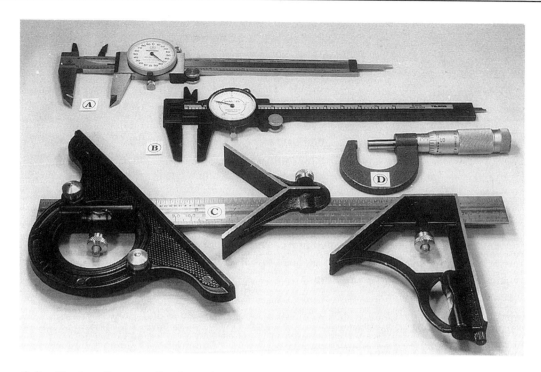

Figure 8-9. Plastic calipers are fine for a classroom workshop, but the dial caliper shown in [A] is a better quality instrument with 0.001" accuracy. *Hosfelt* 54-176, $29.95.   [B] Fiber composite calipers ($15), have an accuracy of about 0.002", which is sufficient for all hobby work.   [C] A combination square/angle/center finder is one of the most useful machinists' tools. This utility grade tool at $24 is adequate for all amateur robotics work, (a high quality version of this same tool costs more than $150).   [D] A 1" micrometer costs about $75 and can resolve 0.0001". Such precision is seldom needed for hobby work. However, measuring capability beyond 1" is frequently necessary, so a 6" dial caliper is more useful.

Hammers: A standard straight claw hammer fulfils most robotic needs and can also be used for pulling nails. Wooden handled hammers are less tiring to use, but heads on wooden shafts may loosen and need tightening periodically.

Tinners metal hand snips, aviation compound cutters: If you can only afford one tool for cutting sheet metal, buy a pair of straight-cut aviation compound cutters shown in Figure 8-10B. Compound action makes this one of the easiest hand tools to use for cutting 1/32" steel metal sheet, or 1/16" aluminum stock. Left and right-hand aviation cutters are also made for cutting curved shapes in metal sheet.

All hand cutters bend metal slightly and nothing beats a shear for a clean undistorted edge. To reduce bending when using hand cutters, make a rough cut to within ¼" of the desired line first. When this ¼" selvage is subsequently trimmed off it will curl away, leaving the remaining sheet flat.

A large pair of tinners snips shown in [A] gives the cleanest cut of any hand metal cutter. Cutting is easier when one grip is clamped in a bench vise. One hand may then be used to support and guide a workpiece, while the other hand performs the shearing actions. This model can be used on 20 gage sheet steel (~1/32"), or 1/16" aluminum.

A small pair of compound straight cutters in [C] makes excellent cuts, and can be used on gages up to 1/32". Plastic handled scissors may break when used on metal, but heavy metal scissors in [D] will cut steel sheet up to 1/64".

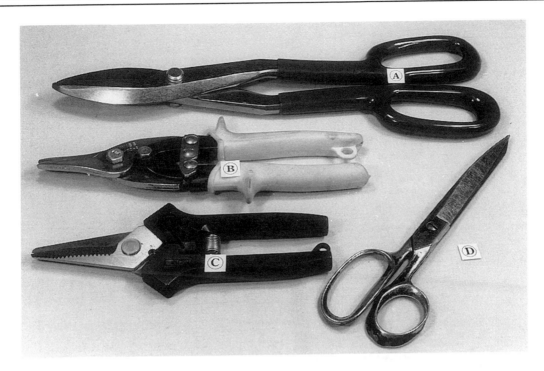

Figure 8-10. [A] One of the best hand tools for cutting sheet metal, these 12" tinners snips with 3" blades from *McMaster-Carr* 3902A6, cost $16.73. [B] Straight-cut aviation snips with compound action have corrugated blades, and will cut 1/32" sheet steel without tiring the hands. [C, D] Light duty compound cutters or heavy scissors are handy for making clean cuts in 1/64" sheet steel.

Files and filing technique: Filing is necessary at times, especially if milling machines, lathes etc. are not available to give a machine-tooled fit and finish. A collection of basic files is required for a classroom, and these should range from needle size to 1" width. Coarseness should vary from bastard to smooth, and shapes in flat, round, half round, rectangular and triangular are required. A wood rasp cuts wood faster than any metal file, and scalloped teeth on one side of these files removes material at a great rate.

Good filing needs slow even pressure strokes, which removes material most rapidly, while reducing file wear. Files are designed to cut on a forward stroke, so maintaining pressure on a back stroke wastes energy and decreases file life. Although metal files work well on wood, wood files should not be used on metal or they will wear out rapidly.

Utility knives, razor blades, chisels, wood planes: Snap-blade plastic handled utility knives with 17.5mm wide blades are adequate for classroom and hobby use. These knives are inexpensive and their useful life can be extended considerably if blades are resharpened on fine abrasive paper, instead of continually discarding snap-off sections. Single edge razor blades are sold in plain steel or stainless steel and are very useful for fine work - these can be sharpened too. Razor blades can be used to make optical slits, and may be used as wedges, producing powerful forces when hammered into small gaps. This is a useful method for separating parts when even a thin screwdriver blade is too large.

When used carefully, chisels can be used for some tricky jobs. For centuries craftsmen made excellent tongue and groove joints with only primitive hand tools. Four chisels from ¼" to 1" are sufficient for a classroom, but individual workers can manage with a utility knife if funds are tight. Chisels are readily sharpened on a grinder or a sheet of carborundum paper backed by a glass plate. Despite a mystique surrounding chisel blade sharpening angles, if a blade is sharp, its precise angle is of minor importance for hobby work.

Wood planes are seldom needed for robotics work, and their use can often be avoided by checking the straightness of all lumber before buying. Always place an eye close to a piece of wood, plastic or metal stock, sighting along its length - any non uniformities or bowing will be readily apparent. Small razor planes are sold at hobby stores for trimming and planing small softwood or balsa pieces.

Plastic sheet cutters: Acrylic and other types of plastic sheet are very useful for building robotic apparatus. Although this material can be cut with a saw blade, a cutter specifically designed for the task is better. An Olfa P-800 cutter with a tungsten steel blade is shown in Figure 8-11 - this model is sold at hardware stores for $7, and replacement blades are three for $2. As with other tools, cutting blades for an Olfa P-800 are sharpened on a hone or carborundum sheet. Refinish the cutter's face, (not the blade's wedge angle).

Curved cuts can be made clean through 1/16" plastic sheet with several passes of an Olfa blade, and examples of sections made with this tool are shown in [A]. Straight cuts in thin or thick acrylic sheets up to ¼" are made with a single score line on one side of a sheet, which is then snapped apart by bending along the line, using a supporting board and clamps if necessary. If a metal bender in Figure 8-2 is available, it can be used to snap prescored ¼" acrylic sheet with ease.

Lexan® and similar polycarbonate sheet used in 'unbreakable windows' is very flexible, and difficult to snap apart even with a deep score. Lexan can be cut with a jig saw blade, operated at a slow speed to avoid melting the material.

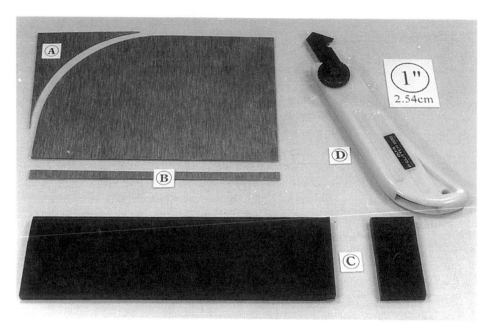

Figure 8-11.   [A, B] Curved or straight cuts in 1/16" hard plastic sheets are made using a cutter shown in [D]. Quarter inch acrylic shown in [C] was sectioned after scoring lines on opposite sides of the material, but on larger sheets scoring from one side is usually sufficient.

Making 'perfect' holes in balsa wood, rubber and soft materials: A method for producing excellent holes in balsa wood, was described for making the rotation sensor in Figure 5-56, and hole quality using this technique with hobby tubing or cork borers, is shown in Figure 8-12C.

Hand held cork boring tubes can be used for making holes in paper, cardboard, Styrofoam®, balsa wood, sponge, rubber, soft plastics, and not surprisingly in cork too. A set of cork borers with 4,6,7.5,8.5,12,13,15,17,18.5,19, 20.5mm outside diameters is sold by *Cole Parmer* H-06298-90, $51. Cole Parmer does not list a sharpener for refurbishing brass cork boring tubes shown in Figure 4B, but this tool may be purchased from *Canadawide Scientific* 32650-11, $38.

Home made boring tubes can be fabricated from brass hobby tubing, spun in a drill chuck and chamfered to a sharp edge with a fine file or emery block. When boring by hand, use a cloth pad and push the tube through with a screwing motion, employing another soft pad to protect the tubing's sharp edge on breakthrough. Take care not to cut your hands when using cork borers (it is easily done!).

Boring tubes made from hobby tubing do an excellent job on thick specimens, if a tube is held in a drill press chuck running at a few hundred rpm. With this method it is easy to drill neat holes through thick books so they can be secured with a cable.

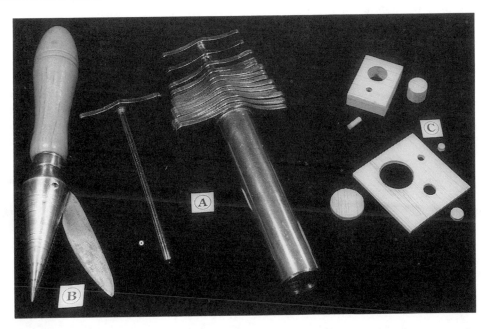

Figure 8-12. [A] This fifteen component set of cork borers from 0.2" to 0.95" outside diameter, can be used to make excellent quality holes in all kinds of soft or spongy materials. [B] Sharpening knife for refurbishing brass boring tubes. [C] Holes cut in balsa also provide mating plugs of the same quality.

Glass cutting: Cutting glass sheet is a useful and readily acquired skill, making custom fabrication of mirrors, beamsplitters and windows, an in-house capability. Purchasing larger sheets of mirror and special glasses is also more economical than buying an equivalent number of smaller individual elements.

Flatness of glass sheet (an inexpensive surface plate):

Most glass plate is flat to better than 100λ/inch (one hundred wavelengths at ~550nm over a one inch length), or about 0.0015" per inch. So almost any sheet of glass makes a good surface for figuring and polishing small objects or whenever an inexpensive flat is required. Glass plate is also very useful as a cutting board for paper, cardboard, tape or similar materials, because glass does not appreciably dull a sharp knife edge. A sheet of graph paper attached beneath the glass makes a useful rectangular guide.

'Float' glass is made by pouring molten glass onto liquid tin, producing the flattest glass from any rapid commercial process (only grinding and polishing produces better surfaces). Float glass is about ten times flatter than most ordinary glass (~7λ/inch or 0.00015"/inch). It is inexpensive compared with a granite surface plate, and equals the quality of a grade 'A' machinist's inspection plate.

Home made cylindrical lenses and beamsplitters: Glass rod, or glass tubing filled with water or a transparent gel (eg General Electric RTV615) is useful for fabricating home made cylindrical lenses. Clear acrylic or polycarbonate rods and tubing can be used for the same purpose, but these materials are softer than glass so are scratched more easily. Thin glass microscope slides or microscope cover glasses, can be used as optical beamsplitters or even as polarizers when stacked together. More information on these elements is given in a future book.

Cutting glass at home: If you are apprehensive about cutting glass, look in the Yellow pages for companies providing this service. However, a little practice on old pieces of glass sheet and tubing will develop the necessary skills required to make many useful items from glass at home. Basic glass cutting tools for a classroom are shown in Figure 8-13, and these handle most robotics requirements.

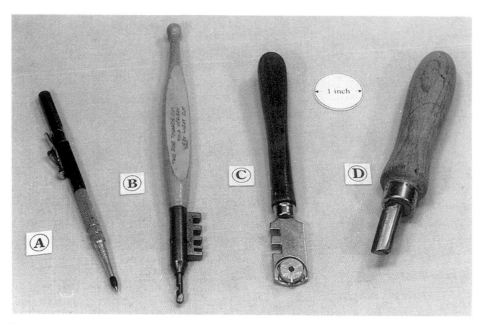

Figure 8-13. [A] a carbide tip scriber can be employed for cutting circles or straight cuts in glass plate ≤0.020" thick. [B, C] Diamond and standard wheel type glass cutters.   [D] Carbide tipped blade for cutting glass tubing or rod.

Glass cutting procedures: Single-wheel glass cutters are inexpensive and available from hardware stores. A six-wheel version shown in [C] allows a new wheel to be rotated into position as required. Wheel cutters are adequate for glass or ceramic from about 0.1" to 0.25" thickness.

When cutting plate glass (⅛" to ¼"), press the cutting wheel at the far glass edge, then draw it toward your body with a bold even continuous motion, leaving a visible trace that is not deeply gouged. Position a ½" length of pencil body or a piece of ¼" wooden dowel beneath the glass, centered on the trace end closest to your body. Place gloved hands on the outer edges at each side of the glass closest to you, then press down simultaneously on both sides, snapping the glass apart.

Try to remember a glass cutter does not cut glass - it makes a scratched channel with local stresses that weaken glass around the scratch. A very deep cut is therefore counterproductive, because maximum stress concentrations occur when a scratch is sharp, not necessarily deep. This subject will be dealt with in a future book, when techniques for cleaving silica fiber optic cables are discussed.

Some glaziers dip their wheels in kerosene and this may help weaken glass at a scratch. It does help to wet a scratch with saliva when cutting glass tubing, and micro cracks on a pristine glass surface, grow larger in a humid environment. So there is probably some merit in wetting a wheel, but the effects are marginal since perfectly good cuts can be made with a dry wheel. Thin glass (0.040 to 0.125") ordinarily needs a diamond cutter, but it can be cut with a wheel cutter after some practice.
     Only light pressure is needed with a diamond cutter, and practice is the best way to learn the technique. Hold a cutter upright, using an even pressure stroke that leaves a fine scratch. It is important for this scratch to extend completely from edge to edge across a plate. Normally thumbs are placed close together on either side of the scratch, and forefingers used underneath the glass for support. Downward and sideways pressure with the thumbs is used to part the two glass sections. Local glass and mirror shops will usually cut glass for a nominal charge if you have trouble with any of these procedures.

Cutting circles from glass sheet: Small circular disks can be cut from thin glass sheet using a circle template as a guide. Use a carbide scriber or a diamond cutter held at a constant angle to the circle. Additional straight line scribes radiating outwards from the circle permit unwanted glass to be broken away. Figure 8-14A shows a 1" circle with radial lines (blackened for contrast), scribed on 1mm photographic plate - a finished circle is shown in [B]. Thin beamsplitters (reflection ~8%, transmission ~92%) made from 0.2mm microscope cover slide and, other shapes are shown in Figure 8-14 [C]. These were made with the carbide tipped scriber from Figure 8-13, using a drafting template as a guide.

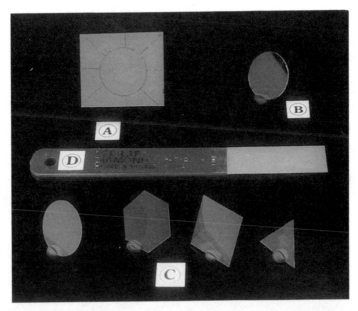

Figure 8-14.   Circles can be cut from a thin glass sheet with a carbide tipped scriber shown in Figure 8-13. [A] A circle is first marked using a light touch. Radiating lines are added so surplus material can be broken away leaving a finished circle shown in [B] (lines have been blackened for contrast).   [C] Various shapes may be cut from thin glass plate, microscope cover glasses, or microscope slides using techniques described in the text.   [D] Diamond hones are superb for smoothing glass edges or sharpening carbide tipped tools.

Cutting glass rod, tubing or bottles: Glass tubing in diameters from 0.05" to 0.5" can be cut with a carbide tool similar to that shown in Figure 8-13A. Buying this special blade for glass cutting (~$40), is not necessary, because satisfactory scratches can be made with any inexpensive straight edged carbide tool available at hardware stores.

Glass rod or tubing up to ⅜" diameter is given a deep scratch about ¼" long perpendicular to its long axis, then snapped apart. Wet the scratch with saliva, then press both thumbs close together and on the opposite tube side to the scratch, flexing the tube with a pulling motion until it parts.

Large diameter glass tubing or even glass bottles are cut by first marking the complete circumference with a carbide cutter or scriber. This mark is subsequently heated at one point on its score, using either a micro-tip flame or a red-hot piece of ⅛" diameter glass rod. A crack will propagate from the hot spot and follow the line. With very large tubing or bottles, an initial crack may not completely encircle the object, so just repeat the hot point application procedure. An alternate crack initiation procedure employing a red-hot nichrome wire is popular with glassblowers, and commercial devices incorporating heated wires are made specifically for this purpose.

Drilling holes in glass: Diamond core drills are fastest for making holes in glass or ceramics, but with prices starting at about $70 they are seldom used by hobbyists. If you can spare the time, a ¼" diameter hole can be drilled in glass at a rate of about one inch an hour, using brass hobby tubing and auto valve grinding compound. Make a dam of modelling clay around the proposed hole, and cut four notches at 90° in the tubing to carry slurry around. Run the drill press at its slowest speed, backing off frequently to carry fresh compound to the drilling area. Light oil, or Varsol added to valve grinding compound reduces its viscosity to a workable level.

Diamond hones: When glass edges need smoothing or carbide tipped drills require sharpening, it is not necessary to use special equipment. Diamond hones can easily handle such tough jobs and they need no cooling water.

Excellent diamond hones shown in Figure 8-14D are made by EZE-LAP Diamond Products, and sold by *McMaster-Carr* in four grit sizes (superfine, fine, medium, coarse) 43545A51/52/53/54, $4.70, $4.70, $7.54, $9,45. Abrasive sections 2" x ¾" are attached to a 6" x ¾" stick, making a very convenient hand-held or table hone. A few well-stocked hobby shops and better machine tool stores also carry these items, as they gain popularity with model builders and hobbyists.

Pipe and tubing cutters: All metal and plastic pipe can be cut with a suitable saw, so a pipe cutter is not an essential tool. A small rolling blade cutter designed for hobby use is shown in Figure 8-15, and cuts copper or brass pipe/tubing up to ½" diameter, with clean ends perpendicular to the pipe axis. Cutters for larger pipe diameters are sold at hardware stores.

Tubing cutters produce a square cut end but always with some squeezing at the cut location, so ends may need finishing with a fine file or abrasive paper. Small diameter brass hobby tubing (≤⅛") can be vee notched through the wall in one spot with a sharp file, then broken apart by bending, leaving a good end finish.

The difference between most tubing and pipe is primarily one of stiffness - tubing is soft and is more easily bent, pipe is harder and will occasionally crack if bent. Copper tubing that has been repeatedly flexed work hardens. It can be resoftened (annealed) by heating with a flame, then allowed to cool slowly.

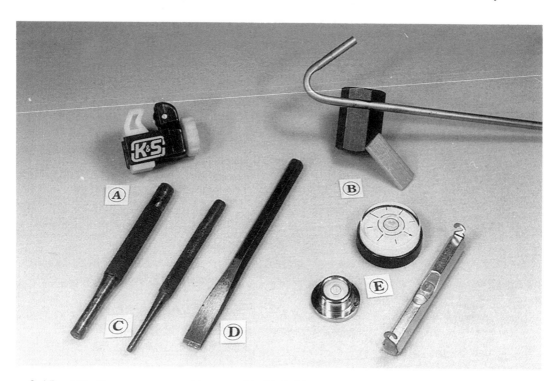

Figure 8-15.   [A] Brass or copper tubing cut with this hobby tubing cutter will have square ends, requiring only minimal finishing.   [B] Both hard and soft 3/16" tubing can be bent without buckling in this hobby tubing bender. Designed for use with a bench vise, it sells for about $4.   [C, D] Pin punches and chisels are very useful for those occasions where no other tool will suffice.   [E] Bubble or line levels are lightweight and small enough to be carried on robotic equipment when inclination is important.

Bending metal tubing without kinking: During bending, an outer tube wall stretches and its inner wall is compressed, therefore a tube must be supported to prevent walls from collapsing. Bending tubing without kinking requires a tubing bender, a bending spring that slips over the tubing, a machined channel, or sand fill and stoppers. Perhaps the most economical tool for hobbyists is a fuel line bender suitable for use on 3/16", ¼", or 5/16" copper, aluminum, or steel tubing. This tool will also bend aluminum or copper up to ⅜" diameter, and is available from *McMaster-Carr* 2478A12, $11.70, or at some auto supply stores.

A machined groove channel bending jig for 3/16" hobby tubing is shown in Figure 8-15B. It is clamped in a vise when used, and will bend even hard brass hobby tubing with only slight distortion. Each pipe size requires a different size groove with this type of bender. If a lathe is available suitable aluminum stock can be channeled with multiple machined slots to make a more versatile and less expensive bender.

Punches and chisels: Even classrooms do not need a complete punch and chisel set, but a few pin punches and a chisel or two shown in Figure 8-15 come in handy for driving out cotter pins, stuck shafts, or cutting where a hacksaw blade cannot reach.

Levels: Standard carpenters levels are very useful but too large to carry on most robotic assemblies. Bubble levels shown in Figure 8-15 are small enough to install on robotic platforms, have omnidirectional sensitivity, are lightweight, inexpensive, and may be obtained at hardware stores.

Most inexpensive bubble levels have a resolution of 0.010"/ft, so a 1 foot level can detect a tilt of 0.048° or about 3 minutes of arc. A 1" diameter bubble level detects tilts of about 0.6° and a line level shown in [D] about 0.2°. Precision 12" levels have a resolution of 10 seconds (0.0027 degs) but cost several hundred dollars. Bubble levels are the least expensive, and one of the most reliable methods for determining tilt, unfortunately they are difficult to interface to a computer. Any bubble level is easily checked for consistency by reversing it end for end on a flat surface - a true level shows the same bubble displacement, regardless of orientation.

Heavy soldering equipment: Soldering is one of the more important skills for practical robotics, not only for making electrical connections (covered in Chapter 7), but for a variety of other tasks. Many jobs are greatly simplified by judicious use of solder. Soldering is primarily under-used because of poor technique.

Fittings and tubes made from copper or brass are readily soldered with rosin cored flux and a propane torch flame. Thin metal sheet or galvanized steel sheet is often rendered unsolderable if overheated by a flame, so is best soldered with a large soldering iron tip attached to a propane nozzle shown in Figure 8-16. Large electric soldering irons (>100W) are excellent for soldering heavy items, but prices for a good 200 watt iron start at $75.

All kinds of mechanical attachments, supports, brackets, stands etc. can be quickly fabricated using brass, copper or sheet metal, so time spent learning how to solder expertly will be amply rewarded.

Solder and fluxes: Acid core solder, or plain wire solder with a liquid soldering flux (usually containing hydrochloric acid and zinc chloride), is required for soldering iron, steel, or stainless steel. Galvanized iron, brass, copper or tin are readily soldered with an acid-based solder paste.

Solder itself is quite weak, and this can be confirmed by pulling apart a length of solder - it withstands very little force before breaking. Therefore if possible, any joint should be seamed or overlapped for extra strength, before adding solder.

Techniques for soldering heavy items: Soldering heavy items requires the same care and attention used for electrical components, but some additional points have been added below to basic guidelines given in Chapter 7:

[1]   All items must be free from grease or oxidation for proper bonding

[2]   Apply either liquid or paste flux to areas being soldered.

[3]   Where possible, pre-tin (cover with a wetted solder layer), items before they are joined.

[4]   Allow sufficient time for a soldering tip to reach its proper operating temperature, then reduce the propane flame to avoid burning a hand while moving the torch around. Clean the solder tip on a damp soldering sponge before soldering.

[5]   Pieces are soldered together only after they have been overlapped then clamped, wired or otherwise supported, so they cannot move during soldering. Take care that too much heat is not carried away by clamps, or soldering may be impossible.

[6]   Apply solder tip and solder simultaneously to the workpiece. This is essential, otherwise heat cannot be efficiently transferred to the work. Neglecting this simple step, is the most common mistake made by beginners.

[7]   When soldering tubing or solid pieces with a flame, do not heat the soldering area directly - this burns the flux. Move a soft flame continuously over other areas close to the solder junction, while touching solder on the soldering area to test for melting. When solder starts to melt, cover the joint with solder, allowing it to flow into all crevices. Then remove the flame and allow to cool.

[8]   Once all joints have been completed on a workpiece, flux paste can be removed with a Varsol soaked rag *(be sure no flame is around when doing this cleaning, because Varsol is flammable)*. Liquid flux can be washed off in warm sudsy detergent water.

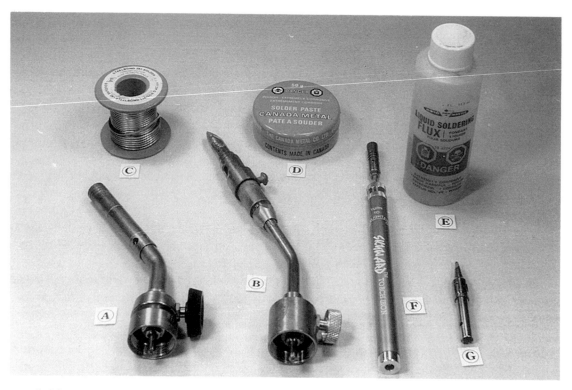

Figure 8-16.   [A] Brazing head fits 14oz or 16oz propane cylinders.   [B] Regular propane flame head with a soldering iron tip attached.   [C] Acid core 'steel bond' solder or low temperature silver solder is much stronger than regular soft solder   [D] Acid solder paste   [E] Liquid flux.   [F] Small butane torch with flame head.   [G] Soldering tip for butane torch in [F].   Items [F, G] are sold at hobby stores $7.50. *Radio Shack* sells a Portasol butane powered flame/soldering kit 64-2182, $34.99.

Soldering aluminum: Perhaps an aluminum flux is manufactured that really works, but those tested by the author on aluminum sheet or heavy stock, have all failed to meet advertised claims.

Soldering aluminum is difficult because an oxide layer immediately reforms no matter how frequently or aggressively the surface is scraped. However, if a puddle of *plain solder* is made on aluminum, its oxide layer can be removed by scraping the aluminum surface with the iron's tip, while poking it through the molten puddle. Oxide formation is inhibited during this procedure, because a solder puddle excludes air from an aluminum surface.

This technique requires a little practice and only works with a heavy duty soldering gun/iron, or a propane torch with solder tip attachment. Be sure to apply enough pressure to cut through the hard oxide layer. Once solder has stuck to the aluminum then other pieces of brass, copper, aluminum etc. can be attached in the normal manner. Do not add any flux, and use only *plain* solder - rosin core solder will not work.

Joints can be made on ⅛" aluminum, or even thicker stock, providing sufficient heat is applied. Connections are as strong as any normal soldered connection, and this may be tested by pulling a joint apart. Both workpieces will retain a solder covering, indicating breakage has occurred at the joint and not at an aluminum-solder interface.

Using the method just described, it is easy to win wagers from 'know-it-alls' who insist *"aluminum can't be soldered"*. However, because soldering other metals is so much easier, only rarely is the extra effort warranted.

Brazing: Copper, brass, cast iron, Nichrome, cold and hot rolled steels, nickel and many other metals can be brazed or silver soldered, providing they can be heated adequately. A brazing head in Figure 8-16A, propane cylinder, brazing rod and flux are necessary for making hard soldered joints.

Brazing procedures are similar to those for soft soldering with a flame, and all parts must be thoroughly cleaned and clamped before heating. Workpieces get very hot when brazing, so work should not be done where a fire could start if a red-hot piece of metal falls on the floor. When pieces being joined are properly heated, high temperature silver solder, or brazing rod flows freely. Brazed bonds are extremely strong and suitable for joining wheelchair axles etc.

Silver soldering and brazing are not synonymous. Some 221°C silver solders may be used with a regular soldering iron, and are only a few times stronger than regular soft solder.

Using a heat gun to bend hard plastics or undo epoxied joints: A heat gun or hot air paint stripper is ideal for bending acrylic and other plastics. Heat guns can also be used to ease seized bolts, accelerate paint drying, or shorten epoxy cure times. They also work well on heat shrinkable tubing discussed in Chapter 7. When gently heated, vinyl and other flexible plastic tubing, can be stretched to embrace much larger fittings than possible at room temperature.

Chemical epoxy debonders work very slowly on closely fitting parts, and heating is the quickest way to separate metal, glass or plastic parts joined with epoxy. Usually a heat gun is adequate, but very heavy metal parts may require heating with a propane torch. Perform these operations in a well-ventilated area because hot epoxy fumes are noxious and mildly toxic.

A 1200W economy heat gun operating up to 1000°F, similar to that depicted in Figure 8-17, may be obtained from *McMaster-Carr* 32605K44, $28.72.

Bending acrylic sheet: Plexiglas™ and similar acrylics in rod or sheet form, are easily bent with very hot air, and some examples are shown in Figure 8-17. Plexiglas has excellent insulation properties, and is a strong and tough material. With only half the density of aluminum, it is stiff enough for robotic fingers or arms, wheels, brackets, shelves etc. Plexiglas is easily sawn, drilled, filed, or tapped, and straight cuts are quickly made with a plastic cutter from Figure 8-11. Strong joints can be made with an appropriate adhesive, so Plexiglas is a very versatile material for making robotic elements.

Even hard laminates similar to Arborite™ or Formica™ can be bent in a hot air flow without cracking, although Plexiglas can be bent more sharply. Most plastics are thermoformable, but it pays to test a new material to ensure its properties are still acceptable after heat treatment.

Figure 8-17. Most plastics are easily bent or formed into other shapes, by heating with a hot air gun.
[A] ¼"Plexiglas™ specimen after hot-air bending.   [B] Hard laminates such as Arborite™ can be bent with appropriate heating.   [C] 3/64" acrylic bends readily around a cylindrical pipe when heated with a gun shown in [D]

Glue gun, glue pot: Hot melt glue is very handy for quick gluing jobs in a laboratory, and more information will be given later in this chapter when adhesives are discussed. A small glue gun is adequate for hobbyists, and a selection of both high and low temperature sticks handles many gluing needs.

Staple gun, pop rivets: A heavy duty staple gun with long and short staples is a handy tool, but is not essential. Most office staplers open, allowing them to be used as a light duty staple gun, fine for tacking cardboard and plastic sheet.

Pop rivets are useful when many fasteners are required in easily accessibly locations, such as on panels or for some rough auto body work. Pop rivets have the disadvantage of being difficult to remove, and requiring large back clearance for the severed shaft. Nuts and bolts or sheet metal screws are better than pop rivets for most robotics work.

Sander, router: Some disk, belt and orbital sanders are sold with dust bags and these can be used in a communal workshop. Use of regular power sanders or plain sandpaper should be restricted in a classroom, because sanding dust sometimes contains traces of toxic paint, glues and binders. When polishing LEDs etc. use wet or dry paper in a wet state.

Solvents: Most people are aware some plastics are dissolved in nail varnish remover (acetone), yet almost nothing harms polyethylene. Rubbing alcohol (methyl alcohol) removes some stains that acetone will not touch, and water is called a universal solvent because it dissolves all materials to some extent. It is useful to know acetone is a solvent for contact cement, plastic wood, hobby and household cement, and a water cleanup is ok for latex products and white glues. No common solvents are available for hot melt glue, silicon rubber, epoxy or super glue, although special debonders or softeners are sold for the last three items.

When testing a surface for solubility try water first, then alcohol, Varsol (VARious SOLvents), acetone and paint remover (usually methylene chloride, or lye based), in that order. If none of these works then test with muriatic acid (hydrochloric acid), lye (sodium hydroxide) - these are water soluble (but add acid to water - not vice versa). Sulphuric acid is used in some drain cleaners (check the label), and this vigorously attacks certain metals. Lye (sodium hydroxide) is a powerful and potentially hazardous alkali. It rapidly eats wet aluminum, generating considerable heat in the process, and with a dangerous evolution of hydrogen.

All these solvents are either flammable, form explosive mixtures, eat flesh, corrode metals and are potentially dangerous. Nevertheless they are extremely useful to well-informed roboticists, and every one of the items mentioned is sold by local hardware stores.

Instructors must use their own discretion when deciding which chemical solvents should be stored in a workshop classroom. If their knowledge is sparse in this area, it is better to err on the side of caution. Flammable solvents may be kept in red polyethylene gasoline cans sold at auto and hardware stores. These containers meet most building safety codes for small quantity storage.

All containers should be clearly marked with tags or labels that cannot wash off. Plastic containers should be used to store flammables, acids and strong alkalis in a classroom workshop, because glass bottles are too vulnerable. Long-term storage of fuels or other flammables in metal cans may be hazardous, due to leaks caused by internal corrosion from condensation or water contamination. Some plastic containers become brittle with age, even when kept in darkness. Therefore, instructors must be vigilant with all hazardous liquid storage.

Greases, oils, cutting fluid: Vaseline™ or petroleum jelly is one of the safest greases for classroom use, but an all-purpose automotive grease is better at higher temperatures. WD40, fine machine oil, and 10W30 oil will handle other lubrication needs.

Drilling, tapping, threading, flycutting, sawing and other tasks on metal are easier and give better results if a cutting fluid is used for these operations. Tap breakage is reduced, and closer tolerances with a better finish are possible, when proper tapping and cutting fluids are employed.

Polishes: Truncated LEDs, opto emitters, plastic fiber optic cable, Plexiglas™ prisms, cylindrical lenses and conical light cones can all be polished at home. Such items are brought to a satisfactory surface quality using various grades of abrasive papers, then given their final finish with a polishing compound. Brasso™ and Silvo™ applied with a soft cloth produce an excellent finish, and both products are inexpensive and sold in hardware stores. Many polishes, including Brasso and Silvo are flammable, so must be used with discretion in a workshop.

Adhesives: Double sided sticky foam tape, superglue, epoxy, hot melt glue, contact cement and silicone rubber, are very useful for roboticists, not only for temporary assemblies but also in more permanent structures. At one time perfectionists frowned on using glues when constructing apparatus. This philosophy is indefensible today, when even some advanced jet fighters have wings fabricated with adhesively bonded composites. Many other examples could be cited where adhesives are the only suitable fastener.

Typical properties for adhesives often used by hobbyists are given in Table 8-3, this list is not complete because many other useful glues are made, and may be better in certain applications.

Removing adhesives is often much more difficult than applying them. Heating weakens most glue bonds and debonders are made for cyanoacrylates and epoxies. DAP's 'Silicone-Be-Gone ™' softens cured silicone rubber, and solvents for some adhesives are listed in Table 8-3.

Table 8-3. Bonding characteristics and other properties of common adhesives

best = ★★★	white glue	hot melt glue	contact cement	super glue (CA) (cyanoacrylate)	epoxy	silicone seal (RTV rubber)
cardboard, paper	★★★	★★★	★★		★	★★★
wood	★★★	★	★★	★★	★★	★★★
fabric	★★	★★★	★★	★	★	★★
hard plastics			★★★	★★	★	★★
polyethylene, vinyl rubber, Teflon (PTFE)			★★★	★★	★	★★
metals			★★	★★	★★	★★★
glass, ceramic	★★	★	★★★	★★	★★	★★★
water resistance		★★★	★★	★★★	★★★	★★★
tensile strength		★	★		★★	stretches well
flexibility	★	★	★★	★	★	★★★
fills gaps		★★★		★	★★★	★★★
viscosity	★★	★	★★	★★★	★	★
solvent (pre-cure)	water	none	acetone	acetone	alcohol	none
solvent (post cure)	hot water	none	acetone	acetone/debonder	debonder	softener
max temperature	150 °F	+140 °F	150 °F	200 °F	150 °F	600 °F
cure time	> 1h	~60s	~1h	<60s	5min, 12h	24h

Categorizing properties in a table is open to all kinds of criticism, and justifiably so by those manufacturers making leading edge products. For example solder can be melted in a bowl made from high temperature epoxy, and copper silicone seal sold for automotive needs operates to 700 °F. Special epoxies used with selected etchants adhere tenaciously to Teflon™, and others will stick and cure under water.

A specially formulated white filler used by dentists can be molded and formed at leisure, then cured to a hard setting plug in a few seconds with ultra violet light. Similar transparent compounds are used for permanently bonding multiple elements in camera lenses, after they have been optically aligned.

For most robotics applications 5 minute epoxy, cyanoacrylates (CA, Superglue or crazy glue), Weldbond™ space age universal white glue, hot melt glues, contact cement, silicone seal (RTV rubber) and Devcon's Duco cement, fill most needs.

If doubt exists concerning which adhesive to employ, remember *'nothing beats a test'*, so test all candidate adhesives on those materials you wish to join, to find the one best suited for your application.

Cyanoacrylate adhesive debonders: Skin suturing with superglue was one of the first uses for cyanoacrylates, and is still used by surgeons for that purpose. Thin glue layers, slightly moist conditions and heavy pressure are all present during most skin bonding accidents, when a tiny drop of glue seeps between finger pads as they are being used to hold a joint. These same conditions should be provided when bonding other materials with superglue. Always keep cyanoacrylate debonder handy for separating accidental skin adhesions, because cementing fingers together is an occupational hazard for modellers. If debonder is not on hand, try a hot water soak or acetone, moving fingers with a sliding peeling action. Pulling fingers apart directly will tear the skin.

Most superglues work very well on glass, ceramic and other non-porous materials. Others will glue leather, vinyl, carbon fiber, plywood, fiberglass, polycarbonate, and some even work on oily or fuel-soaked surfaces. Rearview mirrors are now stuck to windshields with a bond that is as strong as the glass itself, a testament to the impressive performance of modern adhesives.

Cyanoacrylates tend to wick along exposed fibers on across-the-grain wood sections, and this type of joint is always weaker. Special CAs partially alleviate this problem, but bonds parallel to wood grain are stronger. If wood sections perpendicular to the grain are pre-coated with CA then allowed to dry, they can be subsequently glued with CA to give a high strength bond.

Cyanoacrylates are constantly being improved with added features. These include gap-filling, slow setting, fast setting, better flexibility with higher shear strength, compatibility with a wider range of substrates etc. Keeping track with the latest innovations is difficult unless one reads model magazines or frequents hobby stores.

For those hobbyists interested in fingerprint analysis, CA fumes are used to enhance fingerprints so they may be photographed with suitable illumination. This is now a standard forensic procedure

Cementing Teflon™ and other no-stick materials: Teflon™, polyethylene, vinyl, silicone rubber and glass are problem materials for most adhesives. Surface preparations such as Cyanoprep™ made by GC Electronics 10-125 $5, or 200 Catalyst™ from Measurements Group, improve superglue bonds on these tricky materials, by molecular surface activation. Fluorine-based etchants are sold specifically for cementing Teflon with special epoxies, these products must be used with care because they are toxic.

Proper surface preparation is an important factor, no matter what type of adhesive is being used. Clean, well fitting surfaces that are appropriately roughened give the strongest bonds, and this applies to Teflon and other plastics too. Activation treatments should be applied after surfaces have been abraded with a fine carborundum paper and cleaned with acetone.

Contact cement when used correctly is almost a universal adhesive. Unfortunately many fail to read and follow directions given on the glue container. Contact cement must be applied to cleaned roughened surfaces, allowed to dry until it is no longer tacky, then brought into contact. This sounds simple, but due to contact cement's instant bonding characteristics, surfaces cannot be separated once they touch each other. Accurately positioning surfaces before they are brought together is sometimes difficult, but also imperative with this adhesive.

On flat or intimately conforming surfaces of Teflon or polyethylene, contact cement produces a better bond than most other glues. Contact cement requires no pretreatment other than roughening and cleaning, before adhesive is applied.

Coloring epoxy: Epoxy may be colored by the addition of artists dry tempera powder color during the glue mixing stage. If carbon black (see text following Figure 5-21), is added to epoxy, it almost eliminates deterioration due to light absorption, and does not affect the glue's electrical insulation properties. Black epoxy is ideal for making light tight seals around photodetectors, and is totally opaque to visible light in even thin layers. Epoxy doped with carbon black remains in excellent condition under bright lighting for at least 25 years.

Non corrosive silicone rubbers: Silicone seal is a Room Temperature Vulcanizing (RTV) rubber. It cures in 24 hours for layers of reasonable thickness, leaving a strong, resilient, tenacious, heat resistant bond on most materials. Silicone seal has a vinegar (acetic acid) odor, and its fumes are corrosively harmful to delicate metal parts. Bearings, commutators and open relay contacts are especially vulnerable. If parts coated with uncured silicone seal are placed in a closed container such as a submersible or rocket payload, corrosive action may prevent motors and other items from working. Corrosive RTV rubbers are unacceptable for space instrumentation, and special non-corrosive RTV materials are manufactured for use on such equipment.

All silicone products are difficult to remove both before and after curing. No benign solvents are available, and alcohol followed by a wash with warm detergent water is best for cleaning hands and tools.

Hot melt glue, glue guns, glue pots:

Except for time taken for a glue gun to reach its operating temperature, hot melt glue is one of the most convenient gluing methods. Regular size or mini guns are inexpensive, widely available, and glue sticks for a variety of materials are sold in hardware and craft stores. Special glue stick formulations are made for plastics, fabric, leather, ceramics etc. but do not change performance ratings in Table 8-3. High and low temperature glues are sold, but hot melt glues can burn skin.

Hot melt glues have been used industrially for decades, but are relatively recent additions to consumer markets. They make gluing fabrics a breeze and most hot melt bonds can be undone using a heated screwdriver blade, or a soldering iron tip. Anytime a strong but detachable bond is required, hot melt glue should be given consideration, and therefore is recommended for attaching stretch resistors to gloves in Chapter 5.

Sealing a giant balloon with threads of hot melt glue:

A little experimentation will show that combining different varieties of hot melt glues produces enhanced adhesion on some materials.

It is easy to mix combinations of clear, brown, white or translucent glue sticks, by stirring different proportions in a hot melt glue pot, (pots are sold in craft stores, $5).

A toothpick dipped in a molten mix can be pulled, producing a strand that is laid on a test surface. Place a second sheet of the same material over the first, sandwiching the glue, which can then be heat sealed.

A Teflon coated clothes iron run from a Variac or soldering iron controller, is used to heat the trapped glue strand, sticking both surfaces together. Glued sheets can be pulled apart to find which glue mix gives the best adhesion. Special Teflon coated irons with a variable thermostatic control are sold at hobby stores, or from *Tower Hobbies* TA1383, $14.99. These irons are designed for applying covering film to model aircraft, and are ideal for the tests just described.

A 12ft long 4ft diameter remote controlled blimp fabricated from 0.0005" mylar film, is flown at the University of Toronto's 'Open House Day' every October. Student projects helped to decide many features incorporated in this craft, performing all aerodynamic manoeuvres possible with a blimp. Procedures previously described are ideal for sealing a large leak-tight mylar envelope. It is also easily repaired using the same technique.

Mylar™ is a trademark of the Dupont company and is manufactured as a raw film, or aluminum coated on both sides. Translucent mylar is an excellent electrical insulator and made in various thicknesses for that purpose. Mylar is tough, but will tear from an incipient nick. Translucent mylar is best for a homemade blimp, because glue seal integrity is easily checked by visual examination. Dupont sometimes makes small quantities of mylar available to accredited institutions, in response to an appropriate letterhead request. Aluminized 0.0005" mylar is used to make lightweight thermal reflective shields, and this material is available in a 78" x 53" solar mylar blanket from *Edmund Scientific* B60,636, $5.50, and from some safety supply stores.

Some useful adhesive products:

Double sided sticky tape (carpet tape), and conformable self adhesive foam tapes are both under-used by beginning robotics experimenters. Yet just a little experience will illustrate the utility of these versatile materials. Metals plates stuck together with foam or carpet tape can only be separated with extreme difficulty, and foam tape will stick to irregular surfaces and support considerable loads.

Removable Magic Tape™ manufactured by 3M, is very handy whenever a temporary non-marking tack is required. This tape is especially useful when delicate items such as cleaved fiber optic ends must be held firmly yet moved when necessary. Another useful product from 3M, is their 'Post-it' repositionable glue stick.

Modelling clay, Plasticine™, Blu-Tak™ and similar sticky products, are handy as temporary plugs for leaks, holding oddly shaped items, or as a retrieval tool when attached to a long wire.

Permanent and temporary threadlock adhesive compounds prevent fasteners from working loose with vibration. Reusable or temporary lockers are best for hobby work, since they work on wood, metal and plastics, and unlike serrated or lock washers will not mar the finish on a special surface. These compounds are stocked at auto suppliers and hardware stores. Five minute epoxy on threads is very good as a locking compound, and may be removed with debonder or by heating the screw with a soldering iron's tip.

Construction materials: Most amateur robotic apparatus can be assembled from commonly available materials. This is fortunate because exotic materials usually carry expensive price tags. Classrooms need scrap bins of wood, steel, aluminum, acrylic etc. otherwise students must shop for each tiny piece of material, and this is not economical, desirable or efficient. Large sheets of wood and other major items can be bought by individual groups if this conforms to an instructor's operating protocol. Students are always more careful in handling and using materials they have purchased with their own money.

A few basic physical properties of useful materials are given in Table 8-4. Readers will note only two significant figures have been assigned in this listing, (even from the best reference sources data accuracy is seldom better than a few percent). Product properties and performance also vary with composition, heat treatment, hardening, annealing and from batch to batch even for 'identical' items.

Ultimate tensile strength is quoted here but all materials yield at lower values, so hobbyists should incorporate adequate safety factors in their designs.

Table 8-4. Properties of typical robotic construction materials

	density ρ (lbs/in^3)	density ρ (kg/m^3)	tensile strength (psi)	tensile strength (MPa)	Young's modulus E (psi)	Young's modulus E (GPa)	specific modulus (E/ρ)
aluminum	0.10	2700	10,000	70	12×10^6	80	1.2
brass	0.30	8500	60,000	400	15×10^6	100	0.5
copper	0.32	8900	22,000	150	17×10^6	120	0.5
fiberglass	0.065	1800	40,000	275	4×10^6	30	0.6
glass	0.094	2600	15,000	100	10×10^6	75	1.1
carbon composite	0.054	1500	400,000	2800	20×10^6	100-400*	3.7
Kevlar	0.049	1350	200,000	1400	12×10^6	80	2.5
Lexan	0.043	1200	8,000	55	0.3×10^6	2	0.1
nylon	0.041	1150	11,000	75	0.3×10^6	2	0.1
Plexiglas	0.042	1160	6,000	40	0.6×10^6	4	0.1
steel (mild)	0.28	7850	70,000	480	30×10^6	200	1.1
steel (piano wire)	0.28	7800	400,000	2800	30×10^6	200	1.1
wood (balsa)	0.008	215	2,000	14	1×10^6	7	1.3
wood (oak)	0.025	750	15,000	100	2×10^6	15	0.8
wood (pine)	0.015	410	6,000	40	1.5×10^6	12	1.0

* Modulus for a laminated carbon reinforced epoxy depends on fiber orientations.

(1 lb = 0.4536 kg   1 psi = 6895 Pa   1 lb/in^3 = 27680 kg/m^3)

Young's modulus E is a measure of a material's stiffness, but weight is important too when building aircraft, robotic arms or booms. Therefore, a useful design parameter is E/ρ or specific modulus, which incorporates density ρ. Values for specific modulus in Table 8-4 show why graphite and Kevlar composites are so attractive for building aircraft or a space station.

It is interesting to note a wood flying boat built in 1947 still holds the 'largest wing span' record at 319ft 11". Dubbed the "Spruce Goose" (because spruce lumber was the major building material), this colossal plane can be seen at Pier J, Long Beach, California, and serves as a reminder that natural wood is still hard to beat. Fiber composite wheels, titanium frames, and other exotic doodads are state-of-art on modern bicycles. However, these machines have a hard time competing weightwise with a 15lb wooden bicycle employing bamboo frame and bamboo wheel rims made in the 1890s.

Graphite and Kevlar composites are still expensive, and used sparingly in most amateur systems. Their cost is justified when a spar or boom must be lightweight yet have maximum load carrying capacity. Thin composite skins glued to a rectangular balsa framework can produce remarkable stiffening with only a little added weight as shown in Figure 8-28. Carbon fiber reinforced strips are sold at good hobby stores and can be attached to balsa with superglue. Graphite arrow shafts sold at sports stores, are stiff and suitable as pushrods or for stiffening.

Galvanized metal sheet is widely used for making heating ducts, and local fabricators will often provide material cut to specifications at reasonable prices. Small jobbing shops also keep a scrap bin for offcuts, these are ideal for hobbyists and often available at no charge. Sign makers use large quantities of acrylic sheet. So if you need some tricky cuts made, a particular color, or just a few scrap pieces, check out local establishments. It is always best if one offers to pay for scrap items. If requests are not too exorbitant, you may be pleasantly surprised at the generosity of sympathetic dealers.

Learning about fasteners: Browsing through a hardware store is one of the best ways to get acquainted with commonly available fastening techniques, gadgets and tools. Hobbyists need not be disheartened by the plethora of hardware items made, because these are not the sole keys to success. It is the simple, straightforward, well-thought out design, that invariably works best. However, a knowledge of what is available sows and nourishes the seeds of ingenuity, stimulating our brains to engineer something that works as well as that high priced fancy gadget or tool. Time spent browsing and fiddling with devices is really a form of studying, providing a store of knowledge that can be acquired in no other way.

It may come as a shock to learn screws costing 20¢ apiece in a package of five, may only be 3¢ each when purchased as a box of one hundred. Even nails sold loose in a paper bag cost less than when neatly parceled in a pretty box. Check the cost effectiveness and unit price before buying small quantities, you can probably share, trade, swap or sell left over items from a larger package.

Hook and loop fasteners are often used to mount battery packs or oddly shaped packages that must be serviced or removed frequently. Velcro™ or similar hook and loop fastening material is less expensive when purchased in larger quantities, washers, dots and small strips can be cut with scissors as required. Double sided foam tape attached to hook and loop strips increase their versatility. Tough polyolefin mushroom head fasteners work on a similar principle to Velcro™ but have greater holding power, and can be stapled or glued just like ordinary hook and loop fabric.

Hasps, hinges, catches, hitches, latches, wood joiners (corrugated fasteners), eye bolts, expansion plugs, wing nuts, acorn nuts, flange nuts, square nuts, round nuts, self locking clip nuts, threaded studs, threaded anchors, T-slot nuts, slotted washers, wave washers, corrugated washers, spring washers, push nuts, retaining rings (circlips), and many other items are grist for a roboticist's mill.

Although hardware stores stock many useful items, and will special order even more on request, comprehensive catalogs describing such hardware are only provided to commercial users or large institutions. In a perverse way this may actually be beneficial for beginners, because touching hardware often leaves a more lasting impression than a picture, microfiche or computer screen image.

Useful commercial hardware for robotic projects

Wood screws and nails: A few common mechanical fasteners are shown in Figure 8-18. They are made in a wide range of sizes to suit just about any requirement. For example, foot long nails and foot long lag screws are often stocked by local hardware stores, also tiny ¼" wood screws for finicky robotic assemblies. Most steel wood screws are hard and difficult to cut, but drywall and sheet metal screws are even harder and will quickly ruin a saw blade. Sheet metal and self tapping screws work well in wood and come close to being a universal screw, they do not rust easily and some have Scrulox heads, a definite advantage when driving many screws into hard wood with a power screwdriver.

Hobbyists should be familiar with basic types of nails, such as spiral, finishing, ringed, panelling, concrete, aluminum, roofing etc. These all have specific uses but they are not mutually exclusive, so just choose one that does the job. Concrete nails are hardened and can be hammered by hand into masonry, but beware of chipping, and wear safety glasses. Paneling nails are hardened and make good pivot axles when ground to a taper at each end.

Spot welding: Spot welding enables rapid fabrication of small metal boxes, covers. brackets, ducts, motor mounts etc. from galvanized or plain steel sheet. A portable spot welder available from *McMaster-Carr* 7870A33, $445.54 can be used on steel sheet ≤16 gage, and is ideal for a small workshop. Throat working capacity for this unit can be extended to over a foot with extra long electrodes.

A simple riveting technique: Rivets do a neat job when low profile fastening is required, and aluminum rivets with either flat or round heads protrude least, when installed as shown in Figure 8-18E. These rivets are easily cut to length with a pair of diagonal cutters. They are inserted through a suitably sized hole, then peened over with a hammer for a tight mechanical joint. Only spot welding needs less clearance than a rivet used in this manner. Aluminum rivets in various lengths can be obtained by removing the pin from regular pop rivets sold in hardware stores. A wide selection of aluminum, copper or nylon rivets in solid, tubular, flat or round head styles, are available from McMaster-Carr or similar comprehensive hardware suppliers

Blind or T-nuts: Blind or T-nuts remove the complication of trying to position a nut in an inaccessible location while inserting a screw. They are useful clawed devices that can be installed from the rear of a surface (usually wood), Figure 8-18D. These nuts are sold in good hardware stores in sizes from 6-32 through ⅜-16. Hobby stores stock 4-40, 6-32 and 8-32 size T-nuts.

Wiggle nails: Corrugated fasteners, or wood joiners are sometimes used for joining corners as in Figure 8-18D, but they are difficult to remove. Therefore these fasteners are seldom worth the extra expense, since nails, screws or glue work just as well for hobby construction.

Wall anchors: Project elements must occasionally be sturdily mounted to a classroom wall. Expansion or plug anchors work well for these tasks. A tapered peg sliced from a piece of softwood lumber, then hammered into a drilled wall hole, is also satisfactory for all but extremely heavy mounting applications. Toggle, butterfly or similar fasteners are required for attachments to hollow walls. Lag screws are best for heavy installations, they are less likely to snap when torqued down hard, and their heads are easier to grasp.

When possible drill into mortar joining brick or cinder block walls, this is easier than drilling brick.

Hinges: Continuous strip steel or brass hinges are sold in lengths up to 8ft, and may be cut to a specific length. Hinge material purchased in 24" lengths is most economic for classrooms, or for serious hobbyists. Small nylon hinges are stocked at hobby stores, but leather, cloth or plastic strips also make good light duty low profile hinges, and can often be glued to surfaces without rabbeting.

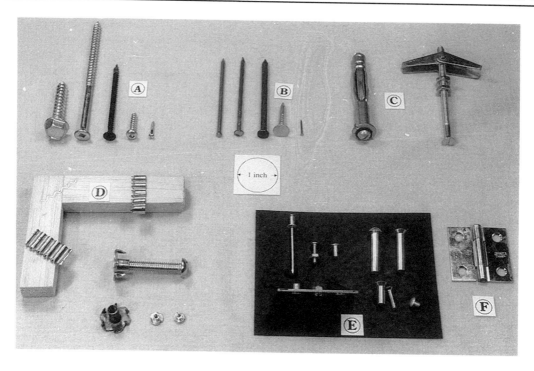

Figure 8-18. All items are listed left to right.   [A] Lag screw, 4" flat head Scrulox wood screw, drywall screw (for cement type sheathing), #8 sheet metal screw, #4 round head wood screw.   [B] 2" finishing nail, 2" spiral nail, 2" concrete nail, 1" aluminum nail, ½" brad (can be pushed in by hand if held in a pair of pliers).   [C] 10-24 expansion anchor, 10-24 toggle hollow-wall anchor.   [D] Corrugated fasteners, ¼-20 round head machine screw with a mating blind nut, ¼-24, 4-40, 2-56, blind or T nuts.   [E] Two plates have been riveted and screwed together to show fastener profiles. Left: pop rivet. Center: 4-40 nut and bolt. Right: pop rivet aluminum head (without pin), hammered flat for a low profile connection. Miscellaneous aluminum rivets.   [F] Brass or steel hinges of various lengths may be cut from a long continuous strip hinge.

Useful screws for roboticists

Whole books are dedicated to screws and fastening techniques, so attempting even a brief summary here is foolhardy. Most amateur needs are met with either round or flat head machine or wood screws purchased at any hardware store. Head styles vary, but slotted screws are more generally useful. Hex heads are superior but more expensive. Steel screws are stronger, but brass screws must often be employed near magnetic fields. Long machine screws can always be shortened with a hacksaw and file, as described below. Try to stick with American size threads because these are more easily obtained, and less expensive.

Nuts and bolts of all sizes are widely used in projects and come in various sizes, head formats, threads, materials etc. Good quality Allen head screws, self sealing screws, and very small screws are only sold by comprehensive hardware or machine tool dealers, and usually by the box (100). Building a collection of machine screws and nuts from salvaged equipment is a good idea, especially if these are stored in marked drawers or containers so they can be efficiently retrieved.

An excellent square head screw developed in Canada in 1908 is gaining acceptance in other countries (*Consumer Reports, November 1995, p695*). Scrulox or square-drive screws are alternative names for the Robertson screw, which comes in five different head sizes. As with hex-head or Allen screws, ball screwdrivers are also made for Scrulox screws, giving them that extra versatility absent with many other screw head designs. In addition, screwdriver slippage on square-head screws is rare, consequently these screws may be torqued at a high level.

Carriage bolts and wing nuts: Carriage bolts in Figure 8-19A are self-anchoring. A square neck just beneath the head digs into wood preventing bolt rotation, a convenience when items must be frequently disassembled. Wing nuts are faster than regular nuts when hand secured connections are adequate, and are available in sizes from 6-32 to ¾-24. Wing screws are also sold, but home made wing nuts or wing screws are easily fabricated by soldering a small bar to a nut or bolt.

Figure 8-19. Listings are from left to right: [A] A carriage bolt's head will not rotate once its square head section is embedded in wood. Wing nuts are handy for quick change installations. Two nuts jammed together on a shaft will act as a lock nut.   [B] Nylon insert lock nut. 1-72 socket screw. 0-80 brass round head screw is also available in socket or flat head style. 0-80 socket set screw needs the smallest Allen key made. ¼-20 stainless steel socket set screw. [C] Self sealing screws have silicone rubber O-rings in their heads, and are suitable for 500°F and 500psi.   [D] Brass threaded inserts are sold in hobby stores. The cylindrical portion from a blind nut is shown on a bolt. This cut-off portion also makes a good threaded insert.   [E] Washers come in all sizes, but home made washers can be fabricated from many materials.   [F] Threaded rod is widely used for robotic assemblies and is available in sizes from 2-56 (shown here).   [G] Screw extractors are made, but only work on #3 or larger screw sizes.

Set screws and motor shafts: Socket set screws are widely employed for locking motor shafts, or anywhere dynamic balance is important or space is tight. Socket head set screws are easier to install than slot head set screws, especially in smaller thread sizes. A complete list of Allen key sizes for hobby work is given in Table 8-2 and the smallest key manufactured (0.028" across flats), is shown in Figure 8-19B.

     Hardened set screws can easily disfigure a motor shaft making it impossible to remove a pulley or other item attached to the shaft. Always file a small flat on a motor shaft, making sure the set screw contacts only on this flattened area. Not only will this prevent shaft seizure, it gives a more secure bite and needs less screw force to prevent slippage.

Coupling nuts and threaded couplers: Plain steel threaded rod in 36" lengths is sold by hardware stores, and is usually stocked in sizes from 8-32 to ¾" diameter. Hobby stores sell fully threaded 12" steel rods in 2-56, 4-40 sizes, or 30" lengths, threaded at both ends for use with threaded couplers. Threaded couplers are especially handy because they can be soldered to piano wire or thin rods, eliminating the necessity for cutting threads. Comprehensive hardware supply houses (eg McMaster-Carr), stock threaded rod in stainless steel, aluminum, fiberglass, brass and nylon in lengths up to 12 feet, from 2-56 to 1½" diameter. Steel couplers for extending threaded rod are sold at local hardware stores for ¼"-20 and larger sizes, depending on a particular store's inventory.

Self sealing screws for submersibles: Submersibles often need sealing fasteners and hobbyists should be aware that both self sealing washers, or O-ring types are manufactured. Stainless steel pan-head machine screws fitted with O-rings, are sold by *McMaster-Carr* in 4-40, 6-32, 8-32 and 10-32 sizes from ¼" to 1" lengths, approx $40/100. These screws make excellent seals on smooth flat surfaces and are good for both pressure and vacuum applications. Silicone rubber O-rings in grooves just below the screw head, seal to more than 500psi and from -150°F to 500°F, (Figure 8-19C).

Hobbyists can make their own seal screws using washers cut from neoprene or Viton sheet with sharpened hobby tubing or cork borers, shown in Figure 8-12. These home made washers work best when used with large pan-head screws.

Washers: Only a few of many washer styles are shown in Figure 8-19E because hobby work rarely needs square beveled, bonded neoprene, nylon, fiberglass, fiber, leather, Kapton, nickel, aluminum, spherical, slotted, copper, felt, shoulder, wave, spiral or other special washers, though these are widely used in industry. With a little ingenuity hobbyists can make special washers when required, or buy them when large quantities are needed.

Thread lockers: Most locking nuts use nylon inserts, an integral serrated washer or a jam feature. Wire tie nuts are used on critical installations, and these ensure a nut can never spin loose to cause catastrophic damage. A dab of epoxy on threads before a nut is finally tightened gives a permanent lock, or two nuts may be tightened against each other for security as shown in Figure 8-19A.

Fixing loose screws in wood and metal: Flat toothpicks packed into a wood screw hole and then clipped off at the surface, enable a screw to bite and take hold. If a hole is in really bad shape cut a conical wooden peg from softwood, then hammer this peg into the hole and cut off flush with the surface. Glue may be added for a more permanent fix in both these remedies.

Screw slippage is common when screws are repeatedly inserted and removed from thin sheet steel or aluminum. Threaded clinch nut inserts are manufactured to correct this problem, but these are sometimes impossible to install when only one side of a plate is accessible.

If a short length of steel wire is slipped through a sheet metal hole and held while a screw is inserted, this will make a tight connection. Clothesline wire is ideal for this fix, because it is tough and several wire sizes are twisted together in each cable. Do not use copper or other soft metals since these are totally unsatisfactory. Softer wire is easily cut by hardened sheet metal screws, and provides no support.

Gently squeezing bolt threads with pliers deforms them so they will often take hold in a tapped hole where threads have been partially stripped.

Inserts for stripped threads: Special metal inserts are available for tapped holes completely stripped of their threads. These helical coil inserts are self locking, non-corrosive, providing a professional fix stronger than the initial holding capacity. Helical inserts are made for all screw sizes from 2-56 to 1½-12, and mid size inserts are often stocked at better auto service repair shops. Helical Screw-lock tools and inserts are sold by *McMaster-Carr.* Typical repair on a stripped hole involves re-tapping, then installing a spring coil with a special tool.

Although helical coil inserts are an ideal solution in many cases, insert tools and taps are not cheap, so amateurs might like to consider the following alternatives. Steel and brass threaded inserts (from 4-40 to ¼-20), are sold by hobby stores. Designed for use on wood, these inserts require a little modification but can then be inserted into metals. Even easier is using the threaded portion of a blind or Tee nut, after its flange has been removed with a hacksaw (Figure 8-19D). Cyanoacrylate, epoxy glues or solder are usually best for installing these home made inserts into metal.

A 4-40 brass threaded insert sold by hobby stores (4/$1), is shown in Figure 8-19D. Intended for use in wood, it may be glued into metal if its coarse outer threads are first removed. Blind or T-nuts shown in Figures 8-18D and 8-19D, are stocked by hobby stores in sizes from 2-56 through 8-32 (4/75¢). They require less space than a regular threaded insert. 'All Threads' inserts are sold for 2-56 to ¼-20 screws, *Tower Hobbies*, eg. OHIQ3050/3055, $1.60 for eight.

Cutting screws to length: A screw cutter is included on some crimping or wire stripping tools, (eg. *McMaster-Carr* 7007K91 $10.64). These cutters will shorten screws and bolts from 4-40 through 10-32. A screw is inserted into a threaded hole, then severed. Subsequent screw removal performs a thread-chasing action. These tools are handy but any size screw can be shortened by hand with a vise, saw and file. First wind a nut onto threads close to the screw's head. Cut the screw slightly longer than the desired length while clamping on unneeded threads. Mount the screw so it is securely held, using masking tape on vise jaws to prevent thread damage, then file off excess material carefully. When the nut is unscrewed, any swarf will be removed leaving a clean thread.

Removing seized or broken screws: When removing seized or rusty screws it may be necessary to apply more torque than can be supplied by a hand alone. In such circumstances a pair of Vise grips® clamped to a screwdriver shaft produces impressive torque. The Vise gripping hand should both push a screwdriver into the screw head, and simultaneously provide extra torque required, taking care not to break the screw apart.

Broken screws are difficult to remove without making a mess of surrounding material. Screws larger than 10-32, can sometimes by removed with a screw extractor shown in Figure 8-19G (the smallest size available). Brass screws can be drilled out, but usually drill wobble restricts this operation to short screws. Drilled out holes are usually filled with an epoxied plug of the host material. Removing steel screws is always a problem. But if material can be removed from around the remaining screw shank, a pair of mini Vise grips is the best tool for extracting a problem screw.

Screwing into hardwood: Screwing wood screws into oak or other hardwoods can be a challenge. If an appropriate size pilot hole is not provided, it is easy to tear the head off a screw or snap its body when excessive torque is applied. In hardwoods, always drill a pilot hole with a drill matching the screw body diameter, leaving wood only for screw threads.

"What load can this screw hold?"

If you are apprehensive about using a screw for supporting a large weight, it is straightforward to calculate its load bearing capability. High strength steel has a tensile strength of about 200,000 psi, and a ¼" diameter rod has a cross sectional area of about 0.05 square inches, so a ¼-20 steel screw can hold approximately 10,000lbs. Tensile strengths for commonly used robotic materials are listed in Table 8-4

Naturally, if a bolt is over-tightened it may be pre-loaded to the point where it will snap with only a small additional stress. Nevertheless, even larger bolts are subject to pre-stressing failure of this kind. Shocks or side loading can also decrease a bolt's load bearing capacity, and roboticists must use discretion and appropriate safety factors when designing for critical load bearing.

Flexible tubing: Flexible plastic tubing is used for many pneumatic and hydraulic experiments. Maximum safe operating pressure for any tubing depends on tube diameter, wall thickness, operating temperature, tube material and filling media. Usually pressure ratings are specified for air, because wall integrity on some tubing is compromised when carrying certain solvents or reactive gases. Table 8-5 gives data for flexible tubing commonly used for robotics experiments. Values may be interpolated if reasonable adjustment is made for tube diameter and wall thickness.

For safety all tubing should be properly connected, clamped, and used within its prescribed pressure rating. Latex, silicone, neoprene and similar rubber tubes are not designed for pressure applications, and consequently burst easily. All tubing slips over fittings more readily with a little saliva or soapy water, heating with a heat gun improves flexibility too on most plastics.

Tubing quality varies and is usually proportional to price. For example an economy garden hose rated for 100psi costs about 10¢/ft, but a high quality hot water hose engineered to withstand 400psi at 200°F may sell for over $1/ft

Clear, food-grade PVC flexible tubing is adequate for most hobby work. It is inexpensive and sold by the foot at hardware stores in sizes from 3/16" ID. This tubing has excellent resistance to most chemicals and gases, but slowly hardens and discolors with age or exposure to daylight. Small bore PVC tubing is easily stretched with a heat gun to fit over larger fittings. It can also be stretched by forcing needle nose pliers into a warmed bore, then opening the handles.

Silicone rubber sold in hobby stores is very flexible, does not harden, and can be used to 350°F. With nylon braid reinforcement, silicone tubing pressure ratings increase more than ten times over those listed in Table 8-5.

Neoprene tubing is also sold in some hobby stores, it weathers well, tolerates fuels, oils and greases, but can only be used for low pressures. Latex rubber tubing is excellent for making good high stretch springs, and holds securely when slipped over a rod larger than its bore. It is available from hobby stores, and sports stores sell latex tubing for slingshots.

Table 8-5. Maximum operating pressures for flexible tubing with air at 70°F

Silicone rubber			Neoprene			Latex rubber			PVC (food grade)			Braided PVC		
ID	OD	P_{max}	ID	OD	P_{max}	ID	OD	P_{max}	ID	OD	P_{max}	ID	OD	P_{max}
1/16"	9/32"	10							1/16"	3/16"	65			
1/8"	3/8"	10	1/8"	3/8"	10	1/8"	1/4"	10	1/8"	1/4"	60	1/8"	5/16"	250
3/8"	5/8"	7	3/8"	1/2"	6	3/8"	1/2"	10	3/8"	1/2"	45	3/8"	5/8"	220
1/2"	7/8"	6	1/2"	3/4"	10	1/2"	3/4"	25	1/2"	3/4"	45	1/2"	7/8"	200

Rigid hobby store tubing: Brass hobby tubing can be used to make artificial fingers or joints, and for other applications outlined elsewhere in this book. Brass tubing is easily soft soldered, and because it is made to high tolerances can be used for bearings, axle shafts, vacuum or low pressure fittings and many other tasks. Outside diameters of commonly stocked round brass tubing are: 1/16", 3/32", 1/8", 5/32", 3/16", 7/32", 1/4", 9/32", 5/16", 11/32", 3/8", 13/32", 7/16", 15/32", 1/2", 17/32", 9/16", 19/32", 5/8", 21/32". Consecutive sizes interface with a sliding fit.

Examples of both circular and rectangular interfacing tubes are shown in Figure 8-20. They have been alternately blackened for contrast. A few larger sizes and different shapes of rigid tubing are also sold at hobby shops and some of these are shown in the same figure. Copper and aluminum tubing with similar tolerances to brass is sold, but is available in only a few sizes.

All hobby tubing can be cut with a razor saw or jewellers saw, then finished with a file. A rotary cutter specifically designed for circular cross section tubing up to 1/2" diameter is shown in Figure 8-20E.

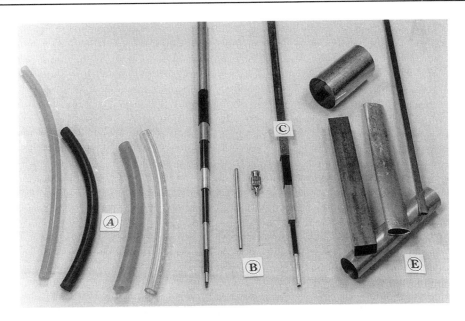

Figure 8-20. Listings are left to right: [A] Silicone, neoprene, latex and PVC food grade flexible tubings are sold by hardware or hobby stores.   [B] Cylindrical hobby tubing gives sliding telescopic fits between consecutive sizes, (alternate tubes are blackened for pictorial contrast). Stainless steel hypodermic tubing (shown above the letter [B]) may be soft soldered and bends readily in a flame.   [C] Consecutive sizes of rectangular hobby tubing   [E] Some other tubing shapes and sizes sold by hobby stores.

Minibore tubing: Small diameter plastic tubing and fine bore stainless steel hypodermic tubes are handy for making miniature pneumatic sensors and transducers. Very small bore flexible plastic tubing (~0.010" ID), is made to fit over hypodermic needles, and is sold by laboratory equipment or specialty plastics suppliers. Thin wall 304 stainless steel hypodermic tubing from 0.010" ID (0.003" wall) to 0.199" ID (0.010" wall) is sold in 12" lengths, $4.40 by *McMaster-Carr*. Hypodermic needles shown in Figure 8-20B are also useful for hobby purposes. Sharpened thin wall stainless steel tubing is excellent for cutting 'perfect' holes in soft materials, and small diameter steel tubing can be bent without collapsing when heated to red heat in a flame.

Music wire, cutting hard materials: Piano or music wire has the highest tensile strength of any readily available wire. It is very hard, and should only be cut with an abrasive disk or specially hardened cutters - a hacksaw blade or file is quickly ruined on this material. Music wire is available from good hobby stores in 36" lengths with the following diameters: 0.015", 0.020", 0.025", 1/32", 0.039", 0.047", 1/16", 0.078", 3/32", 1/8", 5/32", 3/16", 7/32", 1/4". A few samples from 0.004" to 1/4" are shown in Figure 8-21A.
   Large diameter music wire must be cut with a high speed abrasive disk. Small size piano wire (<0.025") may be cut with linesman pliers or Vise Grips®. Electronic wire cutters are designed for copper only and will be permanently damaged if used on music wire.

Piano wire wheel axles: Heavy music wire is extremely stiff and fits neatly inside mating sizes of hobby tubing. It has been used to fabricate a simple axle/bearing combination in Figure 8-21C. Plated brass collars made for 1/16" through 1/4" piano wires, are sold at hobby stores, and a matching collar is shown attached to this 5/32" axle.

Soldering piano wire: Piano wire can be brazed, or soft soldered with acid core flux, and it may be bent in a vise or with special hardened bending jigs from hobby stores. All plain steel rusts, so music wire must be reoiled if its thin protective oil coating is removed by cleaning, soldering or brazing. Piano wire makes excellent coil or leaf springs, and a simple technique for making commercial quality coil springs will be described in a future book in this series.

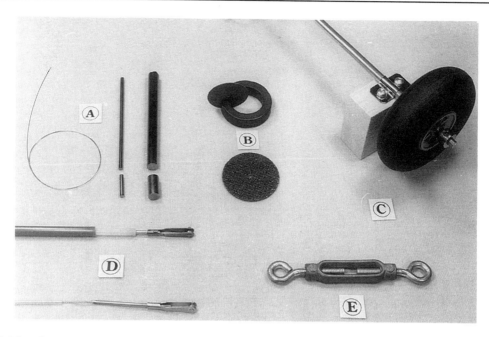

Figure 8-21.   [A] 0.004" piano wire (shown coiled), can be cut with linesman pliers or Vise Grips but will ruin electronic cutters. Heavier piano wire must be cut with an abrasive disk run at high speed. Specimens of 3/32" and 1/4" music wire are shown after cutting with a small disk (in lower [B]).   [B] Ferrite is extremely hard, but can also be cut with an abrasive disk. A 0.025" Dremel disk is shown after making a cut through a ferrite toroid. To prevent cracking, ferrite must not be overheated when cutting. [C] Wheel axle and bearing made with 5/32" piano wire and matching brass hobby tubing. A plated brass collar attaches with a socket set screw.   [D] Sullivan Gold-N-Rods have lower friction than bicycle brake cables or auto accelerometer cables. The small size turns freely even when bent to a 1" radius. Clevises are supplied with each packaged cable.   [E] Quarter scale turnbuckles available from hobby stores are about half the size of the unit shown here, and units as small as 9/16" are sold.

Flexible control rods: Semi flexible and very flexible Gold-N-Rod® push rod cables from hobby stores, have lower friction than ordinary bicycle cables and are very useful for robotic mechanically linkages. Two types of Gold-N-Rod are shown in Figure 8-21D. A small diameter brass plated stainless steel version turns freely even when bent through a 1" radius. The larger size with a splined nylon shaft has impressive low push or turn friction. Cables from 30" to 48" are supplied with separate clevises, or may be hard coupled with 2-56 threaded rod screwed onto their couplers.

Home made miniature flexible control rods may be fabricated with piano wire sliding inside thin wall Teflon tubing. Alpha Teflon tubing is available in sizes from 0.012" ID with a 0.009" wall thickness TFT200-30-100, *Electrosonic*, $20/100ft.

A tough wire: Steel galvanized clothesline is a tough high tensile strength steel wire, ideal as a tourniquet for securing flexible hose to high pressure air and water lines. Always use barbed fittings when making such installations, because hose slips easily on straight pipe. A multiple strand steel clothesline is ordinarily manufactured from two wire diameters, making it an inexpensive source of high strength wire. Unlike piano wire, clothesline steel is not suitable for making stiff springs, but will tolerate repeated flexing, and solders well with acid core flux.

Flexible wire, line and cord: Fishing line sold by hardware or hunting supply stores, is very useful for hobbyists. Most fishing line is load test rated - a useful factor when selecting line for a specific task. Stainless steel fishing cable solders well with an acid core flux, and is sold in various gages either bare, or sheathed in plastic. Fittings are sold at sports stores for connecting to steel or nylon cables.

Turnbuckles: Quarter scale turnbuckles from hobby stores are one of the easiest methods for tightening control cables, and they will take up about 3/4" of slack. The smallest size turnbuckle sold by hardware stores is shown in Figure 8-21E, and has a take-up of 1¼". *Tower Hobbies* carries miniature turnbuckles that take up slacks from 9/16" to 1 5/16": PRCQ2600 $4.30, PRCQ2707 $6.99.

Shaft couplings and bearings: Many mechanical devices are coupled to robotic elements by either rigid or flexible shaft couplings. Devices for coupling rotating shafts are needed most frequently, and a few examples are shown in Figure 8-22.

Rigid couplings need careful installation, because a misaligned heavy load creates forces that cause premature wear on motor bearings. In general amateurs should use flexible couplings, unless a dial gage is available for checking misalignment eccentricities on a rigidly coupled system.

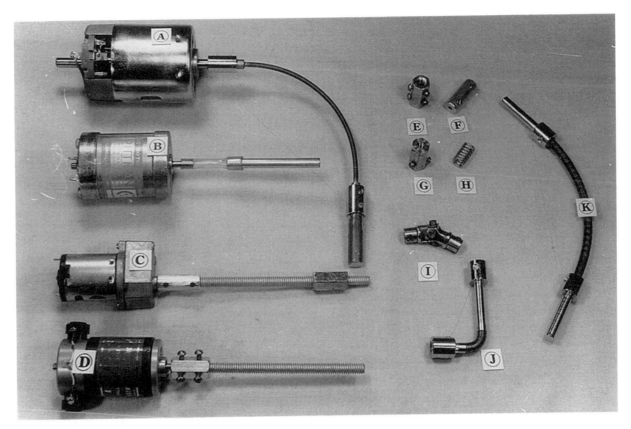

Figure 8-22.   [A] Six inch by ⅛" diameter flexible shaft with ¼" end fittings is useful for applying rotary motion in awkward spots.   [B] A short length of food grade PVC slipped over shaft ends makes an inexpensive flexible coupler, satisfactory for most hobby work.   [C] Rigid couplings are easily machined in brass or aluminum, but a lathe is best for getting different size bores accurately aligned. A nut riding on the threaded shaft is almost impossible to stop by hand, even when driven by this small motor.   [D] Different sized shafts can be accurately aligned with this screw adjustable coupler made from a ¼-20 steel coupling nut.   [E] ¼-20 coupling nut showing screw adjusters.   [F] Rigid aluminum or brass couplings for connecting two different size shafts can be made on a lathe.   [G] A brass bushing in this coupling nut has been tapped to take an 8-32 threaded rod.   [H] Very stiff springs may also be used for shaft couplings, but flexible shafting from A or K is better.   [I] Small universal joints are sold by hobby stores. [J] Flexible bellows made from 0.001" stainless steel can be used at high temperatures or in an ultra high vacuum.   [K] ¼" diameter flexible shaft is made with overlapping spring steel windings and is available in long lengths. End couplings are described in the text.

Flexible shaft couplings: A flexible drive shaft in Figure 8-22A is composed of two overlapping oppositely wound outer springs, on an inner shaft of longitudinal wires. Shafts of this type absorb vibration and can snake around obstacles, providing a simple drive connection normally requiring universal joints or gears. Potentiometers are sometimes remotely operated with these flexible shafts and shafts are sold by good electronics stores, eg. *Electrosonic* #152, 6", $12 for the item shown.

Drive performance from a spring wound shaft is asymmetric, and maximum torque or speeds are obtained when the outer spring is driven so it winds up. Figure 8-22K illustrates a ¼" flexible spring shaft and this type is sold in lengths up to 6ft and operates down to a minimum turn radius of 6", up to 5000rpm at a maximum torque of 10lb-in. Flexible shafts of this type may be shortened after they have been well soldered at the cutting location. Plated brass couplings installed on item K are *Electrosonic* #152, ¼" to ¼", $3. A ¼" to ⅜"coupling #120 costs $2.

Rubber or flexible vinyl tubing couplers are popular with students. They are inexpensive, easy to install, and can be used to connect two different size shafts, as shown on motor B. Free tubing length between shafts should be kept short to reduce backlash. Adhesive and clothesline wire tourniquets may be added if slippage occurs.

Moulded neoprene is used to join two metal collars in some commercial flexible couplings, and three-piece spider couplings with a rubber star insert are made to fit shafts from ¼" to 4½" diameter.

Figure 8-22J shows a thin wall (0.001") stainless steel bellows often used for coupling motor shafts in ultra high vacuum systems, or where high temperatures are anticipated.

Fixed rigid shaft couplings: All elements in a drive train must be concentric when rigid shaft connections are used, and such couplings must be turned on a lathe to achieve the required precision. Rigid couplings are shown in F and installed on geared motor C. This latter coupling has been tapped to accept a ¼-20 threaded rod, and is equipped with a ¼-20 coupling nut rider as described in Figure 6-1D.

Adjustable rigid shaft couplings: If screw adjusters are added to a standard coupling nut as shown on motor D, a plain or threaded rod can be attached to a motor's shaft so it runs concentrically, (4-40 brass screws work well on motor shaft couplings made from ¼-20 coupling nuts). Brass adjusting screws are best because these do not mar a motor's shaft. If steel screws are used, their ends should be filed flat to reduce shaft damage. Always file a small flat on a shaft, and make sure one of the three adjusting screws is aligned with this surface.

Coupling nuts with screw adjusters can be drilled out to accept larger motor shafts, or fitted with a brass insert to accept smaller size threaded rod. A modified coupling nut with an 8-32 insert is shown in Figure 8-22G.

When using a coupling nut with screw adjusters, alignment for true running is made by trial and error. With a little practice shafts can be aligned with negligible wobble in a few minutes.

Coupling nuts: Zinc plated steel coupling nuts are made in 4-40, 6-32, 8-32, 10-24 ¼-20 and larger sizes, and these can be obtained from comprehensive distributors such as *McMaster-Carr*. Hardware stores normally stock ¼-20 (shown in Figure 8-22E), and larger sizes.

Connecting threaded rod to motor shafts: Threaded rod may be attached to coupling nuts, and then driven directly from a motor, Figure 8-22C. For most student projects ¼-20 rod is preferred, because this is inexpensive and readily available. If finer thread is required, a brass screw can be soldered into a larger coupling nut, then threaded with an appropriate tap, Figure 8-22G.

Home made rigid shaft couplings: Rigid couplers shown in Figure 8-22C,F may be made on a drill press from brass or aluminum rod. If different sized shafts must be joined, always drill the smallest hole through the full coupling length first. *Without moving the work* drill a hole for the larger size shaft. This procedure produces two bore sizes with a common axis.

Springs as flexible couplers: Stiff springs Figure 8-22H, soldered to brass coupling collars make inexpensive flexible drives. To avoid misalignment, attach both collars to a common shaft before soldering. It is tempting to use epoxy instead of solder for this task. However, adhesives only withstand modest torques, and will sometimes break under a shock load. Commercial flexible shafting is worth the extra cost in most cases.

Universal joints: Off-axis drive up to 30 degrees is possible with a universal joint. High quality double or telescoping universal joints are expensive (>$30), but economy, or plastic ball and socket joints are available from about $5. A pin type universal joint with brass collars is shown in Figure 8-22I.

Tower Hobbies sells a quality universal joint assembly for marine engines OSMG2944, 22442009, 20FPM, $20.99. Hobby stores specializing in RC boat accessories, often carry universal couplers designed for motor to propeller connections.

Stainless steel bellows couplings: For those applications where backlash cannot be tolerated and drive misalignment is <10 degrees, stainless steel bellows are usually the first choice. These devices are made from very thin material and are acceptable for use in ultra high vacuum conditions, and high temperatures.

A small bellows fabricated from 0.001" stainless is shown in Figure 8-22J. Bellows couplers are used with resolvers, encoders, servo links and anywhere angular motion must be transferred with high fidelity. Stainless steel bellows are not cheap, and the item shown in Figure 8-22 costs about $40.

Shaft extenders and axles: ¼" to ¼" shaft extenders have been attached to a 6" flexible shaft in Figure 8-22K. Made from chromed brass, these elements are equipped with set screws and sold by *Electrosonic* #150, $2.85.

Bolt-on axle shafts and adjustable axles with diameters from 1/8" to 3/16", and lengths to 2" are sold by hobby stores.

Sealed shafts and actuators for submersibles: Water sealed rotating shafts can be made from greased music wire running inside closely fitting brass hobby tubing.

Pushrod seals are convenient for mechanical actuation across a water or fuel interface, and are sold at hobby stores for $3.50.

An easy way to modify a motor's shaft diameter: Sometimes it is necessary to increase a motor's shaft diameter to match a particular fitting. If a piece of brass hobby tubing can be found to match the fitting, it may be attached to the motor's shaft as follows:

[1]   Trim a length of masking tape so its width matches the motor shaft length.

[2]   Wind this tape onto the motor shaft until it is a little larger than the tube's inside diameter.

[3]   Now repeatedly trim *short* lengths from the rolled tape until it just fits the tube.

If this is done carefully, the tubing will be concentric with the motor shaft. When glued, this assembly handles reasonably heavy drive loads. Keep the total tape thickness to <0.040" for best performance, or use additional hobby tubing sizes, one inside the other, to make a better initial fit

Bearings: For most hobby work oiled phosphor bronze, brass or nylon bearings are adequate. Brass hobby tubing makes a good bearing surface, and is available in a wide range of inside diameters. Ball bearings are expensive and more easily fouled in a gritty environment, but are best when low friction is essential.

Beginners sometimes worry about friction losses, but these concerns usually vanish when cost is given full consideration. For example, a ball screw mechanism has an efficiency of about 90% but is only about twice as good as a simple nut follower in Figure 6-1D. A ball nut lead screw costs at least ten times that of a threaded nut follower, so lower friction comes with a hefty price penalty. *Servo Systems* sells a 5" reciprocating ball nut with 5 threads /inch LPS-119 $49.

For heavy assemblies ½" bronze pillow blocks are the most economical bearings, and these are available in straight or self aligning styles from most hardware sources. Standard ½" pulleys, shafts, collars, belts and other fittings are widely available and less expensive than other sizes. Smaller size bearing blocks and accessories are more costly and sometimes difficult to obtain, because they are used less frequently.

Professionals sometimes use linear ball races in conjunction with a hardened shaft, for low friction longitudinal movements. Hobby needs are adequately met with hobby tubing sliding on a music wire shaft, as shown in Figure 6-21. Brass tubing has the additional advantage that it is easily soldered, making it easy to mount other items.

Mounting small electric motors: Reaction torques developed when a motor drives an external element tend to rotate a motor, and for this reason motors should be firmly attached to a suitable base. Ideally a mount should be simple, effective, and allow for easy installation of a new motor. Some mounting methods are shown in Figure 8-23, and these can be made with basic hand tools.

Figure 8-23. [A] Stainless steel worm-drive hose clamp is easily adjusted to fit different motor diameters. [B] 1/16" aluminum sheet was cut with a hacksaw, filed, drilled, then bent across the fine surface lines (rolling mill marks made during manufacture). Flexible vinyl is used to connect motor shaft to gearbox.   [C] Galvanized steel sheet (~0.025") is formed with fingers around the motor body, then bent at right angles in vise jaws or with pliers.   [D] Four-ply hobby plywood is easy to handle, and makes a strong bracket. Note both screws and glue are used to join wood pieces together.

Worm-drive adjustable stainless steel hose clamps make versatile motor holders, as shown in Figure 8-23A. Clamp sizes start at 7/32"-5/8", increasing by small increments to 5 5/8"-8 1/2". Clamps can be cut with aviation snips or a hacksaw. This dc brush motor is ungeared, and its dual shafts make it easier to install an encoder disk.

     Figure 8-23B shows a 12V dc brush motor attached to a gear box via a flexible vinyl coupling. This gearbox is made in Germany by Richard (Marx) and was widely available in train hobby stores a few years ago. It has five selectable gear ratios from 3:1 to 60:1. A Pittman dc gearhead motor from *Servo Systems* DM-616, $21.50, is clamped with a finger-bent galvanized steel band in Figure 8-23C.

Small gear head motors are in great demand for robotic applications and the style in Figure 8-23D has been carried by various distributors, most recently by *Edmund Scientific* M52,467, $10.95. Surplus motor stocks are continually changing and roboticists must scan catalogs for latest offerings.

Gears and gear boxes:
Very few small gear boxes suitable for robotics are sold on the surplus market. Consequently these items are difficult to obtain and usually expensive for hobbyists. It is nearly always best to purchase electric motors with built-in gear heads, if slow speed high torque motions are required. Such motors are widely available and remove the difficulty of gearing down, which is always a problem.

Metal gear sets made for servos, are sold by *Tower Hobbies* HCAM1059 $14.99. A set of polyacetal resin servo gears is shown in Figure 8-24A, retailing for $3. With care these gears can be assembled to satisfy many gear reduction needs. Electric motor reduction gearboxes with ratios from 2.5:1, 3.0:1 and 3.5:1 are $12.99 from *Tower Hobbies*.

Currently *Edmund Scientific* is offering a plastic gear train with 11.6:1 and 18:1 reductions: A52,405, $17.50. A planetary gear box set with reductions from 16:1 to 400:1, A52,408 is $27.95.

Worm and gears behave in the same way as a screw drive shown in Figure 6-1D, where a nut is driven but cannot drive the screw. Such arrangements are ideal when back-loading might cause a motor to be driven in reverse, (as sometimes happens with a planetary gear train). A 30:1 ratio metal worm and gear are shown in Figure 8-24B, and *Edmund Scientific's* worm-gear A52,407, $17.50 has reductions of 216.1:1 and 336.1:1.

Professionals use metal worm and bevel gears shown in Figure 8-24 but these are expensive, and each worm and gear pair shown in B costs over $100.

Small plastic chains and matching toothed gear wheels are employed in some industrial equipment. Unfortunately these products are not widely used, and consequently unduly expensive. Bicycle chains and gears are suitable for wheelchairs but are too bulky for most other amateur robotic applications.

Figure 8-24. [A] Gears from #4005 servo gear set by JR are sold by hobby stores for $3. [B] Metal worm and gear, 30:1 reduction, (details on less expensive alternatives are given in the text). [C] Bevel gears translate motion by 90 degs.

Home-made gearing systems: Many non-standard techniques can be used by hobbyists for gearing mechanisms. For example, grooved, toothed and timing belt pulleys are useful for altering speeds from a rotating shaft. Even several turns of cord wrapped around different diameter wood or metal capstans, may be used as pulleys for changing rotational speeds. Wooden wheels with contacting rubberized rims work well for rotating heavy equipment on a Lazy Susan (ball bearing spin table). The screw-drive principle enunciated in Figure 6-1 is a powerful technique, suitable for many applications. Nylon and aluminum spur gears with 18,20,21,22,48 and 64 teeth are sold by good hobby stores. Food mixers and other high torque, low speed ac equipment are often made with good gear drives, so these items are worth salvaging.

Jacks, winches, scissor positioning lifts: Although forces as large as those provided by auto jacks and winches are seldom needed for most robotic experiments, their principles may be used for making smaller scale devices. Over 100lbs can be lifted with 2ft.lbs of torque using the jack shown in Figure 8-25A, and total lift capacity is 3000lbs for 12" extension. Car jacks were used for translating over 1000 lbs on greased plates as described in text prior to Figure 6-21, showing jacks are useful in a horizontal orientation too.

Electric winches powered from a 12V battery can pull over 4000lbs with a stall force of over 4500lbs. These 12V dc winches at $150 are one of the most commonly available devices for exceptionally high torque. A $20 hand trailer winch shown in Figure 8-25B has 600lb pull, and employs a one-way ratchet to prevent load runaway.

Scissor positioning lifts in Figure 8-25C are widely used in laboratories. A Jiffy Jack ™ shown has a smooth lifting action, can be set to better than 0.01", and has a load capacity of 100lbs. Current price from Cole Parmer is $175, so this item is not intended for casual hobbyists, but it is an extremely useful device for precisely positioning fairly heavy loads.

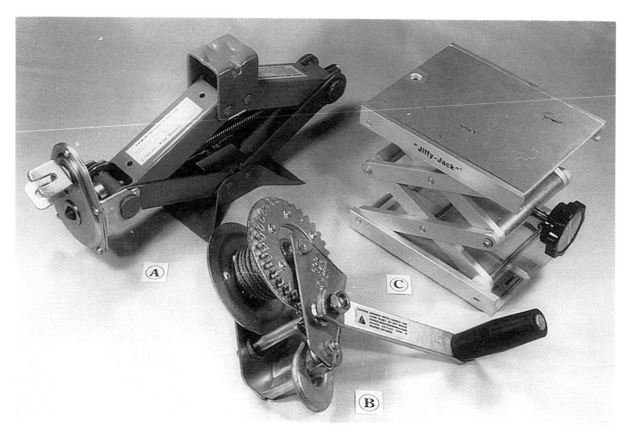

Figure 8-25. [A] 3000lb auto jack uses a well designed screw principle that can be copied by hobbyists. [B] Hand operated trailer winch has 600lb pull and a one-way ratchet to prevent load slippage. [C] A scissor jack can be used to precisely position heavy loads from 3" to 13".

Wheels, axles, collars, brakes, pulleys (sheaves):   Students are sometimes surprised how quickly weight escalates, when building a moving robot with wheels. Batteries, motors, framework and other hardware quickly increase total weight to a point where model car balloon wheels are flattened and will not turn, even when driven by a large motor. Foamed filled or inflatable tires may suffice but often solid resin casters, in-line skate wheels, or wooden wheels are the answer.

Wheels: Hobby stores stock inflatable and foamed filled model aircraft wheels from 1½" to 6" diameter, at prices from $3 to $25 per wheel. A 3½" smooth tire is shown in Figure 8-26A, treaded tires are also available. Model car wheels can be useful too, and Hayes solid neoprene rubber racing wheels in 1.5" and 2.25" dia. are sold by *Tower Hobbies*, HAYQ1113/1114 $1.99, $2.39.

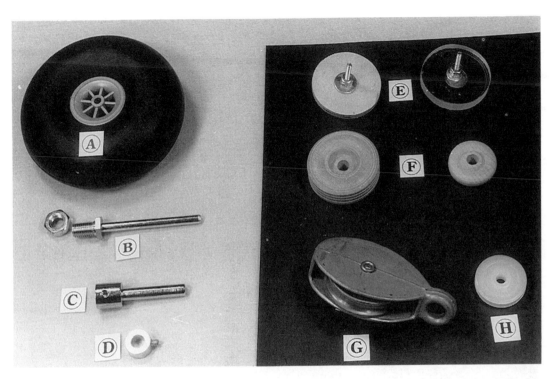

Figure 8-26. [A] Smooth 3½" diameter model aircraft balloon wheel is also made in foam-cored or inflatable styles, and with treads.   [B] Straight spring-steel wheel axle. Bent axles are also sold at hobby stores.   [C] Potentiometer shaft couplings may be used for connecting motors to wheels.   [D] Collars with set screws are made for axle sizes from 1/16" to ¼".   [E] Home made wheels in wood or acrylic, can be fabricated on a drill press, lathe or a clamped hand drill.   [F] Small wooden wheels are sold in craft and hardware stores.   [G] Smaller pulley blocks than this are available at hardware stores.   [H] Nylon pulley wheels are sold by some hardware outlets. They can also be made from acrylic or wood with a file, and an electric drill.

Wheel axles, collars and brakes: Bolt-on spring-steel plated axle shafts are made in straight and bent styles, to fit model wheels. A straight axle is shown in Figure 8-26B. A modified shaft axle is required when wheels are directly driven from a motor, and Figure 8-21C shows one way this may be done. Figure 8-26C shows a ¼" to ¼"shaft coupling from *Electrosonic* #152, $3. This can be attached directly to a motor's drive shaft. Plated brass collars shown in Figure 8-26D, are made to fit the following size axles: 1/16", 3/32", 1/8", 5/32", 3/16", 7/32", and 1/4". These collars may be obtained from hobby stores.

Tower Hobbies sells a servo actuated wheel brake ROCQ1104, $5.79 for two. Even better braking is produced with a multi-disk braking system designed for model cars: *Tower Hobbies* DTXC2650, $25.99.

Home made wheels: Wheels for very small robots can be salvaged from toys or made with a lathe or hand drill. Examples of wheels made from acrylic and four ply hobby plywood are shown in Figure 8-26E. Home made wheels are first cut out carefully with a fine blade saw, then supported in a drill's chuck with a screw, washers and two locking nuts as shown. On-size finish dimensions are obtained by carefully filing a rotating wheel.

Unlaminated wooden wheels up to about 2" diameter, are sold in craft and hardware stores. These wheels are not as tough as either plywood or acrylic wheels, but work well for light loads.

Pulleys: Students have made excellent translation stages, based on cord and pulley designs. Simple designs work best, and are usually made with basic hand tools. Small pulley wheels are required for computer controlled yoyos, and many other robotic mechanisms.

Pulley blocks and wheels in Figure 8-26G,H are sold at hardware depots, but any plastic or wooden wheel spun in a chuck can be grooved at its rim with a suitable file.

Materials for robotic elements:
Wood, plastics and sheet metal, are materials most favored by amateur roboticists. These are good choices, because they are easily worked and fairly inexpensive. Commercial equipment contains more machined parts, and uses a much wider range of materials. Some readily available tubings, bars and fittings useful for assembling robotic elements and other structures are shown in Figure 8-27,

A robotic hand on the front cover of this book is a good example of excellent engineering design requiring only hand tools and basic materials. Half inch diameter copper tube and fittings were used for the wrist mechanism, with rotation force provided by planetary gearing driven by a small dc gear motor. Finger actions were controlled via brake cables operated from motor-driven capstans. Finger actuating motors were placed at the elbow location, thereby reducing the hand's weight.

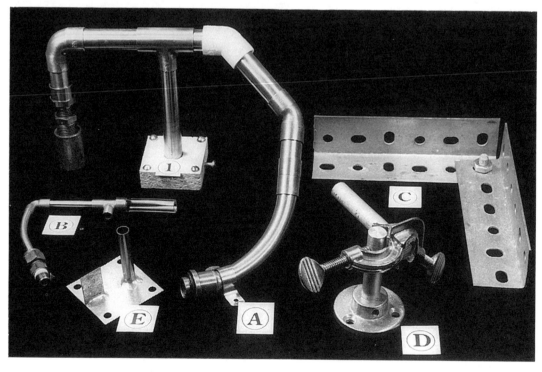

Figure 8-27. [A] Structures assembled from ½" copper pipe and standard fittings have many robotic uses. The robotic wrist shown on the front cover of this book was fabricated from ½" copper pipe.   [1] A wooden flange will hold pipe securely and may be fabricated as described in the text. Copper tubing can be bent into smooth curves. [B] ¼" copper pipe and fittings are also widely available.   [C] Slotted angle bar sold at hardware stores makes a very strong framework. [D] Aluminum laboratory flange, post, and clamp.   [E] Brass tubing and other compatible materials may be soft soldered to give strong joints. Use 'Steel bond' or silver solder for extra strength

Pipe and tubing: Figures 8-27A and B show a few useful pipe and tubing products sold by hardware outlets. Half inch copper pipe is stiff and can be assembled into complicated structures with standard connectors, tees, 45 and 90 degree elbows, laterals, unions, crossings, reducers, plugs or bushings. Some of these fittings are illustrated in Figure 8-27, but many more are carried by good hardware stores.

This figure also shows a short length of ½" tubing bent to form a smooth curve. This can be done with ¼" tubing too, as in Figure 8-27B. Hobbyists are not restricted to copper pipe, although this is easily worked and solders readily. PVC, ABS, aluminum, stainless steel, iron and other pipes and tubings are also sold. Connections can be made with cement, glue etc., and preliminary setups may be tested by just taping joints until a satisfactory model is completed.

Item [1] illustrates how a home made wooden flange may be used to support ½" pipe. A 5/8" hole is first drilled with a wood bit, then a #10 drill (0.101" dia.) is drilled through one edge. Use a 10/24 screw to hold this pipe securely, but do not tap the hole first. For comparison, a commercial base and clamp are shown in Figure 8-27D.

Slotted steel angles: Slotted steel angle and flat slotted bar is sold in 4 and 8 foot lengths at hardware depots, and is shown in Figure 8-27C. This material is best cut with either a jigsaw or cut-off wheel. It is hard, tough, very strong but quickly assembled with 5/16"-18 bolts and nuts.

Slotted steel angles cost about $1.5/ft but *McMaster-Carr* sells 'Erecto' slotted steel angles in either green painted steel or plain galvanized finish. Sizes range from 1½" x 1½" 4664T31 $11.33 to 1½" x 3½" 4664T35 $20.61. These prices are for a quantity of 10 eight-foot bars, plus 40 nuts and bolts. Six inch square reinforcing gusset plates are $14.67 for ten, 4664T22.

Soldered brackets: Various clamps, brackets or attachments can be soldered to galvanized plate, as shown in Figure 8-27E. Follow procedures outlined in text preceding Figure 8-16, when making these items.

Stiffening structures with carbon fiber:

Data presented in Table 8-4 shows graphite composite has a high specific modulus. This means carbon fiber materials can be used by roboticists to reduce flexing in beams, arms and booms, without adding too much extra weight.

Carbon fiber has undergone extensive testing for military and commercial applications and has some remarkable properties. It has good vibration damping characteristics, and deforms elastically over a wide load range. Fatigue tests show carbon composites outperform many other materials, it has low creep and is resistant to most corrosive compounds. This last feature may be a boon for bridge builders in cold climates, where salt used for snow melting corrodes steel reinforcing rods in bridges and similar structures.

Although the thermal expansion coefficient of a single carbon fiber is usually only one twentieth that of aluminum, thermal expansion of a carbon fiber matrix can be reduced to zero. This is achieved by crossing fibers at specific angles during 'lay-up', so their combined expansions cancel.

Carbon fiber has been used to cover foam wings, although this is an expensive proposition. For hobbyists, ½" wide fiber tape or strip is sold by *Tower Hobbies*: DAVR2000 5220 carbon fiber tape 12 feet, $4.59, or DAVR2030 5230 carbon fiber strip 5230, 5.5 feet, $6.69.

Modellers use carbon fiber for reinforcing bulkheads, spars, ribs, wire to wood connections, landing gears and fuselages. Raw carbon fiber is flexible, conforming to almost any shape, it may be bonded to surfaces with epoxy or superglue. Once bonded, most carbon composites are good insulators, and they do not provide significant electrostatic or magnetic shielding.

Tower Hobbies sells Sullivan composite flex pushrods tapped for 2-46 or 4-40. They are available in 36" or 48" lengths: 48" rods are SULQ3281, 581, 2-56 $4.79, and SULQ3285, 585, 4-40 $6.99. Some aeromodellers claim these composite pushrods have better thermal characteristics than nylon, and consequently introduce less offset in ailerons or elevators on hot days.

Kevlar thread has higher tensile strength than kevlar cloth, and its breaking strength may be as much as five times that of steel. Bonding procedures for kevlar are similar to those for carbon fiber and *McMaster-Carr* sells a 375yd spool of 0.008" diameter kevlar thread 8800K81, for $13.50.

Boron and PEEK (Poly Ether Ether Ketone) are products currently being evaluated for various high tech applications. A one foot square sheet of PEEK only ¼" thick costs more than $200, so hobbyists must wait some time for prices to reach the affordabilty level.

Figure 8-28 shows how effective carbon fiber strips can be when used to clad a ¼"x ¼" balsa beam. End deflection is reduced more than 3.5 times if a single 0.007" carbon fiber strip is attached to each side of the beam.

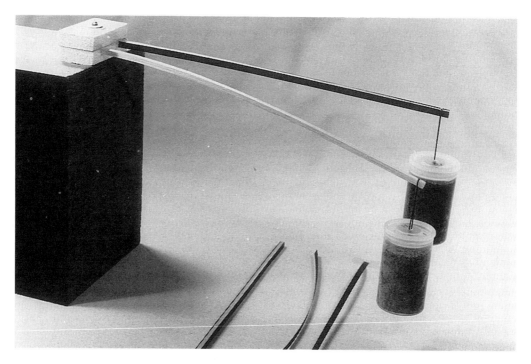

Figure. 8-28. Both beams are made from ⅛" x ⅛" balsa wood, and have identical loads. Very thin (0.007") carbon reinforcing strips glued to the rear beam reduce its tip deflection by a factor of 3.6. Carbon strips shown in right foreground are attached with CA adhesive. A clad beam is shown to the left of these strips. This demonstration is even more dramatic for thicker beams (ie larger values of t in Figure 8-29).

Fiber strips are often attached to hobby structures using a 'suck-it-and-see' approach. However, it is not difficult to calculate the expected improvement when carbon fiber reinforcement is added. Pertinent data for beams in Figure 8-28 are given below, and beam geometry is specified in Figure 8-29. Calculations for *two* carbon fiber strips are shown in Equation 8-1.

Carbon fiber strip	***Balsa wood***
P=0.09x9.81 = 0.883 N	P=0.09x9.81 = 0.883 N
L=0.225 m (~9")	L=0.225 m (~9")
E=250x10^9 Pa (mean value from Table 8-4)	E=7x10^9 Pa (from Table 8-4)
b'= 1.778x10^{-4}m (0.007")	b= 3.175x10^{-3}m (0.007")
t= 3.175x10^{-3}m (⅛")	t= 3.175x10^{-3}m (⅛")
ρ=1500kg/m^3	ρ=215kg/m^3

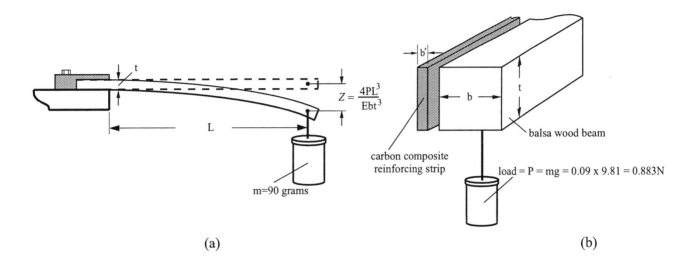

Figure 8-29. Beam deflection for a plain or carbon fiber clad balsa beam, may be determined using variables defined in this figure. Calculations are done in Equation 8-1, using data from Table 8-4.

$$(8\text{-}1) \quad z = \frac{4PL^3}{Ebt^3} = \frac{4 \times 0.883 \times (0.225)^3}{250 \times 10^9 \times (2)^* \times 1.778 \times 10^{-4} \times (3.175 \times 10^{-3})^3} = 0.0141m \quad (1.41cm)$$

* *Two* reinforcing strips are used.

A similar calculation for an unclad balsa beam predicts a deflection $z = 5.7$cm. Measured and calculated values for both beams are compared in Table 8-6.

Table 8-6. Comparison of deflections for beams shown in Figure 8-28.

	calculated deflection (z)	measured deflection (z)
carbon fiber beam	1.41 cm	1.3 cm
plain balsa beam	5.66 cm	4.7 cm

These results are in reasonable agreement because exact moduli for both balsa and carbon fibers employed in Figure 8-28 are unknown. If Young's modulus for both materials is known accurately[4], an agreement of ±5% may be expected.

---

[4] Determining E is straightforward. Deflection z is measured for a known load, then E is calculated using measured values for other variables from the equation in Figure 8-29.

Screw type linear dovetail slides: Screw driven dovetail slides have low friction movement, and are widely used for precisely positioning optical components. Although not quite as good as roller bearing stages, dovetail slides provide an adjustment sensitivity of about10μm. They are simple rugged devices and usually manufactured with either a lead screw as in Figure 8-30A, or a rack and pinion drive, Figure 8-30B.

Residual machining inaccuracies may be corrected if a shim bar is incorporated as in Figure 8-30A. This bar's position is adjusted by three set screws to take up mechanical slop, and is a worthwhile feature to add in any homemade slide. Many top-of-the-line linear translation stages use adjustable shim bars, attaining positional sensitivities <1μm.

Precision screw slides are expensive, but *C & H Sales* presently lists a single axis stage with 2" total travel #GR9556, $50. A top quality slide with 3" movement and repeatability of ±1.25μm (0.00005") is available from *Servo Systems* LPS-117 $69.

Rack and pinion slides: Although their movements are not as precise as screw slides, rack and pinion movements are less expensive to make in long lengths. For example, *McMaster-Carr* stocks four and six foot lengths of rack in 3/16" to 1/¼" widths with tooth pitch varying from 5 to 32 teeth per inch. Price for a 4 foot length of 3/16", 32 tpi track, 6295K242, is $27.45, and a matching spur gear 6325K89 is $5.19. Racks and gears are available in steel and nylon, but prices are comparable for both materials.

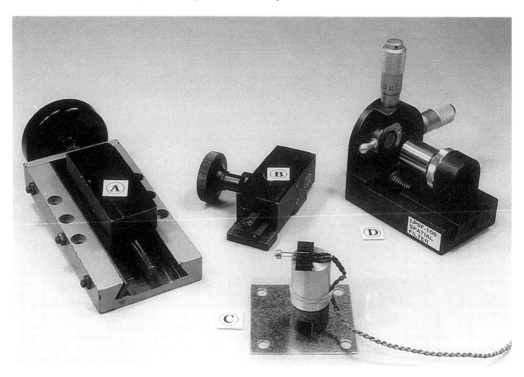

Figure 8-30.   [A] 28 threads per inch screw drive dovetail slide with anti-backlash mechanism, resolution ~0.0002" (5μm).   [B] Rack and pinion dovetail slide with 1" total travel and setting precision of ~0.002".   [C] Plastic coated steel plate works well as a sliding surface for magnetic holders.   [D] Optical spatial filter uses Teflon coated sliding surfaces and magnetic attraction for positioning a 5μm pinhole to ~1μm.

Novel movement systems: Flat sliding surfaces are very useful for positioning purposes, and work well with greased or non-stick coated plates. Students have used flat surfaces for translation stages, and these are sometimes lightly oiled or dusted with talc. If smooth plastic sheet is attached to a steel base, a magnet can be used for holding and positioning items as in Figure 8-30C. Plastic overhead transparency sheet is satisfactory as a plate covering material.

Visible light laser beams can be focussed to spot diameters equivalent to a laser's wavelength ($\sim 0.5 \mu m$). Focused laser beams are passed through tiny pinholes (also with diameters of a few micrometers), to 'clean up' a beam. A microscope objective lens is used for focussing, and an x,y,z translation stage is required to move a pinhole to its optimum position.
      One commercial manufacturer attaches their pinholes to Teflon-coated toroidal magnets, which are attracted to a steel Teflon coated surface. Pinhole movements are controlled by steel micrometer heads, which are attracted to the toroidal magnet. A spatial filter (by Jodon Laser) employing this technique, is shown in Figure 8-30D.

Good quality drawer slides have very little side play when correctly installed, so these may be used for translation purposes. Most drapery tracks are too flimsy for use as monorail translation stages. However, bifold door trackset trolleys have four wheels, and run in a sturdy hanger. A 60lb capacity 4 foot length of door track set, costs about $20, and tracks with 250lb capacity are manufactured. Door specialists listed in telephone directory Yellow Pages, are good sources for these products.

Adjustable x,y stages: A nifty mechanism shown in Figure 8-31A can be duplicated or scaled up for hobby use. Two knurled screw adjusters are used to translate a plate on a plane surface. A socketed coil spring supplies radial restoring forces, staying in contact with a flat on the movable plate. Planar movement is ensured by restraint provided from a two-arm wire leaf spring attached to the outer frame. Adjusting screws have 40 threads per inch, giving adequate fine control.
      Figure 8-31B shows a well-made precision x,y microscope stage with vernier scales. When controlled by hand, each movement is resettable to $100 \mu m$ from the scales, but if an external reference is used uncertainty can be reduced to $\sim 10 \mu m$. This stage is handy for calibrating stretch resistors from Chapter 5, and has been employed by students for microprofiling with a strain gage probe shown in Figure 5-10e.
      Top-of-the-line micropositioning x,y translators used for centering fiber optic cores and other tasks, have movement resolutions of $0.1 \mu m$ or better.

Rotating platforms: Ball bearing rotating platforms or 'Lazy Susans' can support very heavy loads yet still turn with little effort. They are low profile turntables with precision circular ballraces, available in 3", 4", 6", 9" and 12" square format with load capacities of 200, 300, 500, 750, 1000lbs. Lazy Susan in Figure 8-31B, is usually mounted with self-tapping screws, and screw holes are offset to eliminate screw-head interference. If loading is symmetrical, large decks may be employed. Circular decks can be rotated at their edges by a motor driven rubber capstan.

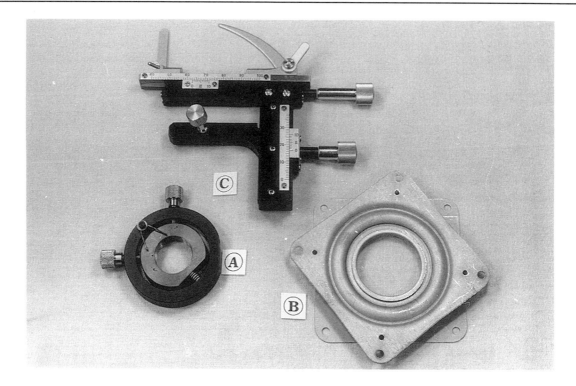

Figure 8-31. [A] x,y translation stage suitable for small displacements.   [B] Rotating ball bearing turntable or 'Lazy Susan' shown here can support 200lbs. If loads are balanced this 3" turntable may be used to support a 12" diameter deck. A quality 12" diameter Lazy Susan has a load capacity of 1000lbs.   [C] x,y microscope stage, *Edmund Scientific,* A30,058, $69.

Multipurpose holders: Two methods for holding small items are shown in Figure 1-6, but many more ideas spawned from fertile minds have been translated into commercial hardware. Professionals employ a variety of nifty stands, clamps, holders and positioners, when breadboarding new apparatus. Such equipment saves time, but many of these items are too costly for an amateur's wallet. However, with a little ingenuity commonly available inexpensive items or homemade gadgets can be very helpful.

'Extra hands' is a handy holder with two alligator clips plus an attached magnifier for close work. Sold by *Radio Shack* 64-2063, $9.99, this useful tool has ball joints allowing objects to be held at just the right orientation.

Figure 8-32 shows a few gadgets that help when small bits and pieces must be jockeyed into the right spot. An articulated flexible spine in Figure 8-32A is easily shaped to any configuration, then locked into a stable form by the large knob. Objects are held firmly in a screw adjustable 'finger'. A sliding clamp on an upright post allows even more adjustability on this versatile, well designed holder with a magnetic base. A similar tool is sold by *McMaster-Carr* 8667A31, $53.18. Many well-designed magnetic base holders are stocked by tool suppliers, and utility grade devices are available for less than $10.

Ball joints and mirror positioners: Camera tripods are great for positioning objects at just the right height and attitude. Three-way pan and tilt action heads are handy, and a ball joint from an old camera tripod shown in Figure 8-32B is useful for mirror tilting and similar applications. A fancier (and more expensive) commercial tilting mechanism in Figure 8-32C, employs magnets for two-degree of freedom (DOF) mirror movements. Six DOF positioning stages with sub micron resolution and seconds of arc rotation capability, may cost more than $10,000.

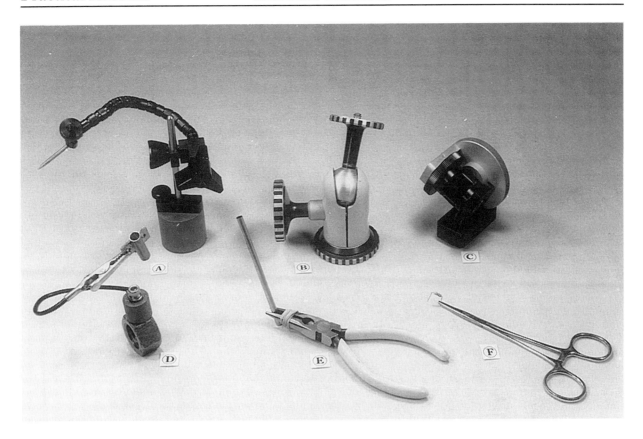

Figure 8-32. [A] Articulated spine positioner with magnetic base. A screw clamp 'finger' holds any small object firmly. [B] Camera tripod ball joints are useful for omni-directional tilting purposes. [C] Precision two-axis tilting mirror holder. [D] Wire solder can be repeatedly bent without hardening, making a handy universal positioner. [E] An elastic band wrapped tightly around plier tips gives a firm grip, even on round tubing. [F] Locking forceps are convenient for holding small objects.

Home made 'third hands': Because of its high lead content, solder can be repeatedly bent without work hardening or breaking. When equipped with an alligator clip or clamp and a magnetic base, a length of solder makes a useful third hand or positioner, as shown in Figure 8-32D. Always encase solder in shrink wrap tubing as shown, to avoid repeated direct skin contact with solder and its lead content.

Figures 8-32E,F show two simple tools for holding objects. Pliers should be on every hobbyist's bench, and stainless steel locking forceps are $2 at flea markets. Manicure scissors, nail trimmers, dental picks, and a variety of other stainless steel tools are sold at flea markets for the same price.

Weigh scales: Robotics teaching laboratories need accurate scales, capable of measuring milligrams to at least 200lbs. Four reliable scales shown in Figure 8-33, are recommended, but many equivalent weigh scales are available.

Digital scales: Item [A] in Figure 8-33 is a Pelouze-4040, 400lb/180kg capacity scale, with digital readout in 0.5lb/0.2kg increments, display module may be positioned up to 9ft from the weighing pan. This unit is rugged and operates from 9V battery or ac adapter. Available from *Office Max* 0702-2759, $119.99 (including adapter)

Weighing is always faster on a digital scale and a Pelouze PE-5, measures 5lb/2.2kg in 0.1oz/2g increments, Figure 8-33B. Available from *Office Max* 0701-8933, PE5R, $84.99, an ac adaptor is sold separately 0702-1876, ADPT2, $6.99.

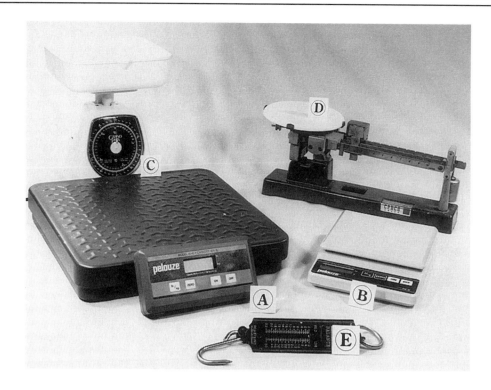

Figure 8-33. [A] Pelouze 400lb digital scale measures in 0.5lb increments and its display unit may be remotely mounted. [B] 5lb scale also made by Pelouze has 0.1oz resolution. [C] Spring type scale shown is best for calibrating force sensing resistors and capacitors. [D] Triple beam balance, 2kg capacity with 0.1g graduations. It can weigh to 0.01g when used with care. [E] 50lb (23kg) hanging spring scale is very useful for rough weighing.

Spring scales: Spring-type food or postal scales are best for calibrating force sensing resistors shown in Figure 5-16, or capacitive sensors in Figure 5-44. A $5 food scale with 10lb/5kg capacity in Figure 8-33C is adequate for many weighing tasks. It has a resolution of ~0.2oz, but repeatability is only ~1oz due to hysteresis. A hanging spring scale is shown in Figure 8-33E.

Beam Balance: Figure 8-33D shows a rugged triple beam balance with 2kg capacity and measurement resolution of 0.01g. A newer model with 20kg capacity and magnetic damping is sold by *Cole-Parmer* H-01045-00, $107. Mechanical balances are simple, basic, very reliable, and ideal for classroom or home use.

Home made weigh scales: Designing and building a simple weighing scale is a good project for beginners. Mechanical principles of beam balances are straightforward and discussed in physics texts. Added features such as digital readout, auto-zero, tare-set etc., allow designs to encompass various levels of sophistication. A homemade microgram electrobalance described in *Scientific American* June 1996, is a worthwhile project for someone with no previous experience in mechanics or electronics.

Hobby-store parts for roboticists: Beginners are often disappointed to find very few parts at hobby stores to solve their problems. This is understandable because hobby stores are commercial enterprises and must make money to survive. Consequently, their product line is specifically tailored to meet modellers' needs. Unfortunately, ready-made kits are now a dominating force in modelling, so building from scratch is experiencing a continual decline.

Larger hobby stores are still interesting places, and a catalog from Tower Hobbies shows many goodies seldom seem on local store shelves. Roboticists should accumulate a good collection of catalogs. Tower Hobbies, Hobby Shack, Hobby Lobby, Ace and similar catalogs make one drool. Buy a copy of RCM (Radio Control Modeler) - a monthly magazine. Its advertisements are a guide to suppliers of many useful products. Other useful magazines, books and journals are listed in Appendix B

Hobby stores sell special tools, propellers, gasoline engines, electric motors, servos, radios, pneumatic retracts, batteries, chargers, rubber bands, tubing, latex rubber, adhesives, voltmeters . . . You may already have some idea of their product lines, and many useful hobby items have already been mentioned or depicted in this book. In closing this chapter a few more useful hobby store items are shown in Figure 8-34, (prices are approximate).

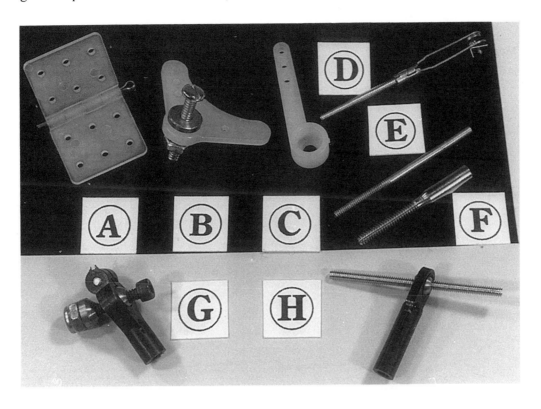

Figure 8-34. [A] Low profile, lightweight nylon hinge, 25¢.　　[B] Nylon bell crank assembly with brass bushing. $1.50.　　[C] 1¼" straight nylon arm, $1.　　[D] Brass threaded clevis with locking tab, $1.　　[E, F] Wire or rod is soldered into these threaded couplers. Couplers are then used to attach wires to ball links or threaded clevises, $1. [G,H] Low friction ball link sockets may be used for tie-rod ends, or anywhere a self aligning swivel connection is required, $2.

Future Books

Much more remains to be documented on Practical Robotics, and a rough outline of planned content for future books in this series is given below. Working circuits, operating protocols, specific part assignations and prices will be included.

Motors

Motors:	Size, torque, power and speed ranges. ac, ac/dc (universal motors).
DC brush motors:	Direct drive, bi-directional drive, slave servoed systems.
Stepper motors:	How to run *any* stepper motor. High power operation, microstepping.
Gasoline engines:	Model aircraft, and small two stroke motors.
Current consumption:	Voltage drop, back EMF.
Torque measurement:	Prony brake, spring scale and strain gage methods. Torque vs back EMF.
Speed measurement:	Encoder disks, Hall sensors, LED strobes, back EMF, tachometers.
Rotation direction:	Quadrature methods.
Power curves:	Power vs rpm.
Gear boxes:	Home made, commercial, larger gear boxes.
Driver circuits:	H bridge, pulsed bridge, ESC devices. Power MOSFET drives
Servos:	Types, torques, circuits model aircraft servo, project applications.
Buying motors:	New motors. Surplus stores, auto parts and junk yard offerings.

Optical devices

Basic optics:	Normal and cylindrical lens action, Axicons, Lensacons, Fresnel lenses, ball lenses, GRIN lens, plastic lenses, home made lenses, collimation, resolution, Kohler illumination, light dumps, Rayleigh horns, Brewster angle dumps, razor blade dumps. Black glass, opal glass, $BaSO_4$, MgO. Integrating spheres and cubes. Color, CIE diagram, XYZ and RGB filters. IR Ektachrome filters. Interference filters. Glass and gelatin filters. Eye sensitivity and color mixture functions. Radiometry, spectrophotometry. Mirrors total and partial reflection, front surface. Home silvering methods.
Emitters:	Incandescent lamps, black body radiation, color temperature, standard lamps. Xenon, mercury, argon, Vortec light sources. Linear and coiled flash lamps.
LEDs:	IR, Visible, superbright. Collimating LEDs. Metal vs plastic LEDs. Opto isolators, opto couplers. High pulsed current operation. Seven segment displays etc.
Lasers:	HeNe, argon, ruby, NdYag, Laser diodes, James Bond's lasers (power and energy).
Detectors:	Spectral sensitivity, relative sensitivity. Linearity, time response. Photodiodes, phototransistors, photodarlingtons, photomultipliers. Reticon arrays. CDs, planar silicon diodes, Avalanche photodiodes. IR cards. Position sensitive detectors and

	home made PSD's. Home made quadrant detector. Star tracking. Motion detector. Total darkness scanning. Very large area detectors (12" x12" or larger).
Opto devices:	Linear finger bend optical sensors. Alphanumeric LCD operation. Kerr cells, Pockels cells, Faraday effect, bleachable dyes. Home made linear optoisolator. Absolute position encoders.
Fiber optics:	Single mode fiber (SMF) Multi mode fiber (MMF). Practical sensors from SMF and MMF. Couplers and efficiencies. GRIN lens coupling. Home made fiber optic pigtails. Cleaving silicon fiber at home. Plastic fiber uses.
LDV:	Practical Laser Doppler Interferometry for amateurs.

Pneumatic, hydraulic and vacuum techniques

Basic theory:	Pressure, vacuum, degassing, pumping speeds, throughput. Overview of pumps and compressors. Glass, stainless steel and home made systems.
Gages:	Manometers, classical vacuum gages, Flowmeters. Solid state and capacitive. gages. Barometric switches.
Seals:	O-ring sizes and materials, make your own O-rings. Back-up rings. Gaskets, Seal screws. Greases and lubricants. Leak detection with bubbles, gages and Tesla coil.
Connectors:	Inexpensive components for vacuum and pressure applications.
Valves:	Taps, solenoid valves, air cylinders.
Hydraulics:	Rotary and linear dashpots. Hydraulic motors. Auto jacks and hydraulic car hoists.
Pneumatics:	Air tools. Levitation with air jets. Coanda and Bernoulli forces. Pyrotechnic actuators, bellows, bolt cutters etc. Air tracks, air tables, air pucks. Hovercraft equation and home built machines.

Computer interfacing and control

Interface chips:	A/D, D/A, 8 and 16 channel multiplexers and demultiplexers. PPIs. Switch debouncing. Latches as temporary memories.
PC interface:	Board design and checkout procedures in BASIC.
BCC52:	Using this 12MHz microcontroller
Timing:	Using a 1μs hardware timer under computer control.

Useful procedures and devices

Encoders: Slot shape and rise time for fast response. Home made rotary and linear encoders.

Miscellaneous sensors: Using bend sensors, touch sensors, thermistors, thermoelectric coolers.

Miscellaneous devices: Electromagnetic tuning forks and vibrating beams. Fiber optic accelerometer. Measuring speed of 400mph paper darts launched with a home made pneumatic gun. Levitation methods. Strength testing with strain gages.

Students projects: Outlines of many interesting student projects.

Teaching aids: Proven methods for teaching and operating a robotics laboratory

Miscellaneous tips: Novel ideas, circuits and procedures for students, roboticists and hobbyists.

References: Books, catalogs and useful sources for roboticists.

Resources: Where and how to get equipment for projects.

Miscellany

Physics: SI units and conversions, useful equations, math formulas, binary and hex conversions, specific mass (density) tables, dielectric constants, electrical conductivities, strength of materials, magnetic equations, PZT constants, permeabilities, wet and dry bulb temperatures.

Electronics: Resistor and capacitor codes, wire gages, coil inductance formulas. Transistor, diode, op amp, and MOSFET data

Circuits: Dozens of additional practical circuits complementing text material

Mechanical: Screw and thread sizes, pipe sizes, O ring sizes, properties of materials.

Optical: Glass and plastic filter transmission curves. Refractive indexes. Encoder disks for photocopying.

Classroom equipment list

a.c. adapters
a.c. power bars
acrylic, cement
adhesives, various
alligator clips
aluminum sheet
amplifiers, op amp
balance, weigh
balance, spring
banana plugs
batteries, various
battery checker
bit, countersink
BNC connectors
books
brackets, angle
breadboards,
solderless
buffers, IC
calipers
capacitance meter
capacitors
capillary tubing
catalogs
chucks, keyless
cigarette lighters
clamps, C and bar
clips, binder
components,
electronic
computers and
accessories
connectors, various
cord, string
cork borers
couplers, shaft,
flexible
coupling nuts
cutter, fly
cutter, carbide
cutter, expansive
cutter, glass
debonders, various
desoldering pump
desoldering braid
diodes
DMM
drafting instruments
drafting film
drill press
drill bits, metal and
wood
drill, cordless
dust pan and brush

elastic bands
electronic components
extension cords
fasteners, various
files, metal and wood
filter, active IC
first aid kit
flashlight
flux, solder
fly cutter
foamboard, IC storage
function generator
galvanised metal sheet
gears
generator blox
glass sheet
glass tubing
glass cutting tools
glues, adhesives
greases
grinder, bench
hammers
headers, Berg type
heat sinks
heat shrink tubing
heat guns
hinges
hobby parts
IC pin straightener
IC extractor
ICs, misc
inductance meter
IR detector card
jack, lab
jacks and plugs
jigsaw, power
knives, utility
lacing cord
lamps and bulbs
LEDs, misc
lenses
lighters, cigarette
linesman pliers
logic probe
logic pulser
loupe, magnifier
lube, oil grease
lubricants, cutting
fluid
magnet wire
magnets
magnifiers
metal punches
metal shear,bender

metal sheet and rod
metal cutters
microcontrollers
micrometer, calipers
microphones
microscope
illuminator
microscope slides
microscope
miniclip probes
mirrors
MOSFETs
motors
multicore solder
multimeters
multivibrators, timers
Mylar film
nails
needles and thread
optical bench
optical components
oscilloscopes
photodetectors
pins, Tee
plastic, sheet and rod
pliers, linesman
plugs and jacks
potentiometers,
various
power supplies
propane torch
pulleys
punch, hand
punch, center
punch, lever
PZT elements
razor saw
rectifier bridges
rectifiers
relays
resistors
rheostats
rubber bands
rulers, scales
sandpaper
saw, hacksaw
saw, hand
saw, power
saw, razor
saw, circular
scissors, various sizes
screwdrivers
screws, socket
screws, flathead

servos, model A/C
sewing kit
shaft couplers
shaft extenders
shear, metal
side cutters
soldering, misc.
solderless boards
solenoids
solvents, various
speakers
spring balance
squares
stapler
storage scope
string
switches
tape measures
tape, misc types
tapping fluid
taps and dies
Teflon tape
test probes
threaded rod
timers, multivibrators
tools, various
transistors
tubing, vinyl, silicone
tubing, metal hobby
utility knives
valves, aquarium
valves, pressure
Velcro
vise, bench
vise, drill press
vise, mini
Visegrips
washers, variety
weigh balance
weights
wheels
wire, piano, music
wire, clothesline
wire, electrical
wire, mechanical
wire strippers
wire wrap, tools, wire
wire electrical
wire mechanical
wood
workbench
wrenches, various

Robotics information sources

How to find nifty ideas

It is probably obvious that to keep abreast of new developments one must constantly scan new issues of magazines and journals. It is perhaps not so evident, that finding some jewels means it is equally important to check old issues too. Computer searches may not reveal these hidden treasures, especially if an archivist failed to recognize their significance when allocating tracer keywords.

Beginners may be disappointed to learn many good technological notions appear in print only once. Sometimes these ideas are buried in a major article and keyword references do not mention the nifty idea. After all it was just a footnote to the main theme.

This is not a rare occurrence, and partially explains why many inventors reinvent devices that have been 'invented' several times previously. Patent assignations often involve a search fee of $10,000 or more because of this phenomenon.

How does this affect hobbyists? Well, *Review of Scientific Instruments* is unlikely to republish that terrific idea which appeared first in 1966. However, a list of some of the more useful magazines, journals books and catalogs will at least give you a place to start.

It would be presumptuous to suggest the following short lists are a representative distillation of many publications containing useful robotic information. Necessary restrictions do a grave injustice to numerous fine books, magazines, journals etc. omitted.

In classrooms, specimen copies of electronics magazines, scientific journals and technical articles may be displayed for student perusal. These will convey the flavor and progressive levels of sophistication characterizing the robotics field, far more eloquently than words.

Magazines

Over the past few years it has become increasingly difficult to find not only good technical books, but also magazines in the same genre. Only the best bookstores now stock all the items listed below. Vanishingly few public libraries carry even one of these publications, although their periodical shelves are filled !

Buying *Electronics Now* or *Popular Electronics* from newsstands should not be difficult since they have huge circulation bases. Other magazines may require subscriptions if they are not available locally.

Advertisements in magazines and other periodicals, are windows to a part's agora, reflecting contemporary commercial developments in electronics and computer fields.

Popular Electronics ($3.99 monthly, P.O.Box 338, Mount Morris. IL 61054-9932)): Articles range from very simple to a medium level of sophistication. One of the best magazines for beginners, who will learn a great deal by constructing projects described. Electronic, robotic and computer offerings are covered in detail. Hardware kits to build specific projects are sometimes available from an article's author.

Electronics Now (formerly Radio Electronics) ($3.99 monthly, Box 55115, Boulder, CO 80321-5115): Projects are slightly more advanced than those from *Popular Electronics*. Tutorials for novices, excellent advertisements, new product reviews and a column by Don Lancaster are monthly highlights.

If something is interesting, new, nifty or puzzling (and not necessarily electronic), it will probably be mentioned by the editorial staff, or in a reader's letter.

ETI Electronics Today International (£2.25 monthly, Nexus House, Boundary Way, Hemel Hempstead, England HP2 7ST): Ideal for beginners and those with some electronics background. English magazines do a good job of providing background related material in their publications.

Electronics (£2.25 monthly, Maplin MPS, Box 777, Rayleigh, Essex, England SS6 8LU):   Good articles on electronic gadgets, physics and new products. This magazine is sold by better bookstores such as Barnes and Noble.

Electronics World (£2.25 monthly, L333, Electronics World, Quadrant House, The Quadrant, Sutton, Surrey SM2 5AS):   High caliber magazine with well written technical articles and substantive explanations of basic phenomena. Good letters from well-informed readers help to keep this magazine among the best.

Circuit Cellar Ink ($3.95 monthly, P.O. Box 5333, Pittsfield MA 01203-9982):   Excellent magazine founded by well-known electronic's guru Steve Ciarcia, at one time a top writer for *Byte*. First rate articles on robotics, computers, electronics. Very useful advertisements and well-written articles make this a first rate magazine. Older issues of *Byte* contain interesting and useful material by Steve Ciarcia. Byte has found no suitable replacement since Steve started publishing his own *Circuit Cellar Ink.*

Microcomputer Journal ($4.95 six times a year, 1700 Washington Ave. Rocky Ford, CO 81067-9900):   This magazine was formerly *Computer Craft* also *Modern Electronics.* Good contributing editors and a knowledgable editor-in-chief make this a high quality product that has published some of the best articles in its field. One of the best sources for information on microcontroller boards and ancillary products for amateurs.

Radio Control Car Action ($4.95 monthly, P.O. Box 427, Mount Morris, IL 61054-9853):   Worth scanning for developments in ESC (Electronic Speed Controls). Latest information on batteries, chargers and excellent advertisements for hobby items make this magazine a good read.

Model Airplane News ($4.95 monthly, P.O.Box 428, Mount Morris, IL 61054-9859):   Interesting articles cover new servos, construction techniques, batteries, chargers and gadgets. Excellent advertisements and interesting letters from readers.

Scientific American ($4.95 monthly, 415 Madison Avenue, New York, NY 10017-1111):   Excellent popular articles in all areas of science and technology by the world's top scientists. 'The Amateur Scientist' column appears sporadically, and excellent 'Amateur Scientist' articles may be found in older issues of *Scientific American*. These are worth seeking out.

Electronic Design (free to qualified readers, monthly, 1100 Superior Avenue, Cleveland, OH 44197-8132): Publishes useful practical circuits in its 'Design Applications' and 'Ideas for Designs' columns.

Sensors (free to qualified readers, or $5.50 monthly US, Helmers Publishing Inc. 174 Concord St. P.O. Box 874, Peterborough, NH 03458-0874):   A journal of applied sensing technology aimed at industrial users. Excellent advertisements and interesting articles. A good way to keep up with state of the art products.

Technical magazines are currently having a rough time. Overheads are high and readership is ephemeral, consequently many minor publications have either succumbed or changed hands. Many other magazines are published but only those with apparent staying power have been listed above.

Scientific Journals

Engineering libraries in universities have subscriptions to many major journals, and these are usually available for outsiders to read and photocopy. Special contractual arrangements permit single copies of most material.

Review of Scientific Instruments (subscription only, monthly, 500 Sunnyside Blvd. Woodbury, NY 11797-2999):   Excellent practical information in all areas of technology. References at the end of each article enable readers to backtrack through older issues and follow topic developments from their inception. Most articles can be understood by anyone with a second year university education.

Measurement Science and Technology (subscription only, monthly, Techno House, Redcliffe Way, Bristol, BS1 6NX, UK):   Previously *Jour. Phys.* E (name changed January 1990). European equivalent of Review of Scientific Instruments. Covers basic scientific measurements, scientific instrumentation, electronics, optics. Interesting easily understood articles.

IEEE Transactions on Instrumentation and Measurement (subscription only, monthly, 445, Hoes Lane, P.O.Box 1331, Piscataway, NJ 08855-1331):   Articles are specific, condensed, and more difficult to follow without some knowledge of the topic. Fewer useful articles for hobbyists, but worth reading for an occasional gem.

Applied optics (subscription only, monthly, 2010, Massachusetts Ave. NW, Washington DC 20036-1023): Older issues were more 'applied', but some current issues still have useful ideas. Thumbnail sketches of recent patents make for enjoyable and thought provoking reading.

Robotics Research (subscription only, monthly, 55 Hayward St. Cambridge, MA 02142): Content is highly specialized and of little use to amateur roboticists, despite the journal's appealing title.

Books

Books already cited in this text are repeated below for convenience. The utility of these texts may be placed in context by referring to information in specific chapters (check index under 'books'). Two books listed below are ideal for those with no previous electronic experience

   Getting Started in Electronics Forrest Mims III *Radio Shack* 276-5003 $4.99

   Engineer's notebooks: (*Timer ICs, Op amps, Optoelectronics, Basic semiconductor circuits, Digital logic circuits, Communications projects, Formulas tables and basic circuits, Schematic symbols*). Forrest Mims III *Radio Shack* 266-5010, 5011, 5012, 5013, 5014, 5015, 5016, 5017 $1.99 each

Books for classroom use

The following books are practical texts providing pin allocations and other parameters for many devices. Hobbyists can preview these books in large bookstores such as Barnes and Noble, Chapters, Borders or Crown Books. Major electronic component stores such as Active Components, stock useful technical texts, but technical books are no longer found in shopping mall bookstores. Prices listed below are approximate.

title	author	publisher	ISBN #	price
CMOS Cookbook	Don Lancaster	H.W.Sams	0-672-22459-3	$25
TTL Cookbook	Don Lancaster	H.W.Sams	0-672-21035-5	$25
Active-Filter Cookbook	Don Lancaster	H.W.Sams	0-672-21168-8	$30
OP-AMP Circuits and Principles	Howard M.Berlin	H.W.Sams	0-672-22767-3	$25
The Art of Electronics	Horowitz and Hill	Cambridge Univ. Press	0-521-37095-7	$80
Analog electronics for microcomputer systems	Goldbrough, Lund and Rayner	H.W.Sams	0-672-21821-6	$50

Additional useful books for hobbyists and roboticists

Physics for Scientists and Engineers by Serway. ISBN 0-03-004854-0 ~$75, Saunders College Publishing: Excellent well-illustrated text, covers all the physics needed for building robots. Many other similar texts are published, but this is one of the best. A good physics text is essential for amateur roboticists, and a second hand book is adequate.

The Best of Steve Ciarcia's Circuit Cellar by Steve Ciarcia. ISBN 0-07-011025-5 ~$30 McGraw-Hill: Written by a master of computer 'how to', this book is full of useful nuggets and has no equal.

The Forrest Mims Circuit Scrapbook by Forrest Mims III. ISBN 0-07-042389-X ~$25 McGraw-Hill: Probably no one makes a beginner feel more comfortable, or explains a subject as well as Mims.

Forrest Mims Circuit Scrapbook II by Forrest Mims III. ISBN 0-672-2252-2 ~$30 Howard W. Sams: A sequel to the previously mentioned book and another winner.

The Forrest Mims Engineer's notebook by Forrest Mims III   ISBN 1-878707-03-5 ~$20 HighText Publications: If you can only afford one electronics data book, buy this one first.

The Robot Builder's Bonanza by Gordon McComb ISBN 0-8306-2800-2 ~$20 TAB Books: One of the first books on 'Practical Robotics'. It contains many useful ideas and circuits.

Transducer handbook H.B.Boyle, Newnes Linacre House, Jordan Hill, Oxford, UK OX28DP $43:   Useful overview of sensors and transducers.

Sensors and Transducers Ian R. Sinclair, Newnes Linacre House, Jordan Hill, Oxford, UK OX28DP ~$50: Good coverage of all types of commercial sensors.

The How and Why of Mechanical Movements by Harry Walton, 1968 Popular Science E.P. Dutton and Co. Inc. New York. Library of Congress Catalog Card # 68-31227:   Excellent book for those with no mechanical experience. Fine illustrations make each topic easy to grasp.

Mechanisms and Mechanical Devices Sourcebook, Nicholas P. Chironis ed. McGraw Hill 1991 ISBN 0070109184, ~$70:   A definitive work on industrial mechanisms. Many neat ideas are explained and illustrated with excellent line drawings.

Sensors for Mobile Robots by H.R.Everett ISBN 156881048-2 $60 A.K.Peters, Ltd.:   Excellent source of references for classroom use. Covers all methods used for ranging, obstacle avoidance. This is not an applications manual and no circuits are included. However, it is one of the best readily understandable robotics reference works available.

Navigating Mobile Robots by Johann Borenstein, H.R. Everett, Liqiang Feng ISBN 1-56881-058-X: Excellent survey of navigation techniques, and a good source of references for classroom use. This book nicely complements *Sensors for Mobile Robots*. No practical data for experimenters but an excellent survey text, and very useful for instructors.

Force and Touch Feedback for Virtual Reality by Grigore C. Burdea   ISBN 0-471-02141-5 $77 Wiley Interscience: Good source of references for classroom use. No practical circuits or construction information for amateurs, but describes many novel sensing techniques and lists hundreds of useful journal references

Manufacturers' data books

No single reference work contains comprehensive data for electronic devices cited in this book. The following short list covers those most frequently used linear and digital items for robotics work.

title	manufacturer	book #	supplier	price
Linear and Interface Integrated Circuits	Motorola	DL128	Newark	$12.46
Linear Databook	NSC*		Electrosonic	$15.50
Schottky TTL Databook	Motorola	DL121	Newark	$12.46
Logic Databook	NSC*	Vol. II	Electrosonic	$12.50
CMOS Logic Databook	Motorola	DL131	Newark	$5.32
Discrete Semiconductor Products Databook	NSC*		Electrosonic	$12.50
Optoelectronics Device Data	Motorola	DL118	Electrosonic	$4
Data Book (covers many optoelectronic items)	OPTEK		Newark	$15.68

* National Semiconductor

Manufacturers' data books are necessary for classrooms but are an expensive investment for hobbyists since full device specifications are seldom essential to make a circuit work. Photocopies of specifications are usually obtainable from any large electronic component store, such as Active Electronic Components. These companies are mostly very cooperative, especially if one offers to pay for photocopying or buy parts under consideration.

Semiconductor guides

Three guides containing useful information and pin specifications for thousands of ICs and discrete semiconductors, are listed below. Transistors, diodes, bridges, transient suppressors, opto-electronics, linear and digital, microprocessors, memories, hardware, TTL and CMOS are cross referenced. These guides often give sufficient information for an experimenter to use a particular device. Asian and European equivalent replacement numbers are given for thousands of components.

ECG	**Semiconductor Master Replacement Guide**	#84F596	$3.81	*Newark*
NTE*	**Semiconductors**	NTE-CRM7	$4.95	*Hosfelt*
Sams	**Semiconductor Cross Reference Book**	#61000	$24.95	*Hosfelt*

* available on diskettes

Reference handbooks

Reference works are expensive, but often available in good technical libraries. Sometimes new copies of outdated editions are sold at reduced prices. Older editions are satisfactory, because reference data rarely changes significantly. Two books worth having on your shelf are:

Handbook of Chemistry and Physics, (CRC Press)

Handbook of Tables for Applied Engineeering Science, (CRC Press)

Catalogs

Technical catalogs provide an opportunity to quickly preview numerous robotics related items. Suppliers expend considerable effort in preparing their catalog information. Some suppliers of surplus equipment even perform in-house tests on obsolete equipment, giving enthusiasts data available from no other source.

A novice may be surprised at the wealth of information in the following catalogs. Many nifty ideas and inventions have been conceived after viewing a piece of equipment portrayed on their pages.

Many excellent companies and their catalogs have been excluded from the following list, and readers are urged to seek alternative product sources. Only by accumulating knowledge and experience from multiple sources, will you be confident your designs are based on optimum choices of equipment and materials.

Useful equipment and parts catalogs

company	catalog contents and comments	pages
Newark	Electronic components and related products. Test instruments, power devices, computer and telephone accessories. Company stocks 120,000 products from 285 manufacturers Every catalog item is stocked.	~1500
Electrosonic	Electrosonic has an impressive inventory and most catalog items are stocked. Catalog provides many useful parameters on numerous semiconductors. More than 90,000 items from 183 manufacturers.	~1850
Active Electronic Components	Good stock of semiconductors and smaller items. Although less comprehensive than Newark or Electrosonic, Active has many outlets in USA and Canada.	~400
Digi-Key	Excellent service and good prices on a wide product line	~450
Hosfelt Electronics	Electronic components, lasers, power supplies, motors, test equipment, tools. First rate surplus house providing excellent service and quality at consistently competitive prices. Catalog updated frequently to reflect stock changes.	~170
C & S Sales	Excellent source for hobbyist equipment, oscilloscopes, DMMs, micro-processor trainers, signal generators, tools etc. Good selection of most commonly used semiconductors. Prices are very competitive.	~60
Radio Shack	Many useful items: books, hardware, common semiconductors etc. Slightly higher prices reflect costs of individual packaging and convenience shopping.	~280
Servo Systems	Annual catalog lists accelerometers, motors, encoders, resolvers, multiple axis positioning stages etc. Top quality new and used equipment, very good prices	~150
C and H Sales	Motors, transducers, gears, test equipment. A good nuts and bolts store with large selection of stocked items. Good prices and up to date catalogs.	~120
Edmund Scientific	World's foremost supplier of affordable optics, lenses, lasers, micro-positioners, optical benches, filters, microscopes, motors, springs, magnets etc. One of the most fascinating full color catalogs for the technically minded.	~260
Fisher Scientific	Complete range of instrumentation and supplies for the laboratory. Everything from books, furniture, flowmeters, safety equipment to pumps, centrifuges and gas chromatographs, chemicals.	~1600
Cole Parmer	Complete range of products for the laboratory. Useful catalog for teaching institutions.	~1720
Pro Hardware	Many hardware supply chains have excellent catalogs listing tools, electrical supplies and auto accessories handy for the experimenter.	~ 50
McMaster-Carr	Perhaps the finest one-stop source for hardware, machine tools, raw and finished materials and numerous other products. First rate catalog.	2952
Yellow pages or commercial phone	One of the most useful and under-used sources of information. A few phone calls can quickly locate an expert in any field. Be nice, explain your problem and free advice	
Thomas Register	A mammoth trade directory occupying over 6 feet of shelf space, it is the best directory for products, services, manufacturers. May be found in major libraries	

Resources and Suppliers

Ace Hardware & Hobbies
1863 El Camino Real
Burlingame, CA 94010
415 697-6099 Fax 415 697-6801

Acheson Colloids
1607 Washington Avenue
P.O. Drawer 611747
Port Huron, MI 48061-1747
800 255-1908 Fax 810 984-1446

Active Components/ Future Electronics
41 Main St. Bolton, MA 01740
800 933-5918 Fax 508 779-3050
(stores throughout US and Canada)

Allegro Microsystems Inc.
115 Northeast Cutoff, Box 15036
Worcester, MA 01615
508 853-5000, Fax 505 853-5049

America's Hobby Center Inc.
P.O. Box 829, North Bergen
New Jersey, 07047-0829
800 989-7950

Astro-Flite, Inc.
13311 Beach Avenue
Marina Del Rey, CA 90292
310 821-6242

Belden Wire and Cable Company
P.O. Box 1980-T
Richmond, IN 47375
800 235-3364 Fax 317 983-5294

C and S Sales, Inc.
150 W. Carpenter Avenue
Wheeling, IL 60090
847 541-0710 Fax 847 520 9904

C and H Sales Co.
2176 E. Colorado Blvd.
Pasadena, CA 91107
Mail address: Box 5356
Pasadena, CA 91117-9988
800 325-9465 Fax 818 796-4875

Canadawide Scientific
1230 Old Innes Road, Unit 414
Ottawa, ON K1B 3V3
800 267-2362 Fax 800 814-5162

Cole Parmer Instrument Co.
7425 North Oak Park Avenue
Niles, IL 60714
800 323-4340 Fax 708 647-9660

Dantec Measurement Technology A/S
Tonsbakken 16-18, 2740 Skovlunde, Denmark
Tel: (+45) 44 92 36 10 Fax (+45) 42 84 61 36

Digi-Key Corporation
710 Brooks Ave. South
Thief River Falls, MN 56701-0677
800 344-4539 Fax 218 681-3380
http://www.digikey.com

Driver Harris Co. Inc.
308-T Middlesex Street
Harrison, NJ 07029
201 483-4802 Fax 302 992-5843

DuPont Company
1007 Market Street
Wilmington DE
302 774-0709 Fax 302 992-5843

Edmund Scientific Company
101 East Gloucester Pike
Barrington, NJ 08007-1380
609 573-6250 Fax 609 573-6295

Electrosonic Inc.
1100 Gordon Baker Road
Willowdale, ON M2H 3B3
416 494-1555 Fax 416 496-3030

EZE-LAP
P.O. Box 20469
Carson City, NV 89721
702 888-9500

Fisher Scientific
112ch. Colonnade Rd.
Ottawa, ON K2E 7L6
613 226-8874 Fax 613 226 8639

Future Electronics/Active Components
41 Main St. Bolton, MA 01740
800 933-5918 Fax 508 779-3050
(stores throughout US and Canada)

GC Electronics 5/26
Division of Hydro Metals Inc.
Rockford, IL 61101

General Scanning
500 Arsenal Street
Watertown, MA 02172
617 924-1010 Fax 617 924-7250

Hawker Energy Products Inc.
617 North Ridgeview Drive
Warrensburg, MO 64093-9301
800 730-7733 Fax 816 429-1758

Hobby Shack
18480 Bandilier Circle
Fountain Valley, CA 92728-8610
800 854-8471 Fax 714 962-6452

Hosfelt Electronics, Inc.
2700 Sunset Blvd.
Steubenville, OH 43952-1158
800 524 6464 Fax 800 524-5414

Kelvin Electronics
10 Hub Drive
Melville, NY 11747
800 645-9212 Fax 516 756-1763

Levitron: Fascination Toys and Gifts
19224, Des Moines Memorial Drive Suite 100
Seattle, WA 98148

Linear Technology Corporation
1630 McCarthy Blvd.
Milpitas, CA 95035
408 432-1900 Fax 408 434-0507

McMaster-Carr Supply Company
600 County Line Road
Elmhurst, IL 60126-2081
Mail address: Box 4355
Chicago IL 60680-4355
708 833-0300 Fax 708 834-9427

Measurement Group Inc.
P.O. Box 27777
Raleigh NC 27611
919 365-3800 Fax 919 365-3945

Microdot
190-T W. Crowther Avenue
Placentia, CA 92670
714 870-6650 Fax 714 524-5346

Micromint Inc.
4 Park Street
Vernon, CT 06066
800 635-3355 Fax 860 872-2204

MKS Instruments, Inc.
34, Third Avenue
Burlington, MA 01803
617 272-9255 Fax 617 229-6659

Motorola Inc.
3102-T N. 56th Street
Phoenix AZ 85036
602 952-3248 Fax 602 952-6100

Newark Electronics
Chicago, IL 60640
800 463-9275
http://www.newark.com

National Semiconductor Corporation
Standard Products Group
3875 Kifer Road
P.O.Box 58090
Santa Clara, CA 95052-8090
408 721-5000

Nonvolatile Electronics, Inc.
11409 Valley View Road
Eden Prairie, MN 55344-3617
800 467-7141 Fax 612 996-1600

Optek Technology, Inc.
1215 West Crosby Road
Carrollton, TX 75006
214 323-2200 Fax 214 323-2396

Philips Semiconductor Sensors
Integrated Circuits Division
811 East Arques Avenue
Sunnyvale, CA 94088-3409
800 227-1817 ext 900 Fax 408 991-3581

Polaroid Corporation
OEM Components Group
Cambridge, MA 02139
800 225-1618 800 391-1170 Fax 617 386-3966

Radio Shack
600 One Tandy Center
Fort Worth TX 76102
800 THE-SHACK
http://www.tandy.com

Servo Systems Co.
115 Main Road P.O.Box 97
Montville, NJ 07045-0097
201 335-1007 Fax 201 335-1661

Sharp Electronics Corporation
Sharp Plaza
Mahwah, NJ 07430-2135
201 529-8757

Tower Hobbies
Box 9078, Champaign, IL 61826-9078
800 637-6050 (information)
800 637-4989 (orders)
http://www.towerhobbies.com

Victoreen Inc.
6002 Cochran Road
Cleveland, OH 44139-3395
216 248-9300 Fax 216 248-9301

Index

A

B

D

Darlington transistor, 47
DC coupling, oscilloscope, 11
DC to dc converter, 70-71
Decoupling power lines, 51, 68
Detector
    bubble, 19, 30
    obstacle, 169
Devices, ingenious, 93
Diamond hones, 276
Dielectric constants (Table), 141
Differentiator, 29
Dimmer switch, 4, 232
Diode
    bridge, 45
    checking with DMM, 44
    clamp, 21, 32, 42
    (Table), 45
    voltage drops, 22, 32, 44
DIP sockets, 238
Displacement sensor, 159
Signal distortion, 12
Divider
    capacitive, 21
    resistive, 21
    voltage, 16, 17, 21
DMM
    inductance, capacitance etc., 6
    inexpensive, 5
    switching noise, 27
Dovetail slides, 306
Dremel tool, 263
Drill press, 257, 258
Drills, sharpening, drilling, 264
Drills, for taps (Table), 267
Dry ice, 134

E

Earth, magnetic field, 153
Eddy currents, braking, propulsion, 161, 162
Eddy current strain sensor, 163, 164
Eggs, squeezing, 113
Elastic
    cord, stretch resistors, 120
    cord, extension force, 210
    limit, 98
Electrocution, 17, 18, 27, 28
Electromagnetic
    arm, 198
    brake and clutch, 211ff
Electromagnetics, 179ff

Electromagnets, 155, 199-201
    moving game pieces, 195, 201
Electronic
    anomalies, 253
    circuits, troubleshooting, 250ff
    components, 39
    (list), 62
    construction, 215ff
    hardware and parts, 236ff
    quiz-refresher, 15, 38
    workstation for classrooms, 9
Emergency procedures, 256
Energy
    batteries, fuel cells, 77
    capacitor, stored, 30
    definition, 87
    density, batteries, fuel cell, 77
    disappearing, 20
    food, 88f
    inductive, 37
    runner, car, bicycle, 86ff
Epoxy, coloring, 283
Equipment, classroom list, 317
Experts, 226

F

Faraday cage, 28, 252
Fasteners, mechanical, 286
Faultfinding circuits, 250ff
Ferric chloride, 249
Ferrite cores, 155
FFTs, 10
Files and filing technique, 271
Filter
    active, 59
    high and low pass, 24, 35
    inductive, 24, 35
    LC, 222
    low pass, 29, 30, 35
Finger
    bend monitoring, 113-115
    strain, measuring, 118
First aid procedures, 256
555 timer, 59-60
Flammable storage, 256
Flat magnetic coil actuators, 181
Flexible, rods and shafts, 294-297
Flexure, dual beam, 109
Flow measurement, 121ff
Flycutter, 265
Foam
    conducting, 110ff
    sensor, home made, 116ff

Q

R

S